AF566924

Computer Arithmetic and Self-Validating Numerical Methods

This is Volume 7 in
NOTES AND REPORTS IN MATHEMATICS
IN SCIENCE AND ENGINEERING
Edited by WILLIAM F. AMES, *Georgia Institute of Technology*

A list of titles in this series appears at the end of this volume.

Computer Arithmetic and Self-Validating Numerical Methods

Edited by

Christian Ullrich

Institut für Informatik
Universität Basel
Basel, Switzerland

ACADEMIC PRESS, INC.
Harcourt Brace Jovanovich, Publishers
Boston San Diego New York
London Sydney Tokyo Toronto

This book is printed on acid-free paper. ♾

ACADEMIC PRESS, INC.
1250 Sixth Avenue, San Diego, CA 92101

United Kingdom Edition published by
ACADEMIC PRESS LIMITED
24–28 Oval Road, London NW1 7DX

Library of Congress Cataloging-in-Publication Data

Computer arithmetic and self-validating numerical methods / edited by Christian Ullrich
p. cm.—(Notes and reports in mathematics in science and engineering; v. 7)
Proceedings of an international conference held Oct. 2–6, 1989, Basel, Switzerland.
Includes bibliographical references and index.
ISBN 0-12-708245-X (alk. paper)
1. Computer arithmetic—Congresses. 2. Numerical analysis—Congresses. I. Ullrich, Christian. II. Series.
QA76.9.C62C663 1990
004'.01'51—dc20 90-816
CIP

Printed in the United States of America
90 91 92 93 9 8 7 6 5 4 3 2 1

Contents

Contributors

Numbers in parentheses indicate the pages on which the authors' contributions begin.

E. Adams (199), *University of Karlsruhe, Institute for Applied Mathematics, Kaiserstr. 12, D-7500 Karlsruhe 1, Federal Republic of Germany*

G. Alefeld (55), *University of Karlsruhe, Institute for Applied Mathematics, Kaiserstr. 12, D-7500 Karlsruhe 1, Federal Republic of Germany*

H. Behnke (155), *Technical University of Clausthal, Institute for Mathematics, D-3392 Clausthal-Zellerfeld, Federal Republic of Germany*

G. Bohlender (1), *University of Karlsruhe, Institute for Applied Mathematics, Kaiserstr. 12, D-7500 Karlsruhe 1, Federal Republic of Germany*

L. Collatz (189), *University of Hamburg, Institute for Applied Mathematics, Bundesstr. 55, D-2000 Hamburg 13, Federal Republic of Germany*

G. F. Corliss (91), *Marquette University, Department of Mathematics, Statistics, and Computer Science, William Wehr Physics Building, Milwaukee, WI 53233*

F. Goerisch (137), *Technical University of Clausthal, Institute for Mathematics, D-3392 Clausthal-Zellerfeld, Federal Republic of Germany*

K. Hafner (33), *Siemens AG, Zentralbereich Forschung und Technik, Otto-Hahn-Ring 6, D-8000 München 83, Federal Republic of Germany*

Z. He (137), *Technical University of Clausthal, Institute for Mathematics, D-3392 Clausthal-Zellerfeld, Federal Republic of Germany*

E. Kaucher (269), *University of Karlsruhe, Institute for Applied Mathematics, Kaiserstr. 12, D-7500 Karlsruhe 1, Federal Republic of Germany*

L. B. Rall (73), *University of Wisconsin-Madison, Department of Mathematics, 610 Walnut Street, Madison, WI 53706*

J. Schröder (247), *University of Cologne, Institute for Mathematics, Weyertal 86-90, D-5000 Cologne 41, Federal Republic of Germany*

C. Schulz-Rinne (269), *ETH Zürich, Seminar for Applied Mathematics, ETH Zentrum, CH-8092 Zurich, Switzerland*

H. J. Stetter (171), *Technical University of Vienna, Institute for Applied and Numerical Mathematics, A-1040 Vienna, Austria*

C. Ullrich (115), *University of Basel, Institute for Informatics, Mittlere Str. 142, CH-4056 Basel, Switzerland*

Preface

Computer performance is expanding year by year. Increasingly expensive experiments and development of prototypes can be replaced by mathematical models for the simulation of technical processes on a computer. A good example is the large field called *numerical fluid dynamics*. However, this process shifts the responsiblity for reliable results to physical and mathematical models, and thus the automatic verification of computed results becomes an essential tool. Automatic result verification permits distinction between the effects of a mathematical model and computational inaccuracies. Only when numerical errors are virtually eliminated can physical and mathematical models be developed systematically.

Many workshops and tutorials have been devoted to this area during the last decade. The most recent event, a first international conference on "Computer Arithmetic and Self-Validating Numerical Methods," was held from October 2 to 6, 1989, in Basel.

This volume is a collection of invited papers from this meeting. The contributed papers will be published in a special issue of the IMACS Journal "Mathematics and Computers in Simulation." Previous related volumes published by Academic Press include *Computer Arithmetic in Theory and Practice* [Kulisch and Miranker, 1981], *Introduction to Interval Computations* [Alefeld and Herzberger, 1983], *Self-Validating Numerics for Function Space Problems* [Kaucher and Miranker, eds., 1984], *A New Approach to Scientific Computation* [Kulisch and Miranker, eds., 1983], *PASCAL-SC: A Computer Language for Scientific Computation* [Bohlender, Ullrich, Wolff von Gundenberg, Rall, 1987] and *Reliability in Computing* [Moore, ed., 1988].

In the first part of this book, a number of papers give a tutorial introduction to the following topics: computer arithmetic with operations of maximum accuracy, differentiation arithmetic and enclosure methods, and programming languages for self-validating numerical methods. In the second part, the authors discuss the determination of guaranteed bounds for eigenvalues by variational methods and the guaranteed inclusion of solutions of differential equations. Finally, an appendix supplies the IMACS–GAMM Resolution on Computer Arithmetic, which is intended to influence and put pressure on manufacturers to implement computer arithmetic operations with necessary care.

Acknowledgments

Support for the symposium from the following sources is gratefully acknowledged: the Institute for Informatics, the University of Basel, the Government of the Kanton Basel-Stadt and the Nationalfonds of Switzerland; GAMM (Gesellschaft für Angewandte Mathematik und Mechanik); GI (Gesellschaft für Informatik); IMACS (International Association for Mathematics and Computers in Simulation); F. Hoffmann-La Roche & Co., Basel; IBM Schweiz; NCR Schweiz; Sandoz AG, Basel.

An excellent demonstration of software was made possible by efforts of Daniel Hollenstein and others at the IBM offices at Basel and Zürich.

Many thanks are due to my collaborators Carlos Falcó Korn, Stefan König, Roman Reith, and others from the Institute for Informatics of the Basel University. Last but not least, many thanks to our secretary, Agnes Mathys, who took care untiringly of organizational work of the symposium.

What Do We Need Beyond IEEE Arithmetic ?

G. Bohlender
Institut für Angewandte Mathematik
Universität Karlsruhe
Federal Republic of Germany

Abstract: While the four usual floating-point operations are the basis of real floating-point arithmetic, the scalar product is the basis of the operations in higher numerical spaces, such as matrices, vectors, etc. In addition, an exact scalar product is an invaluable tool for the verified solution of numerical problems by means of enclosure methods. Therefore, computer arithmetic including such an exact scalar product is a significant extension of IEEE arithmetic. In this paper, several algorithms for an implementation are sketched, problems are discussed and solutions are suggested; finally, typical designs and implementations are summarized and illustrated.

1 Introduction

Under unfavourable conditions, the small rounding error which is involved in each floating-point operation can totally invalidate results. This may happen even after only a few operations, as was demonstrated in [Rump 83a, Rump 83b, Ham 89, Ratz 89, Schu 89]. Let us study two simple examples:

1. The sum $(1 + 10^{50})$ - 10^{50} returns the wrong result 0 using ordinary floating-point operations.

2. Consider a linear equation system $A \cdot x = b$ with the matrix

$$A = (a_{ij}) = \begin{pmatrix} 64919121 & -159018721 \\ 41869520.5 & -102558961 \end{pmatrix}$$

and the vectors $b = (b_i) = \begin{pmatrix} 1 \\ 0 \end{pmatrix}$ and $x = (x_i)$. The true solution of this linear system is $x_1 = 205117922, x_2 = 83739041$. But even in IEEE double precision arithmetic, the following completely wrong results are computed

$$\begin{aligned} \tilde{x}_1 = &\quad a_{22}/(a_{11} \cdot a_{22} - a_{12} \cdot a_{21}) &= 102558961 \\ \tilde{x}_2 = &\quad -a_{21}/(a_{11} \cdot a_{22} - a_{12} \cdot a_{21}) &= 41869520.5 \end{aligned}$$

Note that the computation of $\tilde{x}_1$ and $\tilde{x}_2$ involves only four floating-point operations each!

ISBN 0-12-708245-X

These examples show that ordinary floating-point arithmetic does not suffice to compute reliable and precise results. While the first example can evidently be solved by means of an exact scalar product or an exact sum, the second example requires more subtle methods. For this purpose, enclosure methods have been developed which compute a verified interval enclosure of a given problem. These enclosure methods are again based on the exact evaluation of scalar products. For interval arithmetic see [Moo 66, Ale 74, Ale 83], for enclosure methods see [Kul 82, Kul 83a, Kau 84, Kau 87, Kul 88, Moo 88, Kul 89, Ull 90] and other papers in this volume. Enclosure methods are typically programmed in computer languages like PASCAL-SC, FORTRAN-SC, etc. which have special extensions for scientific and engineering computations; see [Nea 84, Boh 86a, Boh 87a, Kul 87a, Kul 87b, Ble 87, Metz 88, Fal 89, Hus 88, Hus 89a, Hus 89b, Hahn 88, Hahn 89]. For an overview over systems which involve enclosure methods see e.g. [Boh 89a].

In the following sections, let us concentrate on the implementation of an augmented floating-point arithmetic and in particular on the optimal scalar product. Directed roundings (which are needed for interval arithmetic) are required by the IEEE standard for floating-point arithmetic [IEEE 85, IEEE 87] and are therefore available on many modern floating-point processors. In section 2, the scalar product for floating-point numbers is defined and its relation to the IEEE norm for floating-point arithmetic is discussed. In section 3, several algorithms are roughly sketched for the computation of scalar products with maximum accuracy. In section 4, some problems are described which exist in the implementation of these algorithms in software or hardware and several solutions for each problem are suggested. In section 5, some typical designs and implementations of the scalar product in scientific research projects and in commercial products are summarized including concepts for vector processors, parallel processors, and VLSI implementations. In section 6, finally these ideas are summarized.

2 Scalar products and IEEE arithmetic

In the past twenty years, a uniform mathematical theory has been developed by Kulisch and Miranker [Kul 76a, Kul 81]. It describes how an arithmetic operation $*$ which is defined on a superset R can be transferred on a subset F. The basic idea is to execute the operation $*$ in the superset R exactly and round the result into the subset F by means of a so-called semimorphism $\square$ (a monotone projection with certain algebraic properties). This method can be used to define operations for all customary spaces of numerical computation, such as real and complex floating-point numbers, intervals, matrices, vectors, etc.

Let us choose the set $R = \mathbb{R}$ of real numbers, and the subset $F = F(b, p, e_{min}, e_{max})$ of floating-point numbers which is characterized by four integer constants: base $b \geq 2$, precision $p \geq 1$, minimum exponent e_{min} and maximum exponent e_{max}. The set F contains floating-point numbers of the form

$$(-1)^s \cdot b^e \cdot (0.d_1 d_2 \dots d_p),$$

where s is the sign, e the exponent ($e_{min} \leq e \leq e_{max}$), and d_i are base-b digits of the mantissa ($0 \leq d_i \leq b-1$). (This representation differs slightly from the IEEE representation where the decimal point is between the first and the second digit.) Let $*$ be any of the customary operations $+, -, \cdot, /$ and let $\Box$ be one of the roundings $\bigcirc$ (round to nearest), $\bigtriangledown$ (round towards minus infinity), $\triangle$ (round towards plus infinity), or $\Box_T$ (truncate, round towards zero). Then the procedure described above agrees exactly with the definition of operations in the IEEE norms for floating-point arithmetic ([IEEE 85] for binary arithmetic, [IEEE 87] for base-independent arithmetic). Nevertheless, the two methods are clearly different. The aims of IEEE arithmetic are standardization and reasonable definition of *real* floating-point arithmetic; a considerable effort was made to define computations with infinities, signed zeros, and "NaNs" (for non-numeric data), and to provide exception handling, etc.

In contrast, the definition by semimorphism is not specialized on real floating-point arithmetic but instead provides a uniform definition of arithmetic operations for all customary *higher numerical spaces* as well. Therefore, it is less specific about data formats, etc. While the term "vector processor" is usually defined in a technological sense, here we can interpret the idea of a semimorphism as a mathematical definition of a vector processor.

All these operations in the higher numerical spaces can be implemented efficiently, if the basic operations for real floating-point numbers are defined by semimorphism (or equivalently according to the IEEE norm), if the directed roundings toward plus or minus infinity are available, and if a scalar product with maximum accuracy is available. For all floating-point operands $a_i, b_i \in F$, such a scalar product shall compute the correctly rounded result

$$\Box(a_1 \cdot b_1 + a_2 \cdot b_2 + \ldots + a_n \cdot b_n)$$

as if it first produced an intermediate result correct to infinite precision and with unbounded range, and then rounded this intermediate result to the destination floating-point system according to the selected rounding mode $\Box$. This contrasts with the traditional computation of a scalar product which involves a rounding in each multiplication and addition. In the following sections, always such a scalar product *with maximum accuracy* is meant, even if this is not explicitly mentioned.

Note that this definition conforms with the IMACS-GAMM Resolution on Computer Arithmetic [IGR 89] about the elementary compound operations "accumulate" for the components of a vector and "multiply and accumulate" for two vectors.

Let us conclude this section with a remark about scalar products and IEEE arithmetic: in addition to the usual numeric values, a floating-point system according to the IEEE norm contains special representations for signed infinity, signed zeros, and not-a-number (which is intended to encode non-numeric information in the floating-point format and is frequently used in connection with exception handling). These special values can be handled in a scalar product as defined in the IEEE norm. Exceptions and traps in a scalar product can again be handled as described in the IEEE norm; of course the exceptions underflow, overflow and inexact can never occur during the computation of the "infinitly precise" intermediate result, but only when this intermediate result is rounded to the destination floating-point format.

3 Algorithms for the scalar product

The scalar product need not necessarily be provided in hardware by an arithmetic processor - it suffices if it can be programmed efficiently using other available operations. For example, the set of exact floating-point operations suggested in [Boh 89b] could be used; each such operation computes a rounded result and a remainder which together represent the exact result. These operations can in turn be simulated in software by means of e. g. IEEE arithmetic operations - but through several levels of software simulation much efficiency is lost [Dek 71, Lin 74, Hahn 88, Hahn 89].

Another way how a scalar product could be implemented is by means of a floating-point arithmetic with dynamic precision (see e.g. [Lor 71, Krü 86]), which is provided by many algebraic / symbolic manipulation systems or for example in the language Numerical Turing [Hull 85, Hull 87]. But a special scalar product algorithm requires much less effort to control the precision of each operation and is therefore more efficient; e.g. in

$$1 \cdot 1 + 10^{50} \cdot 10^{50} - 10^{50} \cdot 10^{50}$$

100 digits precision are needed to obtain the correct result, but doubled precision suffices for the products and even the sums could be computed with much less effort because one operand is usually much shorter than the other.

These arguments motivate the use of a special algorithm for the computation of scalar products. Usually, in such an algorithm first the products $p_i := a_i \cdot b_i$ of the floating-point numbers $a_i, b_i \in F(b, p, e_{min}, e_{max})$ are computed with doubled precision $2 \cdot p$ and then the products p_i are added by means of a special summation algorithm.

Several algorithms were developed which are intended to increase the precision in repeated sums [Knu 73, Knu 81]. Some of these algorithms use an array of cascaded summing registers [Wol 64, Mal 71], others perform binary summation [Linz 70], change the sequence of operations in order to minimize the error bound [Cap 75], or compute a sum and an estimation of the remainder in each addition [Kah 65, Møl 65].

In [Dek 71] and [Lin 74] the computation of sums with remainder was studied and conditions were found under which the remainder is computed exactly. Let us sketch such an operation in figure 3.1. $s := a \boxplus b$ is the rounded sum of operands a and b, r is the exact remainder, i.e. $a + b = s + r$ (provided that no overflow or underflow occurs).

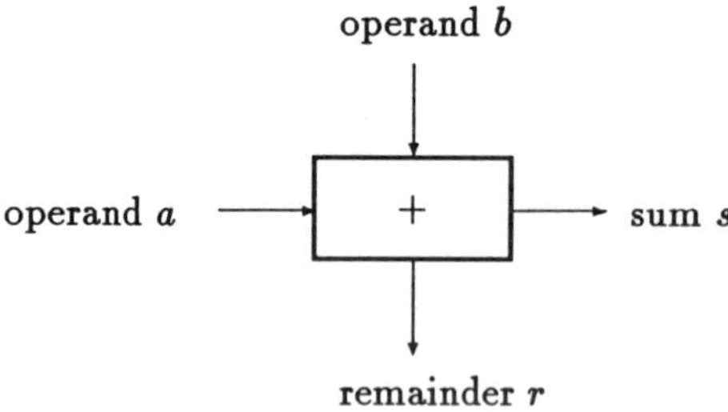

Fig. 3.1: Sum with exact remainder

3.1 Addition with remainder

The first algorithm which could guarantee nearly full precision for the computation of the sum $s := \sum_{i=1}^{n} x_i$ was found by Pichat (see [Pich 72]).

She uses the sum with exact remainder in each addition and can therefore prove that after the execution of the summation no information has been lost, i.e. values x'_i have been computed with the property

$$\sum_{i=1}^{n} x'_i = \sum_{i=1}^{n} x_i$$

This process is illustrated in figure 3.2. It can be interpreted as an algorithm (one adder is available in software or hardware for all operations) or as a circuit consisting of $n-1$ adders. After one pass of this algorithm, x'_n contains an approximation of the sum s and $x'_1, \ldots, x'_{n-1}$ contain error terms. The same process can be repeated by feeding the outputs x'_i back into the inputs x_i which leads to new values x''_i and so on. It could be proved that under certain conditions $x_n^{(k)}$ converges to a value $\tilde{s}$ with an error of less than two units of the last place of the mantissa.

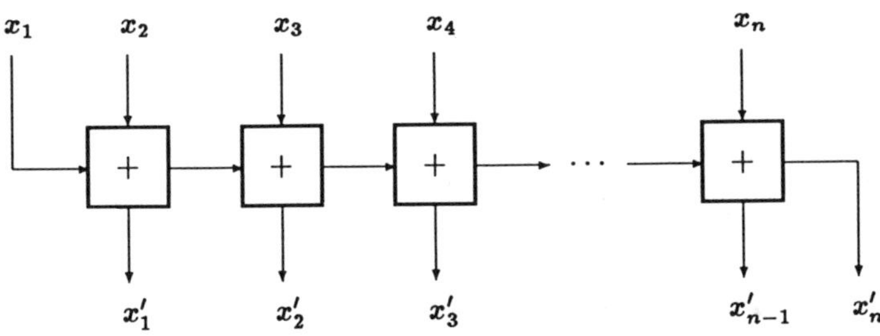

Fig. 3.2: Repeated sum with exact remainder

This algorithm was modified and applied to compute scalar products in [Boh 77a, Boh 77b, Boh 78]: using an appropriate estimation of the sum of remainders $\sum_{i=1}^{n-1} x_i^{(k)}$, the iteration can be terminated as soon as the required accuracy is reached. Secondly, it could be proved that (if one disregards zeros) the operands $x_{n-k}^{(k)} \ldots x_n^{(k)}$ are ordered according to their exponents and if two operands overlap according to their exponents, the mantissa of the larger value contains only digits zero in the overlapping region - see figure 3.3. Therefore, the iteration may be stopped after at most $k = n-1$ steps when all operands have been ordered.

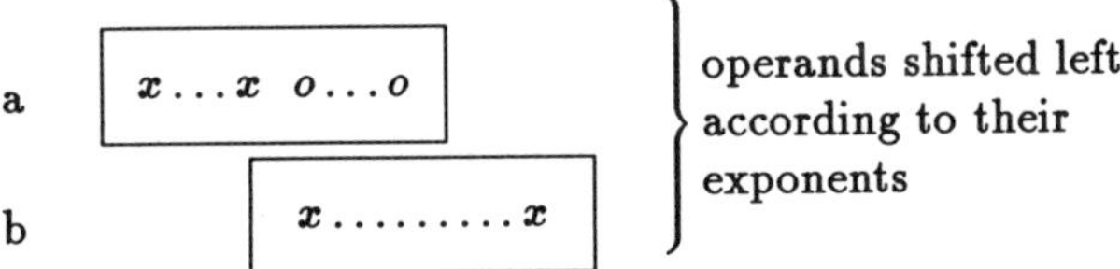

Fig. 3.3: Ordering $a > b$

In the scalar product $s := \sum_{i=1}^{n} x_i := \sum_{i=1}^{n} a_i \cdot b_i$ of the operands $a_i, b_i \in F(b, p, e_{min}, e_{max})$, the summands x_i of the repeated sum are computed in doubled precision $2 \cdot p$, whereas the final result $\Box s$ is only rounded to precision p. Therefore, the approximation $x_n^{(k)}$ of the sum s contains p "guard digits" and it can be proved that (for a certain k and nontrivial p)

$$\Box s = \Box x_n^{(k)} \quad \text{for all rounding modes } \bigcirc, \bigtriangledown, \bigtriangleup, \Box_T \text{ defined above}$$

That means that the result is as precise *as if* an intermediate result was computed with infinite precision and rounded to the destination floating-point format.

3.2 Fixed-point accumulator

A second algorithm for the computation of scalar products makes use of a long fixed-point accumulator A which allows the addition of any product of floating-point numbers without rounding error ([Rump 80, Boh 82, Boh 83]).

The product $x := a \cdot b$ of two numbers $a, b \in F(b, p, e_{min}, e_{max})$ has $2 \cdot p$ digits in the mantissa; this mantissa has to be shifted left or right in the accumulator A according to the exponent of the product which is in the range between $2 \cdot e_{min}$ and $2 \cdot e_{max}$. Therefore, the accumulator has the format which is described in figure 3.4; g guard digits are added in order to prevent overflow, therefore even the square $L \cdot L$ of the largest floating-point number $L = 0.(b-1)\ldots(b-1) \cdot b^{e_{max}}$ can be added b^g times without overflow.

g	$2 \cdot e_{max}$	l	l	$2 \cdot \lvert e_{min} \rvert$

Fig. 3.4: Layout of the fixed-point accumulator A

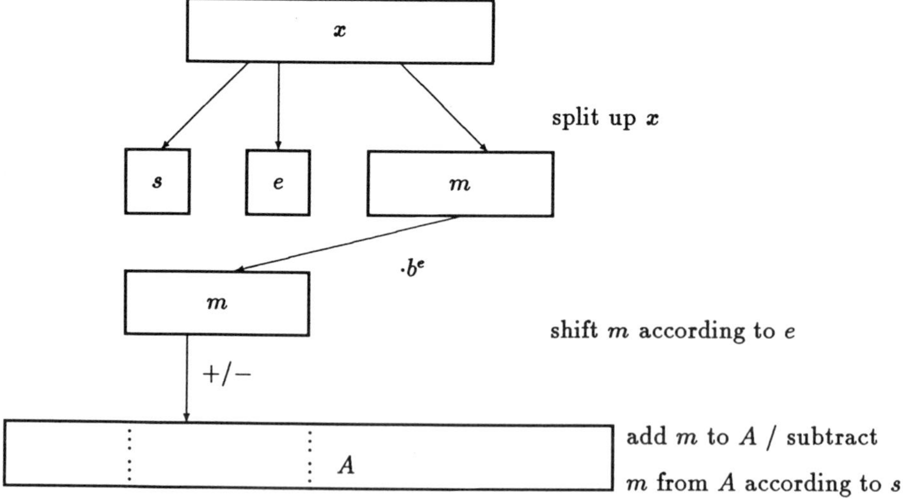

Fig. 3.5: Addition to fixed-point accumulator A

A product $x = (-1)^s \cdot m \cdot b^e$ with sign, s, mantissa m and exponent e can be added to the accumulator A as indicated in figure 3.5.

3.3 Ordered addition

Finally, let us sketch a third algorithm [Kul 76b] which consists of the following steps:

1. compute the products $p_i := a_i \cdot b_i$ exactly

2. order the products $p_i = (-1)^{s_i} \cdot m_i \cdot b^{e_i}$ according to their exponents and add up values with equal exponents; the result is

   $$e_1 > e_2 > \ldots > e_n$$

3. add up double length operands $p_1 \ldots p_k$ starting with the largest operand p_1; add as many terms as can be added without rounding error in an accumulator A_1 for double length operands which has $2 \cdot p$ digits, one guard digit, one rounding digit, one sticky bit, and one carry bit.

4. add up the remaining operands $p_n, \ldots p_{k+1}$ using rounded addition in a second accumulator A_2 of similar layout, starting with the smallest operands

5. add A_1 and A_2 and round the result

Due to the addition from left to right in step 3, no catastrophic cancellation can occur in step 5 and the rounded result can be proved to be correct for all operands and all considered rounding modes.

Let us shortly compare the three algorithms: If n is the dimension of the vectors, the algorithm "Ordered addition" requires at least execution time $O(n \cdot \log(n))$ because of the sorting algorithm; therefore it appears only reasonable for short vectors. The other two algorithms are linear in execution time (in the case of addition with remainder one has to assume that the algorithm is terminated after a few iterations, which is usually true).

The storage space being required in the algorithm with the fixed-point accumulator is proportional to the exponent range but independent from the dimension of the problem; therefore, a fixed-point accumulator can be kept in an arithmetic processor. In the case of some typical floating-point systems, we obtain the following results:

format	bytes in fixed-point accu	base ^{guarddigits}
IBM/370 ACRITH	168	16^{14}
IEEE double	530	2^{42}
BCD, 64 Bit	216	10^{10}

The fixed-point accumulator allows to store intermediate results of scalar products without rounding errors. More terms can be added to it later on with minor overhead (see the following section about dotprecision expressions).

In the other two algorithms, the storage space is proportional to the dimension n of the vectors; in the general case of long scalar products, this is a disadvantage. Additional terms which are to be added to an intermediate result of a scalar product require a considerable overhead.

4 Problems and suggestions

In this section, problems are discussed occurring in the realization of scalar products and possible solutions are suggested. Let us start with two short remarks.

Let us assume that each step in a scalar product consists of a multiplication and an addition of the resulting product to an accumulator. If the multiplication is much slower than the addition, it seems to be more reasonable to compute several partial products (by splitting up one or both operands) and to add these sequentially to the accumulator.

A scalar product unit should not be optimized exclusively for very long scalar products; short scalar products consisting of just two or a few products occur quite frequently (in complex multiplication and division, in interval operations, etc.). This problem can be solved by keeping the number of pipeline stages small. The set-up time (overhead for management, clearing of accumulator, etc.) and the finish-up time (rounding of the result, etc.) should be minimized. Alternatively, short scalar products of the form $a * b + c * d$ could be handled by a special algorithm which does not involve the overhead of the general case.

4.1 Dotprecision expressions

Frequently, the operands of a scalar product are not stored in two contiguous vectors. Instead, products (or simple operands) have to be added to or subtracted from a previously computed intermediate result of a scalar product. Therefore, it should be possible to store intermediate results of scalar products "with infinite precision" in a special data format. A finite storage representation suffices, which is trivial in case of a fixed-point accumulator. For the algorithms "addition with remainder" and "ordered addition" this is less evident; in these cases, the execution of the addition requires much more effort.

However, the algorithm "addition with remainder" can be modified for this purpose: the operations described in figure 3.1 can be cascaded in order to compute a sum, a first order remainder, a second order remainder, etc. as illustrated in figure 4.1. The register which is connected with each adder stores the previous value until it is added in the next time step. If enough adder stages are provided, the intermediate value of the scalar product is contained exactly in the registers R_i. Two similar designs were described in [Win 88] and in [IBM 86b] respectively.

Let us call a variable which can store the intermediate value of a scalar product (or synonymously dotproduct) a "dotprecision variable". The following table lists the "dotprecision operations" which should be provided for such dotprecision variables. These operations are available e.g. in PASCAL-SC [Boh 86a, Boh 87a] and FORTRAN-SC [Ble 87]; see also [Jül 87, Kreu 88].

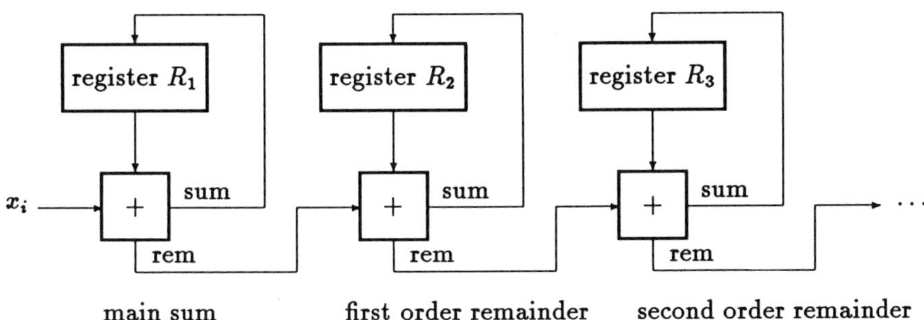

Fig. 4.1: Cascaded adders with remainder

A dotprecision expression may consist of sums and differences of the following terms:

- integer or floating-point variables or constants
- products of two integer or floating-point variables or constants
- products of two floating-point vectors
- dotprecision variables
- sums of dotprecision expressions in the form:
  <u>for</u> iv := ilow <u>to</u> ihigh <u>sum</u> (dotprecision expression)

Such a dotprecision expression can be stored in a dotprecision variable (without rounding error); alternatively it may be rounded to a floating-point number according to the specified rounding mode, or to the smallest floating-point interval that contains the exact value. Example: for real vectors u, v, x, y the expression $\nabla(u \cdot v + x \cdot y - 1)$ computes an optimal lower bound for the scalar product $u \cdot v + x \cdot y - 1$.

Dotprecision expressions can be implemented if the following basic operations for dotprecision variables are available. In this table, $A, B, \ldots$ denote dotprecision variables (called accumulators), and $x, y, \ldots$ denote floating-point or integer variables.

instruction	explanation
$A := 0$	clear accumulator A
$A := A + x$	add operand x to accumulator A
$A := A - x$	subtract operand x from accumulator A
$A := A + x * y$	add product $x * y$ to accumulator A
$A := A - x * y$	subtract product $x * y$ from accumulator A
$x := \Box A$	round contents of accumulator A to a floating-point number x (according to the specified rounding $\Box$)
$A := A + B$	add two accumulators
$A := A - B$	subtract two accumulators

Additional operations might be useful, such as determination of the sign of a dotprecision variable, addition of vectors or products of vectors to a dotprecision variable (for reasons of efficiency), determination of a main sum $x_1 := \Box A$ and remainders $x_i := \Box(A - x_1 - \ldots - x_{i-1})$ in a single operation (for "staggered correction" long-real arithmetic).

Dotprecision expressions can be defined in a similar way for floating-point vectors and matrices. These consist of a scalar product in each component. Example: for floating-point vectors b and x, floating-point matrices A, R, I and an interval matrix B, the following assignments should be possible:

$$\begin{array}{rll} x := & \bigcirc(R \cdot b) & \{\text{round to nearest}\} \\ B := & \Diamond(A \cdot R - I) & \{\text{residue, rounded to interval}\} \end{array}$$

4.2 Carry propagation

In the solution which makes use of a fixed-point accumulator, care has to be taken to prevent carries from propagating over a very long distance. The most obvious representation of negative numbers in the fixed-point accumulator A would be complement respresentation. Without counter measures the simple sequence of operations

$$\begin{array}{rcl} A := & 0 & \{\text{clear}A\} \\ A := & A - 1 & \{\text{subtract one}\} \\ A := & A + 1 & \{\text{add one}\} \end{array}$$

would lead to a borrow propagation over the upper half of the accumulator in the subtraction (changing all zeros into digits $b - 1$) and a carry propagation over the upper half of the accumulator in the addition (changing all digits $b - 1$ into zeroes).

Sign/magnitude representation of the fixed-point accumulator seems not very useful, since subtractions would frequently require large parts of the accumulator to be complemented.

There are several possibilities how these carries can be avoided (or at least made much less probable).

a) Two seperate fixed-point accumulators A^+ and A^- can be used for positive and negative operands, respectively; this method was used in the first software implementation of an exact scalar product in PASCAL-SC and in the first hardware implementation BAP-SC [Teu 84, Teu 86, Boh 86b, Boh 87b]. When the result of the scalar product is rounded to a floating-point number, the two accumulators A^+ and A^- have to be subtracted, which is a minor extra effort as compared to the normal rounding process. Since twice the storage space is needed, the method is only reasonable if just one or a few fixed-point accumulators are available in the system.

b) Bounds can be used to indicate the part of the fixed-point accumulator which is nontrivial; the digits below the lower bound are all equal to zero, and the digits s above the upper bound are all equal to zero or all equal to $b - 1$, according to the sign bit (see e.g. ACRITH [IBM 84, IBM 86a]). The fixed-point accumulator has therefore the layout which is illustrated in figure 4.2.

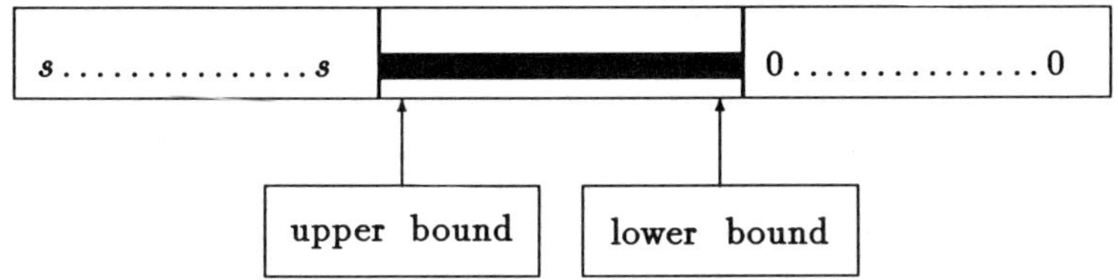

Fig. 4.2: Fixed-point accumulator with bounds, $s = 0$ or $s = b - 1$.

Such bounds are very useful in the rounding: the upper bound indicates the most significant digit of the result, the lower bound can be used in the determination of the sticky bit (which is needed for the rounding). Therefore, this method was used in the first PASCAL-SC implementation on a Z80 processor in combination with two accus A^+ and A^-.

Of course the bounds have to be updated in each operation, which takes some execution time. Therefore, there is a tradeoff between faster rounding (with bounds) and faster addition (without bounds); in the arithmetic processor BAP-SC no bounds were used for this reason.

The digits outside of the bounds may be valid or invalid. If they are invalid, they have to be defined as soon as an operation increases the upper bound or decreases the lower bound.

c) A third method preventing carries over a long distance is the use of an offset bit pattern: the fixed-point accumulator is defined to contain the value zero, if it is filled with a certain bit pattern.

In the 68000 version of PASCAL-SC, version 1, which uses decimal arithmetic, the pattern ...5050... was chosen. In the case of a binary system, the pattern ...0101010... could be used. In principle, any pattern will do which is not ...000... or $\ldots(b-1)(b-1)(b-1)\ldots$

Any addition to the pattern or subtraction from the pattern will generate at most one carry or borrow. Of course, the pattern has to be subtracted from the accumulator when the result is rounded to a floating-point number; but this is only a small extra effort.

d) Carries can be prevented by different methods: the fixed-point accumulator can be split into words that allow a redundant representation of the contents. For example signed words could be used, or each word could be provided with a carry counter (as was suggested by Kirchner and Kulisch [Kir 87, Kir 88]).

Each word can be provided with flag bits which indicate whether all digits in the word are zero or $(b-1)$. With such flags, the carry can be propagated much faster. A scheme which is based on such flags is presently being investigated at the University of Saarbrücken.

4.3 Selection of wordlength

The fixed-point accumulator A can be organized as a sequence of base-b digits. In this case the digits of a product can be added to the accumulator without shifting. But such an organization of the accumulator requires that either all digits have to be accessed and operated on serially (which might be very slow), or that fields in the accumulator have to be accessible according to a digit-based address (which might be very complicated).

Therefore, it appears to be more reasonable, to split the acccumulator A in words of an easily addressable wordlength W - usually a power of two. In this case, the exponent E of an operand (which is to be added to or subtracted from A) has to be split into two fields: the quotient $Q = E \underline{\text{ div }} W$ indicates the index of the word of A, where the mantissa M has to be added; the remainder $R = E \underline{\text{ mod }} W$ indicates how far the mantissa M has to be shifted in order to adapt it to the wordlength of the accumulator. These coarse and fine shift operations are illustrated in figure 4.3. (In practice, these computations are slightly more complicated to account for the size of a digit in bits and for the wordlength in bytes).

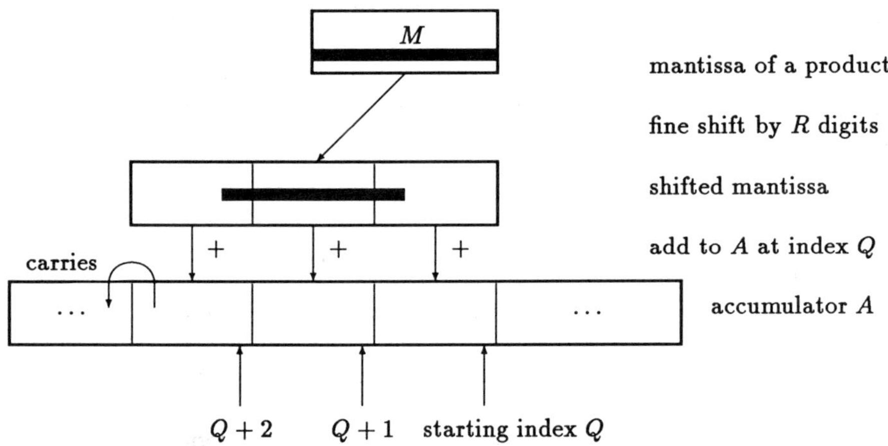

Fig. 4.3 Fine shift and coarse shift

In figure 4.3 it is assumed that the word length of the accumulator is the same as the word length of the original floating-point operands. In the example of IEEE double precision arithmetic this means 64 bits. The mantissa of a product is 106 bits wide and has to be fine-shifted by between 0 and 63 bits. Therefore, the shifted mantissa fits into 169 bits, i.e. in three 64-bit words which have to be added to the fixed-point accumulator A. If a carry beyond the most significant word occurs, one or possibly a few more words of the accumulator have to be modified (see the previous section about carry handling).

The number of accessed words of the accumulator can be reduced from 3 to 2, if the accumulator is organized in double-length words; in the example with IEEE double precision arithmetic this means a 128-bit structure. The number of accesses can only be reduced to 1, if the accumulator is constructed of overlapping words as illustrated in figure 4.4. If the accumulator words are chosen wider than the mantissas which have to be added, a carry handling can be nearly always avoided.

Such a redundant representation has two disadvantages: it needs more storage space and the final rounding becomes much more complicated: the overlapping parts have to be added in order to obtain a non-redundant representation. The idea of overlapping accumulator words is similar to Malcolm's algorithm [Mal 71]; a software solution using floating-point operations was implemented in [Sue 86], but it could as well be realized in hardware or with integer operations.

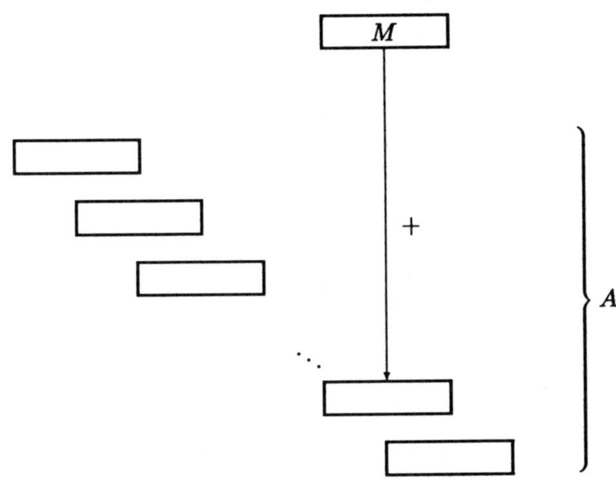

Fig. 4.4: Fixed-point accumulator A with overlapping words

4.4 Interface and context switch

In a hardware realization of a scalar product, a bottleneck exists between the arithmetic unit and main memory which contains the operands. The number of operand transfers is identical with the number of operations - no results can be reused in internal registers. In a well-designed arithmetic processor for scalar products, the operation itself can be performed very efficiently. But still the speed is limited by the narrow bus-bottleneck. In figure 4.5, the situation is illustrated in the case of a fixed-point accumulator, but for the other algorithms the problem is basically the same.

Let us sketch some solutions for this problem. The interface has to be made as efficient as possible. This can be achieved by wider and faster busses, but it requires a redesign of the whole system and increases the costs considerably. An alternative would be to include vector registers VX and VY on the arithmetic unit which can store a certain number of components of the vectors x and y. These vector registers can be filled from main memory more efficiently by means of DMA blocktransfers (in a pipelined design, the transfers are executed in parallel with the arithmetic operations).

Another possible solution is the use of vector registers for vectors which are used several times: In matrix-vector and matrix-matrix products the same row or column vector is reused n times. Therefore in principle, the amount of data which have to be loaded in the arithmetic unit for a matrix multiplication can be reduced from $2 \cdot n^3$ (for n^2 scalar products of length n) to $2 \cdot n^2$ (for two matrices).

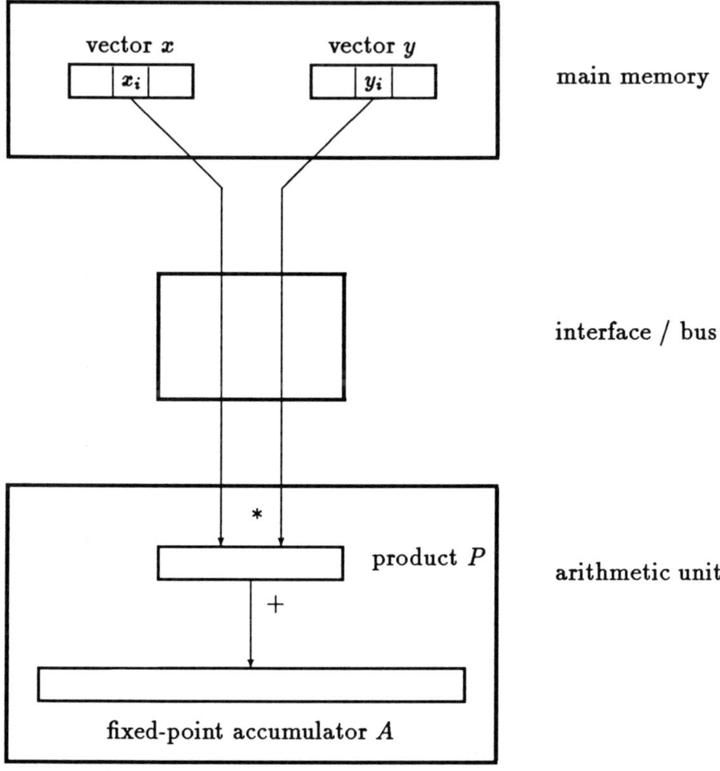

Fig. 4.5: The interface / bus bottleneck

A similar problem which requires different solutions is the context switch in a multitasking system: if the arithmetic unit is required by a different task, the data of the current task have to be saved, including the intermediate value of the scalar product. The same situation occurs in the use of dotprecision variables as defined in section 4.1. This means that as little data should be kept in the arithmetic unit as possible – in contrast with the suggestions above. The following solutions may be suggested (they are formulated for a fixed-point accumulator but may be applied to any realization of the scalar product):

a) the scalar product unit is assigned exclusively to a single task, like a printer

b) the fixed-point accumulator is located in the arithmetic unit, but in a context switch, only the active part of it is transferred to main memory; this includes the values between the upper and lower bounds and some status information

c) the fixed-point accumulator fetches three operands: x, y and the required part of A; the result is written back to main memory immediately; this scheme seems to be reasonable for RISC-based computers, as well as for computers which automatically keep the most recently accessed words in a fast cache memory

d) the fixed-point accumulator is kept in main memory, but a part of it is copied to a special cache in the arithmetic unit (usually about four words plus some status information should suffice); as long as the exponents vary only in a moderate range, the cache does not have to be reloaded from main memory

e) several fixed-point accumulators are available in the arithmetic unit; they are managed and assigned to different tasks by the operating system; one task may request several fixed-point accumulators

5 Designs and implementations

In [Kul 83b], a large number of variants for the implementation of scalar products was listed, and a hardware unit was described basing on a long accumulator and a triple length adder. Finally, parallelism in scalar products and pipelining of the arithmetic operations were discussed.

Let us now study several designs and implementations of the scalar product which were performed in scientific research projects or are contained in commercial products. Of course, this section can not be a complete overview over all available designs, but rather is a selection of some designs which highlight interesting properties. For a more detailed description let us refer to the listed literature.

5.1 PASCAL-SC

The first implementation of an exact scalar product was performed in an early version of PASCAL-SC for the 8-bit microprocessor Z80; it was later adapted to the 16-bit processor 8086/8088 which is used in the IBM/PC [Kul 87a]. PASCAL-SC is an extension of PASCAL for scientific and engineering computations; it supports interval computations and is augmented by libraries for the verified solution of basic numerical problems by means of enclosure methods.

A decimal floating-point system F(10, 12, -98, 100) is used with packed BCD representation; one of the eight bytes is reserved for sign bit, status information, and a "zero" flag which allows a more efficient handling of zeros. The scalar product is implemented by means of a pair of fixed-point accumulators A^+ and A^- for positive and negative operands; bounds are used. Basically, only one pair of accumulators is available in the runtime system.

In the 68000 version of PASCAL-SC [Kul 87b] the floating-point system $F(10, 13, -98, 100)$ is used; the scalar product is implemented by means of a fixed-point accumulator A with an offset bit patterns (see 4.2) and bounds. In principle, several such accumulators can be used – but not in a comfortable way.

Binary floating-point arithmetic which is compatible with the IEEE standard has been developed for PASCAL-SC, see section 5.3

The scalar product was first implemented in hardware on the arithmetic processor BAP-SC [Teu 84, Teu 86, Boh 86b, Boh 87b]. The same BCD-floating-point system is used as in the 68000 version of PASCAL-SC; therefore enclosure algorithms programmed in PASCAL-SC can be implemented on BAP-SC by simply linking the program with a different runtime system.

BAP-SC is basically a 64-bit processor with special extensions for decimal computation and scalar products (see figure 5.1). The arithmetic / logic unit ALU consists of 16 bit-slice processors AMD 29203 with additional carry-lookahead logic. The internal registers and the indirect address logic are used to speed up the computation of products, etc.

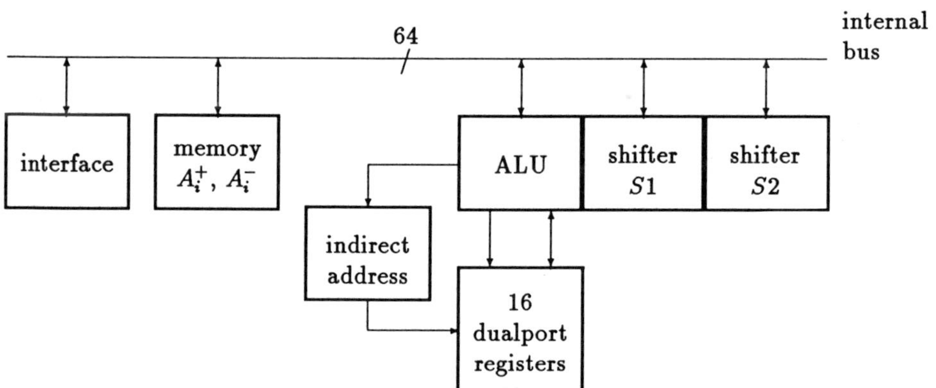

Fig. 5.1: Architecture of BAP-SC

Scalar products are implemented by means of a pair of two fixed-point accumulators A^+ and A^- which are located in the internal memory and consist of 35 words of 64 bits width each. Because of the low number of words, no bounds are needed. Eight such accumulators $A_0^{+/-}, \ldots, A_7^{+/-}$ are available.

The mantissa of a floating-point number is 13 digits wide, i.e. 52 bits. A product is thus contained in 104 bits. A shift of 0,4,8, ..., 60 bits is necessary in order to adapt the product to the 64-bit structure of the fixed-point accumulator, the shifted product is thus contained in a 164 bit word. Therefore the ALU has to be extended by two shifters $S1$ and $S2$ which are both 64 bits wide. The shifted result is added in three steps to the corresponding words of the accumulator A^+ or A^-.

5.2 ACRITH, ARITHMOS, HIFICOMP

The first commercially available implementation of the exact scalar product was contained in the subroutine package ACRITH which is a library for the verified solution of basic numerical problems by means of enclosure methods [IBM 86a]. The IBM/370 hexadecimal format $F(16, 14, -64, 63)$ is used. Scalar products are implemented by means of a fixed-point accumulator with bounds. It occupies 168 bytes and is aligned on a 256-byte boundary (therefore $420 = 168 + 256 - 4$ bytes are needed in main memory) [IBM 84]. On several machines of the 4300 and 9370 series it is supported by microcode or hardware, on all other machines with 370 architecture it is simulated in software. As the fixed-point accumulator is located in user memory, an arbitrary number of such accumulators can be used. Usually the accumulator can be accessed very efficiently because it can be kept in cache memory.

A set of operations for fixed-point accumulators $A, B, \ldots$, floating-point operands $d, e, \ldots$, and floating-point vectors $v, w, \ldots$ is defined in ACRITH. The operands $d, e, \ldots$ and the components of the vectors $v, w, \ldots$ may be double precision operands (64 bits, as described above) or single precision (32 bits), but not mixed. The vectors may be stored contiguously or with a constant stride for each vector (e.g. rows and columns of a matrix). Let us list the relevant operations:

instruction	explanation
$A := 0$	clear accumulator A
$A := A + d$	add floating-point number to accu
$A := A - d$	subtract floating-point number from accu
$A := A + v \cdot w$	add product of two vectors to accu
$d := \bigcirc A$	round A to nearest floating-point number
$(d, e) := \Diamond A$	round A to floating-point interval
	(i.e. d is a lower bound $\bigtriangledown(A)$, e is an upper bound $\triangle(A)$)
$A := A + B$	add two accus
$A := A - B$	subtract two accus

These operations are basically equivalent to the dotprecision operations which are defined in section 4.1; they permit the definition of dotprecision operations in FORTRAN-SC [Ble 87].

The program packages ARITHMOS [SIE 86] and HIFICOMP [Vel 89] contain equivalent sets of operations. ARITHMOS is supported in hardware on a large range of mainframe computers; the development of a VLSI chip was studied [Haf 89]. HIFICOMP works on machines which are compatible with the IBM/370 architecture or the IBM/PC [Vel 89, Mar 89].

5.3 Software implementations

In addition to the software implementations being mentioned above, several others have been developed. Let us discuss only a few of these.

In [Kie 88] a scalar product is described by Nassi-Shneidermann-diagrams and implemented in 8086 assembler. It is based on IEEE single precision arithmetic and makes use of a fixed-point accumulator.

A new version of PASCAL-SC is being developed at the University of Karlsruhe according to the specifications in [Boh 86a, Boh 87a]. In this version the user can choose between decimal arithmetic and binary IEEE floating-point arithmetic. Several implementations of binary floating-point arithmetic including a binary scalar product are being prepared: assembler versions for the Intel 386 and Motorola 68000 families of microprocessors, and a portable version which is written entirely in C. All of these versions provide dotprecision expressions as defined in section 4.1.

Portability is an important aspect in software implementations that are written in high-level languages, even if such implementations are not as efficient as assembler or hardware versions. Portable software implementations of the scalar product have been developed in Modula for the Modula-SC system [Fal 89], in Ada [Erl 88], in APL for APL/PCXA [Hahn 88, Hahn 89], and for the Abacus system [Hus 88, Hus 89]. The APL implementation is based on the addition with remainder algorithm, in contrast with the other implementations which are based on a fixed-point accumulator.

5.4 Designs for pipelined processors

In a pipelined computer, a sequence of operations is fed through a pipeline of specialized processors (multipliers, adders, etc), which may again consist of several internal pipeline stages. This method leads to a considerable speedup of repeated sequences of operations - such as scalar products - because all pipeline stages operate in parallel (on different data).

Pipelining may obviously be applied to the multiplications and additions which are involved in an exact scalar product. Multiplications can be implemented very efficiently in a pipelined processor, e.g. by means of a Wallace tree. A product may thus be computed in each clock cycle. If the addition in the exact scalar product is slower than the multiplication, several adders (e.g. several fixed-point accumulators) may be used; in this case, an adder is responsible for each k-th product.

In a pipelined design, the execution time of each processor has to be constant for all possible operand combinations. In [Kir 87, Kir 88, Cap 88] techniques are studied how this aim may be achieved for the exact addition of products to a fixed-point accumulator. A product has to be added in each clock cycle, even if carries are generated. In all three designs, a considerable amount of additional hardware is required for this purpose.

Two designs are proposed in [Kir 87, Kir 88] which are both based on a fixed-point accumulator A in complement representation. The basic layout of the scalar product units is described in figure 5.2. The shifter is a cyclic shifter: digits which are shifted out at the left are shifted in from the right again. For these digits, the exponent has to be adjusted (using the exponent tag defined below). The rounding unit is responsible for a possible final carry handling, for the determination of the most significant part of the sum, and for the rounding to the desired floating-point result according to the selected rounding mode.

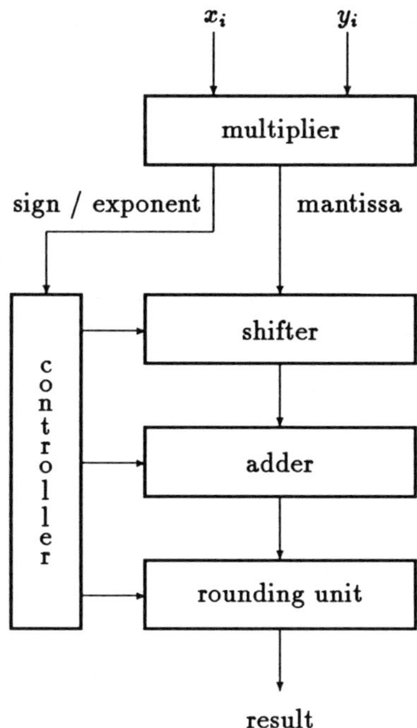

Fig. 5.2: Layout of scalar product unit

Let us consider the operations of the adder more precisely. In the first design this adder has the form of a matrix, as illustrated in figure 5.3. The shifted mantissa fits into a line of the matrix; in fact one has to require that even $(c-1) \cdot d + 1 \geq 2 \cdot p$ in order to prevent the most significant digit and the least significant digit from being shifted into the same block. In particular, the number of columns must be greater than one.

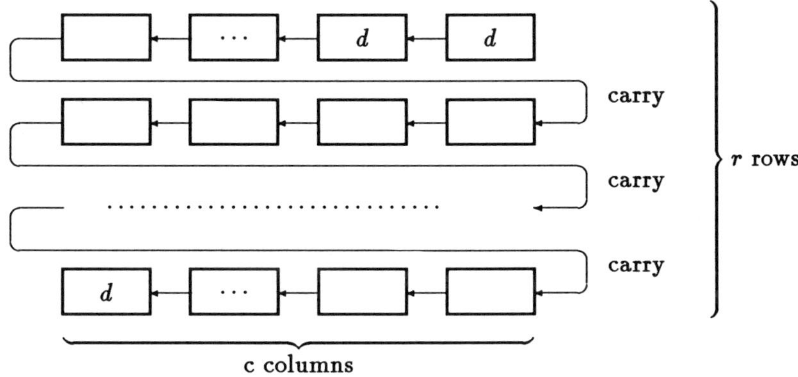

Fig. 5.3: Adder matrix with carry propagation

Each block in the matrix represents a subadder which can add or subtract d digits in a single cycle. Of course, the matrix has to be large enough to represent the fixed-point accumulator, i.e.

$$r \cdot c \cdot d \geq 2 \cdot (p + |e_{min}| + e_{max}) + g$$

for the floating-point system $F(b, p, e_{min}, e_{max})$ and g guard digits. Carries and borrows which are generated in a block are stored in a carry register and passed on to the next block on the left in the next cycle; therefore, carry handling requires no extra execution time. Note that each carry flag consists of two bits (carry and borrow) and that each operation may involve a carry or borrow from the previous operation (which may be different); addition with borrow and subtraction with carry can occur!

The shifted mantissa is split into c blocks M_i with d digits each. Each block is supplied with an exponent tag E_i which indicates in which row of the matrix this block has to be added. The shifted mantissa is fed into the first line of the matrix and is passed on to the second line in the next cycle and so on. If the exponent tag indicates that the correct line has been reached, the block is added/subtracted. This process, being illustrated in figure 5.4, can be fully pipelined: each row contains the data of a different product; a new product can be fed into the matrix in each cycle. For IBM/370 format the amount of hardware which is required for this matrix is approximately equal to a Wallace tree for the multiplication.

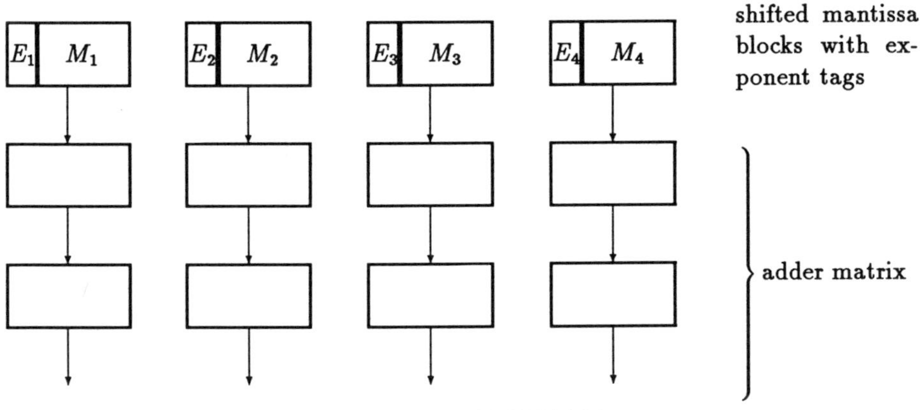

Fig. 5.4: Transfer of shifted mantissa through matrix

After the last product has been fed into the matrix, up to $r \cdot c - 1$ cycles are needed to handle carries, before the final result can be determined.

In the second design, the fixed-point accumulator again is in principle shaped as in figure 5.3 in the form of a matrix. But much less hardware is required in this design because the individual blocks are no adders but only registers. For each column in the matrix, a single adder/subtracter is provided. Carries are not handled during the additions/subtractions; instead in each block a carry register is included which is incremented for each carry and decremented for each borrow. At the end of the summation process, each carry register has to be added to the next block. This requires up to $r \cdot c - 1$ cycles to eliminate the carries - as in the previous design.

Figure 5.5 shows the layout for one column of the matrix. It contains an adder/subtracter and a dualport registerfile (one for each row of the matrix). Adder and registers are $d + y$ digits wide, where d is the width of a block and y is the number of guard digits in the carry counter. In contrast, the slice of the mantissa m is only d digits wide.

An operand (exponent tag e, sign bit s, slice of the mantissa m) is fed into the unit. The exponent tag e determines the register in the register file which m should be added to or subtracted from. In the next clock cycle the operands are transferred to e', s', m' and the register is copied into a temporary register r'. The sign s' determines whether m' (which is extended with y zeros) and r' are added or subtracted. The result is contained in r'' in the next cycle, and the exponent tag is transferred to e''. The result r'' is written back into the registerfile at the write address e''.

This design is pipelined, i.e. while e', s', m' represent one operand, the next operand is entering the circuit at e, s, m. If $e = e'$ or $e = e''$, we have a pipeline conflict because the values in the registerfile have not yet been updated. Therefore, the multiplexer blocks the data path from the registerfile to r', but instead loads r' with r'' or with the result of the addition/subtraction which is presently being executed.

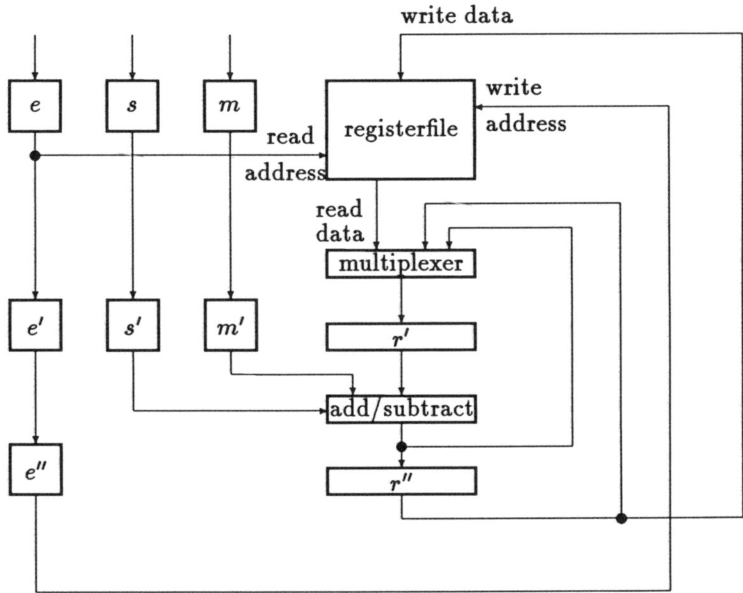

Fig 5.5: Column of the adder unit

In [Cap 88] a systolic super summer is proposed for the scalar product which allows the addition of a product in each cycle. The design being illustrated in figure 5.6 consists of a sinking region, a sieve and an accumulator-pipeline. In the sinking region, the mantissa sinks vertically down until it reaches the correct exponent tag. At this moment it changes its direction and sinks diagonally to the lower right until it reaches the accumulator a_0 where it is finally added. As soon as the last operand has been entered, the accumulator a_0 is fed into a pipeline of accumulators in which the remaining carries are handled and the result is normalized. At the end of the pipeline the result can be rounded.

The design has the advantage that it can be constructed from systolic building blocks which are very regular in structure and which are connected in a very regular way. The disadvantage is that the design requires an extreme amount of hardware: if $L = 2(p + |e_{min}| + e_{max}) + g$ is the size of a fixed-point accumulator, the design requires about L^2 building blocks.

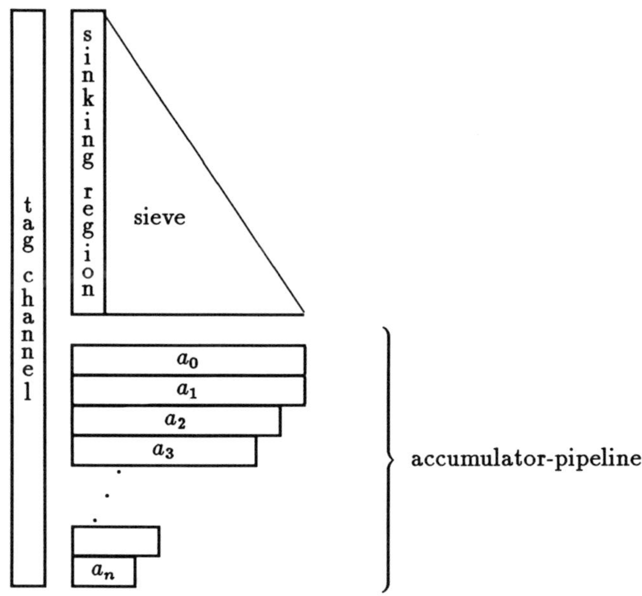

Fig 5.6: Systolic super summer

5.5 Studies for chip design

Several studies about the implementation of the scalar product on a chip were performed. In the context of the ARITHMOS project, one of these studies has already been mentioned in section 5.2 [Haf 89].

A second study was performed at the University of Eindhoven [Yil 89]. A pair of fixed-point accumulators A^+ and A^- is used to compute the scalar product of two IEEE single precision floating-point vectors. The two accumulators are shaped as two concentric rings which continuously rotate. Single bit adder stations are used to add a product serially to the accumulator A^+ or A^-. These adder stations are relatively slow because of their bit-serial operation. But a large number of adder stations can be used which add many products at the same time. The number of adder stations is only limited by the length of the accumulator. In figure 5.7 the design is illustrated, where the two accumulator rings are seen from the side.

The dispatcher finds a free adder and sends the operand to that adder. In the adder the operand is temporarily stored until the correct position of the accumulator reaches the adder. At that time the adder starts to add the operand and any possible carries to the accumulator. After the last operand has been added, the accumulator has to perform a full revolution before all carries have been handled and the start of the result has reached the subtraction station. If the exponent range is large and the vectors short, this could be a disadvantage. In this case, the accumulator and the adders should be structured in blocks of sufficient size.

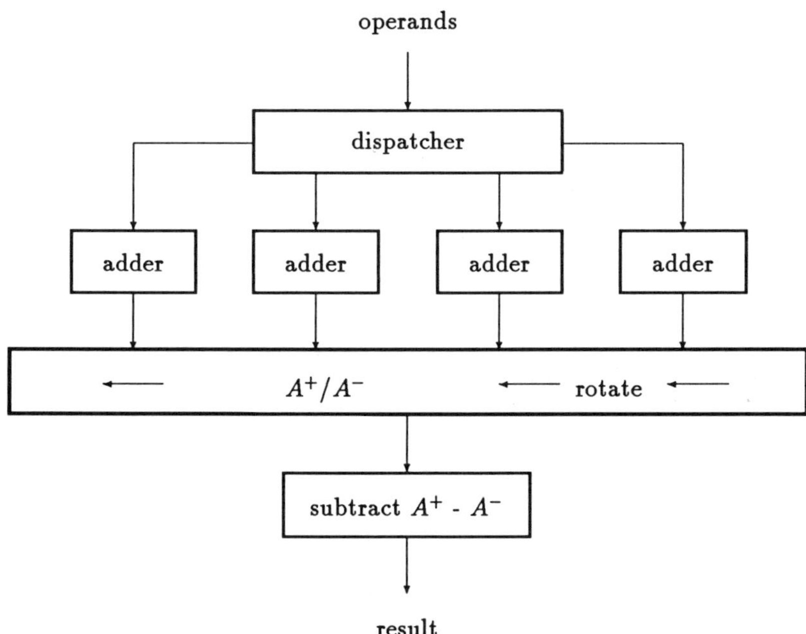

Fig. 5.7: Rotating accumulators

At the University of Saarbrücken, two studies were performed which are based on a fixed-point accumulator A in complement representation [Win 85, Lich 88]. The accumulator is folded on the chip as illustrated in figure 5.8 and can add/subtract products of IEEE single precision operands.

Each basic cell is responsible for a single bit of the accumulator. The accumulator consists of a matrix with 34 rows and 32 columns of such basic cells - similar to the design in figure 5.3. A pair of an odd and an even row is grouped together to form a "doublerow". A product, which fits in two doublerows, is added to the accumulator by a fine shift, a coarse shift and an addition. The fine shift brings the product in the correct position relative to the doublerows. The coarse shift is performed by transporting the fine shifted product over a vertical bus to the correct two doublerows; this bus is four bits wide for each of the 32 columns. The bits below the least significant bit are cleared, and the bits above the most significant bit are set to the sign bit in all remaining doublerows, i.e. the product is extended to full length of the fixed-point accumulator. Finally, the product is added to the previous contents of the accumulator by means of a full-adder in each cell. The carry is saved and can be added in the next cycle with the next product. Since a large number of full adders works simultaneously, the power consumption is relatively high.

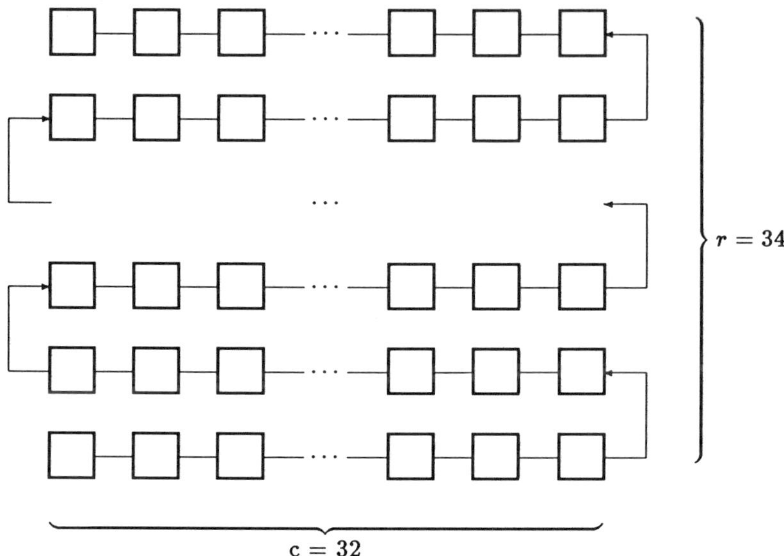

Fig. 5.8: Folded accumulator

At the University of Karlsruhe, a scalar product for IEEE double precision operands was studied [Knö 88]. The design which is based on a fixed-point accumulator A is similar to the matrix described in figure 5.3. It consists of 9 rows with 8 columns of 64-bit processors. The available silicon compiler Genesil could not pack the processor on a single chip; instead, the design had to be cut into eight equal slices. If the exponent range of the operands can be restricted, less slices suffice. The slices can be adapted to individual requirements by means of external control signals.

A product is added to the accumulator after an external fine shift. The shifted result has to be split into 64-bit words which are fed to all rows and columns of the matrix in parallel. An exponent tag is provided which selects the one out of the 72 adders which is responsible for the addition. Negative numbers are represented in complement; therefore, the adders with a larger exponent tag all add sign bits $0 \dots 0$ or $1 \dots 1$; the adders with a smaller exponent tag add zero. All adders handle carries which might be generated by previous operations.

5.6 Parallel concepts

Several algorithms for the exact evaluation of scalar products on parallel processors were proposed in [Leu 82]. These algorithms are all based on the addition with remainder which is described in section 3.1.

a) The algorithm [Pich 72] described in figure 3.2 can be pipelined: one processor computes the sum and a vector of remainders of the original operands $x_i^{(1)}$, the remainders $x_i^{(2)}$ are passed to a second processor which computes the sum of the remainders and a vector of second order remainders $x_i^{(3)}$, and so on. This process, which is illustrated in figure 5.9, requires one processor for each line. It is similar to the algorithm described in [IBM 86b]. Processor #2 can start working as soon as the operands $x_1^{(2)}$ and $x_2^{(2)}$ are available. Therefore, the second line may only start after a delay of *two* time steps and so on, which necessitates $n-1+2\cdot(l-1)$ steps to compute the $l-th$ line. As the sum and all remainders are ordered after at most $l=n-1$ steps (see [Boh 78, Boh 77b]), the parallel algorithm needs at most $3\cdot n-5$ time steps.

In [Leu 82] this process is described in a slightly different way: each processor stores its results (sum and remainder) back into the original array. Processors which become idle at the end of the array can be reused to compute an additional line in this scheme, which is the reason why at most $n/2$ processors are needed. The termination criterion from [Boh 78, Boh 77b] is used.

If less processors are available, the algorithm can be easily adapted. Two processors seem to be a good compromise between average execution time and required hardware.

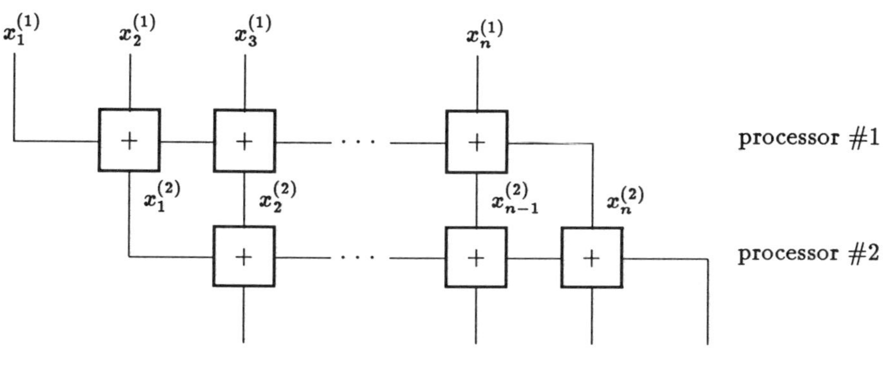

Fig. 5.9: Pipelined addition with remainder

b) If $n/2$ processors are available for the summation of n operands x_i, an algorithm can be described in close analogy to the well-known "sorting by exchanging". In an even time step t, all pairs of operands $x_{2i}^{(t)}, x_{2i+1}^{(t)}$ are added in parallel. The rounded sum and the remainder are stored in $x_{2i}^{(t+1)}$, $x_{2i+1}^{(t+1)}$, respectively. In an odd time step, all pairs of operands $x_{2i+1}^{(t)}$, $x_{2i+2}^{(t)}$ are added in a similar way. This process which can be named odd-even-summation is illustrated in figure 5.10, where the adder cells are rotated by 45 degrees as compared with the previous decriptions. The $n/2$ processors execute one line in parallel and then proceed to the next line.

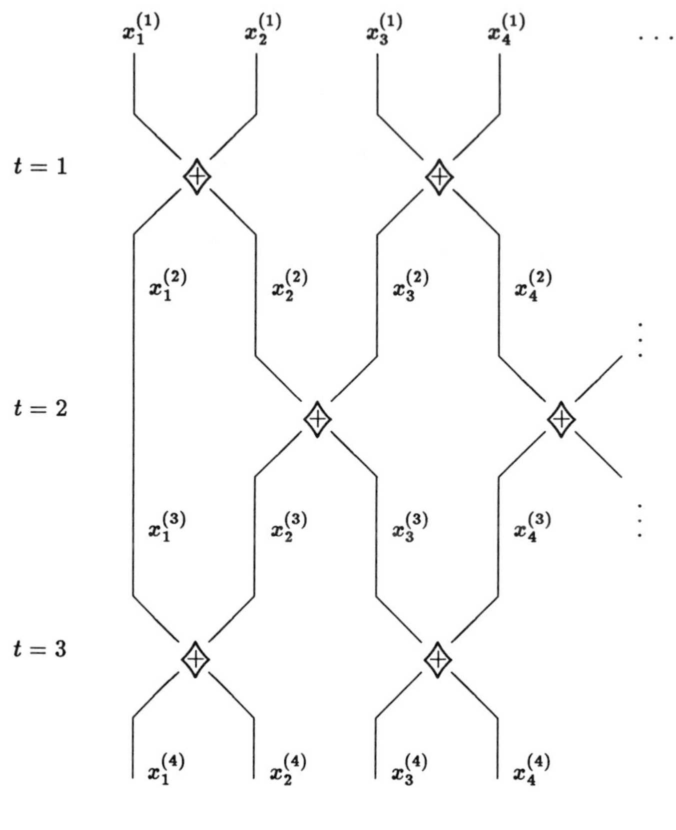

Fig. 5.10: Odd-even-summation

c) Finally an adder tree was described in [Leu 82]. The sum is computed by means of a binary tree. All remainders are saved and added in a second pass by the tree. In a repeated application of this process a sequence of a sum, a first-order remainder, a second-order remainder, and so on is computed. This sequence has finally to be corrected by means of a pipeline-algorithm as described in a).

6 Conclusions

A large number of variants for the implementation of an exact scalar product have been designed and implemented at many different locations. Efficient methods are available for hardware and software solutions. Experience with many implementations of scalar products has shown that frequently, an exact scalar product is even faster than an inexact traditional implementation – provided that they are based on comparable hardware support. The reason for this effect is, that in an exact scalar product less and simpler operations have to be performed: the product does not have to be rounded, composed into a floating-point number and decomposed again for the addition; the addition of a product to a long fixed-point accumulator is in principle no more complicated than a traditional floating-point addition.

Due to its numerous applications in higher numerical spaces and in interval enclosure methods, the exact scalar product has to be considered a basic operation which should be provided in hardware on a modern floating-point processor. It can also be considered as a valuable extension of the IEEE standard for floating-point arithmetic; it agrees with the philosophy of the standard and contains no contradictions with it. If necessary, the exact scalar product can be composed out of simpler operations, such as operations with a rounded result and an exact remainder.

References

[Ale 74] Alefeld, G., Herzberger, J.: *Einführung in die Intervallrechnung.* Bibliographisches Institut (Reihe Informatik, Nr. 12), Mannheim / Wien / Zürich, 1974.

[Ale 83] Alefeld, G., Herzberger, J.: *Introduction to Interval Computations.* Academic Press, New York, 1983.

[All 86] Allendörfer, U.: *Dezimale Gleitpunktsysteme zur effizienten Implementierung der optimalen Rechnerarithmetik.* Ph.D. thesis, Universität Kaiserslautern, 1986.

[Ble 87] Bleher, J. H., Rump, S. M., Kulisch, U., Metzger, M., Ullrich, Ch., Walter, W.: *FORTRAN-SC: A Study of a FORTRAN Extension for Engineering/Scientific Computation with Access to ACRITH.* Computing **39**, 93–110, Springer, 1987.

[Boh 77a] Bohlender, G.: *Genaue Summation von Gleitkommazahlen.* Computing Suppl. 1, 21–32, Springer Verlag, 1977.

[Boh 77b] Bohlender, G.: *Floating-Point Computation of Functions with Maximum Accuracy.* IEEE Transactions on Computers, vol. C-26, no. 7, July 1977.

[Boh 78] Bohlender, G.: *Genaue Berechung mehrfacher Summen, Produkte und Wurzeln von Gleitkommazahlen und allgemeine Arithmetik in höheren Programmiersprachen.* Ph.D. thesis, Universität Karlsruhe, 1978.

[Boh 82] Bohlender, G., Grüner, K.: *Gesichtspunkte zur Implementierung einer optimalen Arithmetik.* In [Kul 82].

[Boh 83] Bohlender, G., Grüner, K.: *Realization of an Optimal Computer Arithmetic.* In [Kul 83].

[Boh 86a] Bohlender, G., Rall, L. B., Ullrich, Ch., Wolff von Gudenberg, J.: *PASCAL-SC: Wirkungsvoll programmieren, kontrolliert rechnen.* Bibliographisches Institut, Mannheim, 1986.

[Boh 86b] Bohlender, G., Teufel, T.: *Demonstration of the Bit-Slice Processor Unit BAP-SC in a 68000 Environment.* Proceedings of the 11th IMACS World Congress on Systems Simulation and Scientific Computation, Oslo, 1985. Appeared in: Computer Systems: Perfomance and Simulation, North Holland, 1986.

[Boh 87a] Bohlender, G., Ullrich, Ch., Wolff von Gudenberg, J., Rall, L. B.: *PASCAL-SC: A Computer Language for Scientific Computation.* Perspectives in Computing **17**, Academic Press, Orlando, 1987.

[Boh 87b] Bohlender, G., Teufel, T.: *BAP-SC: A Decimal Floating-Point Processor for Optimal Arithmetic.* Appeared on pp 31–58 in [Kau 87], 1987.

[Boh 89a] Bohlender, G., Miranker, W.L., Wolff von Gudenberg, J.: *Floating-Point Systems for Theorem Proving.* Proceedings of the Computer Assisted Proof Conference, Cincinnati, March 1989; IBM Research Report RC 15101 (# 67356), Yorktown Heights, 1989.

[Boh 89b] Bohlender, G., Kornerup, P., Matula, D.W., Walter, W.: *A Proposal for Exact Floating-Point Operations in Fortran 8x.* Proposal submitted to ANSI Committee X3.9-198x, November 1989.

[Cap 75] Caprani, O.: *Roundoff Errors in Floating-Point Summation.* BIT 15, 5–9, 1975.

[Cap 88] Cappello, P.R., Miranker, W.L.: *Systolic Super Summation.* IEEE Transactions on Computers, 37(6), 657–677, June 1988, and *Systolic Super Summation with Reduced Hardware.* IBM Research Report RC14259 (# 63831), Yorktown Heights, Dec. 1988.

[Dek 71] Dekker, T. J.: *A Floating-Point Technique for Extending the Available Precision.* Numer. Math. **18**, 224–242, 1971.

[Die 85] Dietrich, Th.: *Realisierung der optimalen Arithmetik in FORTRAN 8x.* Diplom thesis, Institut für Angewandte Mathematik, Universität Karlsruhe, 1985.

[Erl 88] Erl, M., Hodgson, G., Kok, J., Winter, D., Zoellner, A.: *Design and Implementation of Accurate Operators in Ada.* ESPRIT-DIAMOND, Del. 1-2/4, 1988.

[Fal 89] Falcó Korn, C., König, S., Gutzwiller, S.: *MODULA-SC, A Precompiler to MODULA-2.* SCAN-89, Basel (Oct. 2-6, 1989), in [Ull 90].

[Haf 89] Hafner, K.: *Chips for High Precision Arithmetic.* SCAN-89, Basel (Oct. 2-6, 1989), in this volume.

[Hahn 88] Hahn, W., Mohr, K.: *APL/PCXA, Erweiterung der IEEE Arithmetik für technisch wissenschaftliches Rechnen.* Hanser Verlag, München, 1989.

[Hahn 89] Hahn, W., Mohr, K.: *APL/PCXA.* SCAN-89, Basel (Oct. 2-6, 1989), in [Ull 90].

[Ham 89] Hammer, R.: *How Reliable is the Arithmetic of Vector Computers?* SCAN-89, Basel (Oct. 2-6, 1989), in [Ull 90].

[Hull 85] Hull, T. E., Abrham, A., Cohen, M. S., Curley, A.F.X., Hall, C.B., Penny, D.A., Sawchuk, J.T.M.: *Numerical Turing.* ACM SIGNUM Newsletter 20, 3, 26–34, July 1985.

[Hull 87] Hull, T.E., Cohen, M.S.: *Toward an Ideal Computer Arithmetic.* Proceedings of the 8th Symposium on Computer Arithmetic, Como, 131–138, IEEE Computer Society, 1987.

[Hus 88] Husung, D.: *ABACUS: Programmierwerkzeug mit hochgenauer Arithmetik für Algorithmen mit verifizierten Ergebnissen.* Diplom thesis, Institut für Angewandte Mathematik, Universität Karlsruhe, 1988.

[Hus 89a] Husung, D., Rump, S.M.: *ABACUS.* SCAN-89, Basel, Oct. 2-6, 1989.

[Hus 89b] Husung, D.: *TPX, Version 1.0.* Public-Domain-Software. TU Hamburg / Informatik III, Eißendorfer Str. 38, D-2100 Hamburg 90, 1989.

[Jül 87] Jüllig, H.–P.: *Untersuchung zur Implementierung von Skalarproduktausdrücken.* Diplom thesis, Institut für Angewandte Mathematik, Universität Karlsruhe, 1987.

[Kah 65] Kahan, W.: *Further Remarks on Reducing Truncation Errors.* Comm. ACM, vol. 8, no. 1, Jan. 1965.

[Kau 84] Kaucher, E., Miranker, W. L.: *Self-Validating Numerics for Function Space Problems.* Academic Press, New York, 1984.

[Kau 87] Kaucher, E., Kulisch, U., Ullrich, Ch. (eds.): *Computerarithmetic: Scientific Computation and Programming Languages.* B. G. Teubner, Stuttgart, 1987.

[Kie 88] Kießling, I., Lowes, M., Paulik, A.: *Genaue Rechnerarithmetik, Intervallrechnung und Programmieren mit PASCAL-SC.* Teubner Verlag, Stuttgart, 1988.

[Kir 87] Kirchner, R., Kulisch, U.: *Arithmetic for Vector Processors.* Proc. of the 8th Symp. on Computer Arithmetic, May 19 - 21, 1987 in Como, IEEE Computer Society, 256–269, 1987.

[Kir 88] Kirchner, R., Kulisch, U.: *Accurate Arithmetic for Vector Processing.* Journal of Parallel and Distributed Computing (Academic Press), vol. 5, no. 3, June 1988.

[Knö 88] Knöfel, A.: *Entwurf eines Arithmetikprozessors für das optimale Skalarprodukt mit Hilfe eines Silicon Compilers.* Diplom thesis, Institut für Angewandte Mathematik / Institut für Rechnerentwurf und Fehlertoleranz, Universität Karlsruhe, 1988.

[Knu 73] Knuth, D. E.: *The Art of Computer Programming, Volume 1/ Fundamental Algorithms.* 2nd ed., Addison-Wesley, Reading, 1973.

[Knu 81] Knuth, D. E.: *The Art of Computer Programming, Volume 2/ Seminumerical Algorithms.* 2nd ed., Addison-Wesley, Reading, 1981.

[Kreu 88] Kreutzenberger, R.: *Formelauswerter mit E–Verifikation für allgemeine arithmetische Ausdrücke und Einbau in einen FORTRAN-SC Compiler.* Diplom thesis, Institut für Angewandte Mathematik, Universität Karlsruhe, 1988.

[Krü 86] Krückeberg, F.: *Arbitrary Accuracy with Variable Precision Arithmetic.* In: Interval Mathematics (Proc. of the Intern. Symposium 1985, ed. by K. Nickel), Lecture Notes in Computer Science **212**, 95–101, 1986.

[Kul 76a] Kulisch, U.: *Grundlagen des numerischen Rechnens: Mathematische Begründung der Rechnerarithmetik.* Reihe Informatik **19**, Bibliographisches Institut, Mannheim, 1976.

[Kul 76b] Kulisch, U., Bohlender, G.: *Formalization and Implementation of Floating-Point Matrix Operations.* Computing 16, 239–261, 1976.

[Kul 81] Kulisch, U., Miranker, W. L.: *Computer Arithmetic in Theory and Practice.* Academic Press, New York, 1981.

[Kul 82] Kulisch, U., Ullrich, Ch. (eds.): *Wissenschaftliches Rechnen und Programmiersprachen.* Berichte des German Chapter of the ACM **10**, B. G. Teubner, Stuttgart, 1982.

[Kul 83a] Kulisch, U., Miranker, W. L. (eds.): *A New Approach to Scientific Computation.* Notes and Reports in Computer Science and Applied Mathematics, Academic Press, Orlando, 1983.

[Kul 83b] Kulisch, U., Bohlender, G.: *Features of a Hardware Implementation of an Optimal Arithmetic.* In: Kulisch/Miranker (eds.): A New Approach to Scientific Computation, 1983.

[Kul 86] Kulisch, U., Miranker, W. L.: *The Arithmetic of the Digital Computer: A New Approach.* SIAM Review **28**, no. 1, 1986.

[Kul 87a] Kulisch, U. (ed.): *PASCAL-SC: A PASCAL Extension for Scientific Computation*; Information Manual and Floppy Disks; Version IBM PC/AT; Operating System DOS. Wiley–Teubner Series in Computer Science, B. G. Teubner, J. Wiley & Sons, 1987.

[Kul 87b] Kulisch, U. (ed.): *PASCAL-SC: A PASCAL Extension for Scientific Computation*; Information Manual and Floppy Disks; Version ATARI ST. B. G. Teubner, Stuttgart, 1987.

[Kul 88] Kulisch, U., Stetter, H. J. (eds.): *Scientific Computation with Automatic Result Verification.* Computing Suppl **6**, Springer, Wien, 1988.

[Kul 89] Kulisch, U. (ed.): *Wissenschaftliches Rechnen mit Ergebnisverifikation, Eine Einführung.* Vieweg-Verlag; Akademie-Verlag, Berlin, 1989.

[Leu 82] Leuprecht, H., Oberaigner, W.: *Parallel Algorithms for the Rounding Exact Summation of Floating Point Numbers.* Computing 28, 89–104, 1982.

[Lich 88] Lichter, P.: *Realisierung eines VLSI-Chips für das Gleitkomma-Skalarprodukt der Kulisch-Arithmetik.* Diplom thesis, Institut für Angewandte Mathematik und Informatik, Universität Saarbrücken, 1988.

[Lin 74] Linnainmaa, S.: *Analysis of Some Known Methods of Improving the Accuracy of Floating-Point Sums.* Bit 14, 167–202, 1974.

[Linz 70] Linz, P.: *Accurate Floating-Point Summation.* Comm. ACM, vol. 13, no. 6, June 1970.

[Lor 71] Lortz, B.: *Eine Langzahlarithmetik mit optimaler einseitiger Rundung.* Ph.D. thesis, Universität Karlsruhe, 1971.

[Mal 71] Malcolm, M.A.: *On Accurate Floating-Point Summation.* Comm. ACM, vol. 14, no. 11, Nov. 1971.

[Mar 89] Markov, S. Velichkov, T.: *HIFICOMP - a Subroutine Library for Highly Efficient and Accurate Computation.* SCAN-89, Basel (Oct. 2-6, 1989), in [Ull 90].

[Metz 88] Metzger, M.: *Weiterentwicklung von Programmiersprachen für die Numerik unter besonderer Berücksichtigung der Rechengenauigkeit.* Ph.D. thesis, Universität Karlsruhe, 1988.

[Møl 65] Møller, O.: *Quasi Double-Precision in Floating-Point Addition.* Bit 5, 37-50, 1965, and *Note on Quasi Double-Precision.* Bit 5, 251–255, 1965.

[Moo 66] Moore, R. E.: *Interval Analysis.* Prentice–Hall, Englewood Cliffs, N. J., 1966.

[Moo 88] Moore, R. E. (ed.): *Reliability in Computing, The Role of Interval Methods in Scientific Computing.* Perspectives in Computing **19**, Academic Press, San Diego, 1988.

[Nea 84] Neaga, M.: *Erweiterungen von Programmiersprachen für wissenschaftliches Rechnen und Erörterung einer Implementierung.* Ph.D. thesis, Universität Kaiserslautern, 1984.

[Pich 72] Pichat, M.: *Correction d'une somme en arithmétique à virgule flottante.* Num. Math. 19, 400–406, 1972.

[Ratz 89] Ratz, D.: *The Effects of the Arithmetic of Vector Computers on Basic Numerical Algorithms.* SCAN-89, Basel (Oct. 2-6, 1989), in [Ull 90].

[Rump 80] Rump, S. M.: *Kleine Fehlerschranken bei Matrixproblemen.* Ph.D. thesis, Universität Karlsruhe, 1980.

[Rump 83a] Rump, S. M.: *Solving Algebraic Problems with High Accuracy.* In [Kul 83], 51–120, 1983.

[Rump 83b] Rump, S. M.: *How Reliable are Results of Computers?.* Jahrbuch Überblicke Mathematik 1983, Bibliographisches Institut, Mannheim, 163–168, 1983.

[Schu 89] Schumacher, G.: *Genauigkeitsfragen bei algebraisch-numerischen Algorithmen auf Skalar- und Vektorrechnern.* Ph.D. thesis, Universität Karlsruhe, 1989.

[Sue 86] Suess, W.: *Das genaue Skalarprodukt im IEEE-Zahlformat.* Diplom thesis, Institut für Angewandte Mathematik, Universität Karlsruhe, 1986.

[Teu 84] Teufel, T.: *Ein optimaler Gleitkommaprozessor.* Ph.D. thesis, Universität Karlsruhe, 1984.

[Teu 86] Teufel, T., Bohlender, G.: *A Bit-Slice Processor Unit for Optimal Arithmetic.* Proceedings of the 11th IMACS World Congress on Systems Simulation and Scientific Computation, Oslo, 1985. Appeared in: Computer Systems: Performance and Simulation, North Holland, 1986.

[Ull 90] Ullrich, Ch. (ed.): *Contributions to Computer Arithmetic and Self-Validating Numerical Methods.* Special issue of IMACS Journal on Math. and Computers in Simulation, to appear 1990.

[Vel 89] Velichkov, T., Cohen, R., Stoyanov, P.: *Basic Computer Arithmetic Operations.* SCAN-89, Basel (Oct. 2-6, 1989), in [Ull 90].

[Win 85] Winter, T.: *Ein VLSI-Chip für Gleitkomma-Skalarprodukt mit maximaler Genauigkeit.* Diplom thesis, Institut für Angewandte Mathematik und Informatik, Universität Saarbrücken, 1985.

[Win 88] Winkler, S.: *Auswertung von Programmteilen mit maximaler Genauigkeit.* Diplom thesis, Institut für Angewandte Mathematik, Universität Karlsruhe, 1988.

[Wol 64] Wolfe, J.M.: *Reducing Truncation Errors by Programming.* Comm. ACM, vol. 7, no. 6, June 1964.

[Wol 88] Wolff von Gudenberg, J.: *Arithmetische und programmiersprachliche Werkzeuge für die Numerik.* Habilitation thesis, Universität Karlsruhe, 1988.

[Yil 89] Yilmaz, T., Theeuwen, J.F.M., Tangelder, R.J.W.T., Jess, J.A.G.: *The Design of a Chip for Scientific Computation.* Euro Asic, Grenoble, Jan. 25 – 27, 1989.

[IBM 84] IBM *System/370 RPQ, High-Accuracy Arithmetic.* SA 22-7093-0, IBM Corporation, 1984.

[IBM 86a] IBM *High-Accuracy Arithmetic Subroutine Library (ACRITH)*, General Information Manual. 3rd ed., GC 33-6163-02, IBM Corporation, 1986.

[IBM 86b] IBM *Verfahren und Schaltungsanordnung zur Addition von Gleitkommazahlen.* Europäische Patentanmeldung, EP 0 265 555 A1, 1986.

[IEEE 85] American National Standards Institute / Institute of Electrical and Electronics Engineers: *IEEE Standard for Binary Floating-Point Arithmetic.* ANSI/IEEE Std 754–1985, New York, 1985.

[IEEE 87] American National Standards Institute / Institute of Electrical and Electronics Engineers: *IEEE Standard for Radix-Independent Floating-Point Arithmetic.* ANSI/IEEE Std 854–1987, New York, 1987.

[IGR 89] IMACS, GAMM: *Resolution on Computer Arithmetic.* Mathematics and Computers in Simulation 31, 297–298, 1989.

[SIE 86] SIEMENS *ARITHMOS, (BS 2000) Benutzerhandbuch.* SIEMENS AG, U2900–J–Z87–1, 1986.

Chips for High Precision Arithmetic

K. Hafner
Siemens AG
Corporate Research and Technology
Otto-Hahn-Ring 6
D-8000 München 83, FRG

Abstract: The efficient implementation of a computer arithmetic including a maximum accurate scalar product requires specialized hardware. Depending on the implementation constraints very different types of architecture turn out to be adequate. For a typical scenario an appropriate architecture is derived. The kernel of this architecture consists of an application specific VLSI chip which is presented in some detail.

1. Introduction

The accumulation of rounding errors in scientific computations using floating-point numbers can falsify the results entirely. However, rounding errors even in complex arithmetical computations using addition, subtraction and multiplication (e.g. scalar products) can be completely avoided by converting the original floating point numbers into fixed-point numbers of sufficient length (usually called *accu numbers*) and by performing the computations in that accu number format. Maximum accurate results can thus be obtained if the conversion back into the original floating-point format is performed at the end of such a sequence of arithmetical operations only. In that way maximum accurate results for higher arithmetical operations (e.g. sums and scalar products) are achievable [12].

The format of the accus must be long enough that all possible floating-point numbers of a chosen format can be represented exactly. Depending on the particular floating-point format an accu of some thousand bit length may

ISBN 0-12-708245-X

thus be necessary. The elementary arithmetical operations on the accus and the conversions between the formats can easily be executed by software but with respect to performance these computations should be supported by specialized hardware.

Within the ESPRIT project DIAMOND we have studied the existing proposals for hardware support of an arithmetic with maximum accurate scalar product ([8], [9], [10], [11], [14]). In addition we have evaluated them with respect to VLSI realizations. Since none of the published proposals has satisfied our requests we have developed our own concept. Our work includes the problem definition, the development of an architectural concept, and the evaluation of the architecture with respect to feasibility, performance, and design effort.

2. Exploration of the Design Space

2.1. Design Goals

The hardware implementation of an arithmetic with maximum accurate scalar product has to fit in with various design goals. These goals concern – among others – the interfaces to the hardware and software environment, the realization of the required functionality, and the adaption to the applications which are to run on the new system. In our context there are the following alternatives concerning those *design goals*.

Regarding the *interface to the host system* one could integrate the hardware extension as an autonomous coprocessor or as a dumb peripheral, and one could implement a tight or a loose coupling to the target system. In addition the communication between hardware extension and host could be accomplished by rather different operations.

Concerning *functionality* one could execute just elementary operations on accu fragments, or complete accu operations, or even higher arithmetical routines. One could implement various instruction sets, and one could support different data types and a different number of accus ranging from just one to any quantity whatever.

With respect to *applications* one could optimize the design for quite distinct workloads, e.g. small or large sums and scalar products respectively, diverse frequencies of the various operations, and different assumptions on the input data.

2.2. Realization Constraints

The architectural design space is restricted by some *realization constraints*. These constraints mainly concern the splitting of the required functionality into manageable subfunctions and especially the associated partitioning of the system into components.

To cope with the design complexity the specified behavior must be mapped into several distinct modules. Costs are reduced if these modules can be composed of standard components. In addition the design methodology and the design system have to be adjusted to the technical requirements and to the expected production volume.

In demanding projects it may even be necessary to develop chips which are specially tailored to the application at hand. In that case there are some general considerations which have to be taken into account:

- a large number of pins is very expensive and there is a strict limit in the absolute pin count (about 200 pins),
- there is a limitation to the maximum complexity achievable on one chip,
- broad communication paths are feasible on-chip but not across the chip boundary,
- contrary to on-chip information distribution there are large delays of signals passing the chip boundary,

- on-chip memory accesses are an order of magnitude faster than off-chip.

To sum up, depending on the present design goals and realization constraints very different types of architectures turn out to be adequate. The guiding design criterion is a balanced trade-off between chip areas allocated to execution functions and to storage functions.

2.3. Feasibility

In many cases the highest performance – and often the highest costs too – are achieved if all required functions are implemented in one single chip. A first glance at our problem will already clarify the feasibility of that approach and will in addition yield a crude performance estimate.

For that purpose let us focus on one of the most typical accu operations, the multiplication of two long floating-point numbers and the accumulation of the result in the accu. If we denote by k the length of the floating-point mantissa, then a straightforward implementation would require approximately $2\,k^2$ single-bit operations (additions or shifts). If we assume furthermore that every single-bit operation requires 10 elementary gates for its hardware realization then we arrive at 5×10^4 switching operations (SwOps) per accu instruction with long floating-point operands ($k = 54$).

Let us on the other side determine what is achievable on one chip. Today's complexity limit for application specific chips is around 10^5 gates. The switching time for a gate is about 10^{-9} seconds. This results in an upper limit of 10^{14} SwOps/s.

That performance rate however is reduced by a number of reasons. To begin with, information distribution (i.e. signal delays) absorb an order of magnitude. Another order of magnitude is lost by architecture (e.g. modularity, reusability of components, and standardized interfaces) and a further one by the design methodology used (e.g. semi custom instead of full custom design). Finally, an application specific chip with low production volume is manufactured normally with a matured technology only. This is in contrast to system houses which make use of the latest technological evolution in the design of their most advanced standard components. Accordingly the maximum complexity and the minimum switching times in our case have to be downgraded by some factor. To sum up, 10^{10}–10^{11} SwOps/s is the maximum complexity we can achieve today on one chip.

Consequently a *maximum performance* of 10^5–10^6 accu operations per second is obtainable today. That rate is in accordance with corresponding values achieved in standard floating-point units. Due to the higher complexity of arithmetical operations with maximum accuracy in contrast to ordinary arithmetic this is a rather demanding aim which lies, nevertheless, within the bounds of possibility.

3. Architecture of the ARITHMOS Processor

Guided by the general considerations outlined in the last chapter we will now explain our architectural concept in some detail. It is adapted to the following requirements and constraints:

- hardware support for the basic routines of ARITHMOS, the software implementation of a maximum accurate arithmetic from Siemens [1],
- support of the IEEE floating-point formats single and double [5],
- suitable for present VLSI realizations,
- easy to incorporate into various PCs and workstations,
- low design and production costs,
- support of a multi-tasking and multi-user environment,
- support of multiple accus per task,
- performance maximization for typical workloads.

3.1. System Integration

A reasonable solution to the problem of enhancing the arithmetical functions of a computer is to integrate an additional coprocessor. There are a few alternatives for doing this (see figure 1).

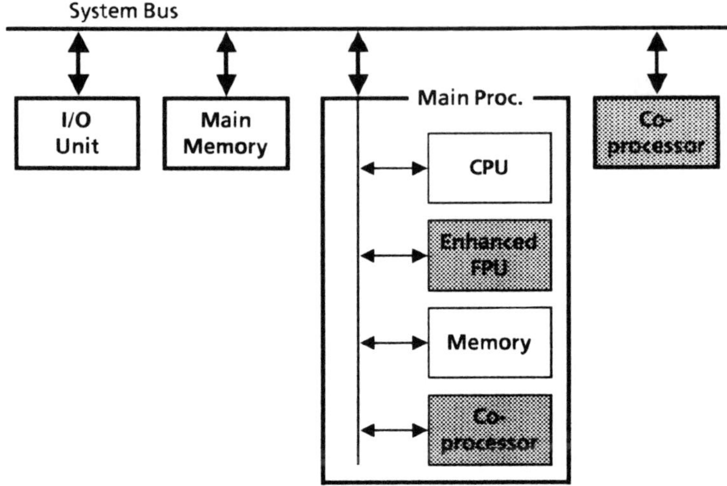

Figure 1: Alternatives for the System Integration of Additional Functions

(1) We could realize it as an *enhanced floating-point unit* which is pin-compatible to the original one. This approach to system integration is followed e.g. in the new Cyrix FasMath 83D87 numerics coprocessor [2], which is pin and software compatible to the Intel 80387 FPU. Because of the tight coupling to the main CPU this solution yields high performance. But there is a rather high price to pay for it: The design effort is rather high and its usage is restricted to one CPU family only.

(2) A related solution uses the coprocessor extension facilities of existing CPUs with which *additional user-defined coprocessors* may be connected to the main processor by the same tightly coupling protocol. Because there is no need to replicate the ordinary floating-point instructions which

are already realized in the floating-point unit the design is much simpler than in solution (1). Nevertheless it has the same serious drawback as being restricted to one CPU family only.

(3) A third alternative is the realization as an additional *coprocessor* which is *loosely coupled* to the main processor via the system bus. This solution frees us from being restricted to one CPU family. It is adaptable to all systems which use a certain system bus. By a proper choice of this bus system it will fit into various PCs and workstations. A modification of the driver software will suffice for the integration into a related system. In addition it is much easier to integrate further powerful functions and there are more possibilities for the realization than there are in a one-chip solution. In view of those benefits we favor the realization of a maximum accurate arithmetic as a loosely coupled coprocessor.

3.2. Programming Model

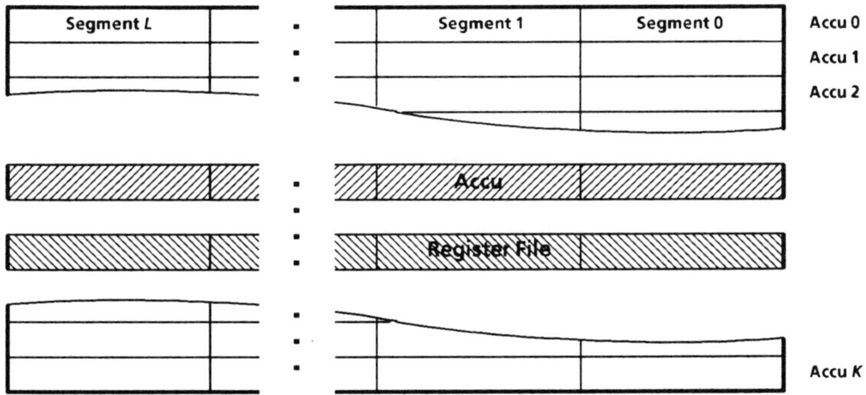

Figure 2: Logical Organization of the *Segment Memory*

The most important aspect of the programming model is the logical organization of the accus. We have decided to support several accus and to divide the accus into segments of equal length. The corresponding data are stored in the *segment memory* (see figure 2) which is organized in rows and columns. Every accu number makes use of one row of the segment memory. To support ordinary floating-point operations and operations with integers as well, arbitrary rows of the segment memory may be used as *register files* for storing floating-point and integer numbers instead of using them as accus.

To raise performance for typical workloads, we in addition have decided to take care of the accu's *living domain*. The living domain of an accu consists of its significant part without trailing zeros and without leading zeros respectively signs (depending on the number representation). We therefore introduce two pointers. The first one, *LB*, points to the left boundary of the living domain, i.e. it contains the index of the most significant accu segment. The second one, *RB*, points to the right boundary of the living domain, i.e. it contains the index of the least significant accu segment. The *administration data* of an accu consist of the accu number, *LB*, *RB*, zero flag *Z* and its sign *N*. The data of an accu number are stored in segment 1 .. *L* of one row, the corresponding administration data are stored in segment 0 of the same row.

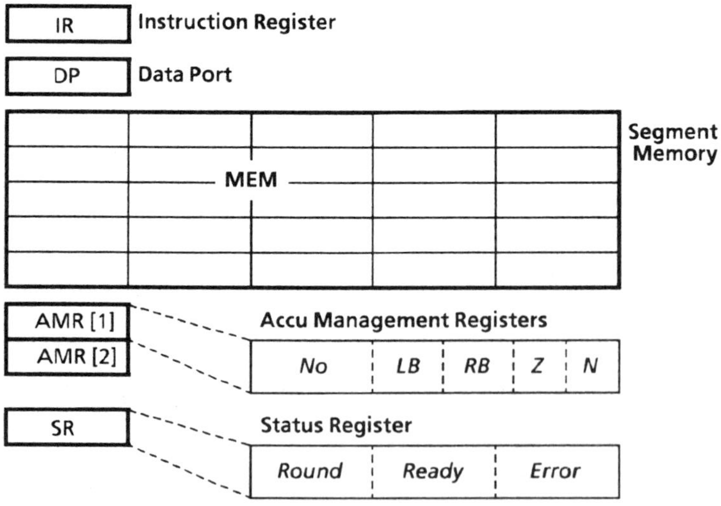

Figure 3: Resources of the Programming Model

We now give a more detailed description of the resources and instructions of our programming model in pseudo Pascal notation (see also figure 3). For sake of simplicity, we only consider the case of a segment length which is compatible with the floating-point format used (e.g. 64 bit segments fit to long IEEE floating-point numbers). In addition we assume that the data types *AccuSegmentType*, *FloatingPointType*, and *IntegerType* are predefined.

```
RoundType : (ToNearest, Upwards, Downwards, ToZero, Outwards);

StatusType :     RECORD
                     Round :     RoundType ;
                     Ready :     BOOLEAN ;
                     Error :     BOOLEAN ;
                     END (* Record *) ;

AMRType :        ARRAY [1 .. 2] OF
                     RECORD
                         No :  [0 .. K] ;
                         LB :  [0 .. L] ;
                         RB :  [0 .. L] ;
                         Z :   BOOLEAN ;
                         N :   BOOLEAN ;
                         END (* Record *) ;

SegmentType :    RECORD
                     CASE SegmentKind OF
                         A :   AccuSegmentType ;
                         F :   FloatingPointType ;
                         I :   IntegerType ;
                         M :   AMRType ;
                         END (* Case *)
                     END (* Record *) ;

AccuType :       ARRAY [0 .. L] OF SegmentType ;
DP :             SegmentType ;
SR :             StatusType ;
AMR :            ARRAY [1 .. 2] OF AMRType ;
MEM :            ARRAY [0 .. K] OF AccuType ;
```

Next we describe the instruction set of the ARITHMOS processor. There are three groups of instructions. Besides the ordinary arithmetical operations on floating-point numbers there are arithmetical operations on accus and data movement instructions. Almost all instructions use indirect addressing of operands. In ordinary floating-point instructions the locations in the segment memory are referenced via accu management register *AMR* [1]. In the accu instructions there is a working accu which is implicitly referenced via accu management register *AMR* [1] while the other input operands (floating-point numbers or a second accu) are referenced via *AMR* [2]. Note, that the operators used in the following description depend on the data types of the operands. The instructions are defined without regard to data types of operands, therefore other mnemonics are used than in ARITHMOS.

- IEEE floating-point operations [5]:

ADD (*R*1, *R*2, *R*3) : Addition of two floating-point numbers with rounding of the result,
MEM [*AMR* [1].*No*, *R*1] : = MEM [*AMR* [1].*No*, *R*2] + MEM [*AMR* [1].*No*, *R*3] .

SUB (*R*1, *R*2, *R*3) : Subtraction of two floating-point numbers with rounding of the result,
MEM [*AMR* [1].*No*, *R*1] : = *MEM* [*AMR* [1].*No*, *R*2] – *MEM* [*AMR* [1].*No*, *R*3] .

MUL (*R*1, *R*2, *R*3) : Multiplication of two floating-point numbers with rounding of the result,
MEM [*AMR* [1].*No*, *R*1] : = *MEM* [*AMR* [1].*No*, *R*2] * *MEM* [*AMR* [1].*No*, *R*3] .

DIV (*R*1, *R*2, *R*3) : Division of two floating-point numbers with rounding of the result,
MEM [*AMR* [1].*No*, *R*1] : = *MEM* [*AMR* [1].*No*, *R*2] / *MEM* [*AMR* [1].*No*, *R*3] .

- Accu operations [1]:

AFADD (*R*) : Addition of a floating-point number to the accu,
MEM [*AMR* [1].*No*] : = *MEM* [*AMR* [1].*No*] + *MEM* [*AMR* [2].*No*, *R*].

AFSUB (*R*) : Subtraction of a floating-point number from the accu,
MEM [*AMR* [1].*No*] : = *MEM* [*AMR* [1].*No*] – *MEM* [*AMR* [2].*No*, *R*].

AFRND (*R*) : Round accu to floating-point number,
MEM [*AMR* [2].*No*, *R*] : = *MEM* [*AMR* [1].*No*].

AFMAC (*R*1, *R*2) : Multiplication of two floating-point numbers and accumulation of the result to the accu,
MEM [*AMR* [1].*No*] : = *MEM* [*AMR* [1].*No*] +
MEM [*AMR* [2].*No*, *R*1] * *MEM* [*AMR* [2].*No*, *R*2].

AAADD : Addition of two accus,
MEM [*AMR* [1].*No*] : = *MEM* [*AMR* [1].*No*] + *MEM* [*AMR* [2].*No*].

AASUB : Subtraction of two accus,
MEM [*AMR* [1].*No*] : = *MEM* [*AMR* [1].*No*] – *MEM* [*AMR* [2].*No*].

CLAC (*N*) : Initialization of an accu,
AMR [1].*Z* : = TRUE;
AMR [1].*N* : = FALSE;
AMR [1].*No* : = *N*.

AIMUL (*R*) : Multiplication of an accu with an integer number,
MEM [*AMR* [1].*No*] : = *MEM* [*AMR* [1].*No*] * *MEM* [*AMR* [2].*No*, *R*].

- Data Transport Instructions:

MOV (*R*1, *N*1, *R*2, *N*2) : Movement of a segment in segment memory. By using reserved addresses this instruction also includes I/O operations on data in segment memory and load and store operations on the status register *SR* and on the accu management registers *AMR*.
MEM [*AMR* [*N*1].*No*, *R*1] : = *MEM* [*AMR* [*N*2].*No*, *R*2].

SWAMR : Swap the two accu management registers,
dummy := *AMR* [1] ;
AMR [1] := *AMR* [2] ;
AMR [2] := *dummy*.

3.3. Block Architecture

The *ARITHMOS processor* is supposed to serve as a coprocessor in existing computers. The partitioning into the main modules according to the guidelines explained in chapter 2 is illustrated in figure 4. The ARITHMOS processor is designed as a plug-in board and is coupled to the target system via a *bus interface* module. By changing this module even couplings to different bus systems and hence various hosts are possible.

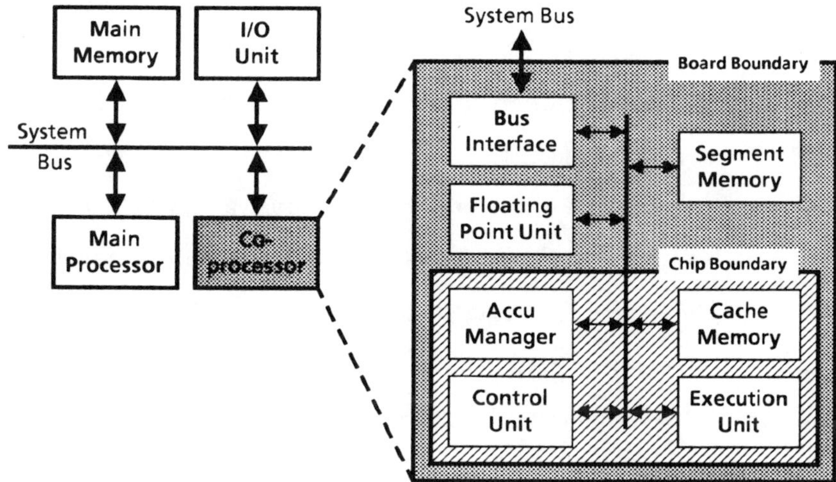

Figure 4: Partitioning of the ARITHMOS Processor

The core of the ARITHMOS processor is an application specific chip (*ARITHMOS chip*) which is surrounded by several standard components. The functions integrated into that chip execute arithmetical operations on accu segments with high performance.

The accus, the corresponding accu management data, and additional registers for floating point numbers are stored in the external *segment memory* which is built up by standard components on board level. If the segment memory is mapped physically into the address space of the host system easy transfer of data is possible between host and coprocessor. If there is not enough space in the segment memory then the host system can be called to store some parts of it in the host's main or even secondary memory. The optimum physical size of the segment memory strongly depends on the expected workloads. If there is only one user of the ARITHMOS processor which needs just a few accus then there is obviously no need for a large segment memory. But if many accus are used at a time and if even several users share the access to the coprocessor then much more memory would be needed.

Substantial performance improvements can be achieved if the living domain of the actual accu is stored in a local memory on the ARITHMOS chip. We have decided to organize this on-chip segment memory as a fully associative *cache* [7]. The first advantage of a cache memory is that the mapping between the logical addresses and the corresponding physical memory locations is quickly done by specialized hardware and in addition is fully transparent to the software. The second advantage is that the locally stored data set is automatically and dynamically adjusted to the actual requirements. This is especially useful if several accus are used at the same time.

Our cache is a content addressable memory (cf. figure 5). Each of its entries can hold one *data segment* and the corresponding *address* consisting of accu and segment number. If there is a request for a certain segment then its

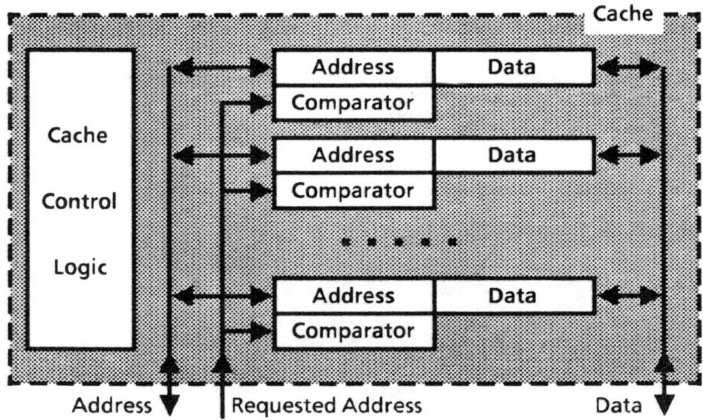

Figure 5: Cache Architecture

address is simultaneously compared with the addresses of all segments actually stored in the cache. If a matching entry is found (cache hit) then the corresponding data field is read out respectively written in.

If there is no match (cache miss) then the segment data are fetched from the off-chip segment memory. The fetched data segment can additionally be stored in the cache to get faster access in the future. If the cache is full and a cache miss occurs then some data has to be removed from the cache. We have studied a least recently used (LRU) strategy as a replacement algorithm. If a data segment is modified and written back to the cache, then there is the question whether the corresponding segment in the off-chip segment memory should correspondingly be updated (write-through cache) or not (store-in cache). We have decided to support both approaches, the corresponding cache control flags can be placed in the status register or accu management segment.

There is a reasonable choice of the granularity of the segment cache. Compatibility with the IEEE floating-point and with the standard integer formats, and the request of limiting the overhead arising from the associative parts (address fields and comparators) lead to a segment length of 32 or 64 bit. The optimum cache size depends on the actual workload. Our simulations show that for reasonable assumptions surprisingly few segments (5–20) suffice already.

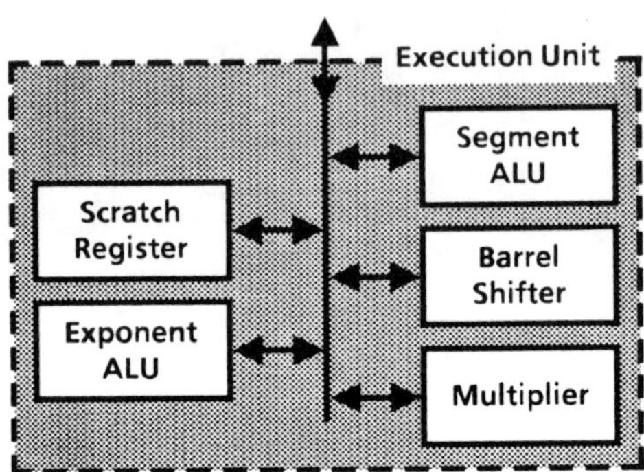

Figure 6: Architecture of the *Execution* Unit

The *execution unit* of the ARITHMOS chip (cf. figure 6) performs the elementary arithmetical and logical operations (e.g. addition, subtraction, shifts, multiplication) on the data segments. Let us look for example at the AFMAC instruction, in which two floating-point numbers are multiplied and added to the working accu.

Here, the two floating-point numbers are fetched and decomposed into sign, exponent and mantissa. The different parts are stored in dedicated scratch registers. The exponent of the product is built by adding the two exponent parts in the *exponent ALU*. Concurrently the product of the two mantissa is computed in the *multiplier*. This multiplication has to be divided into several steps if the multiplier width is smaller than the mantissa length. The resulting product is now adjusted to the segment boundaries by a *barrel shifter*. The exponent is corrected by the *exponent ALU* accordingly. The segments of the intermediate result are stored in the *scratch register file*. They can now be added to the working accu. The resulting accu segments and the corresponding accu flags N and Z are computed in the *segment ALU*.

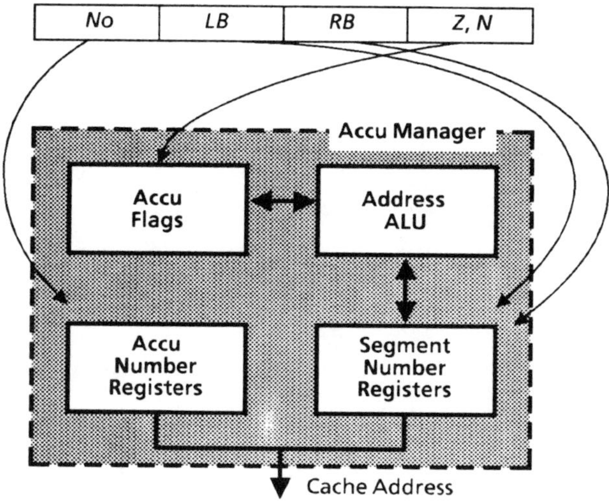

Figure 7: Architecture of the *Accu Manager*

The main purpose of the *accu manager* (cf. figure 7) is address calculation. The accu manager contains the accu management registers and a simple *address ALU*. Let us again consider the AFMAC instruction, in which two floating-point numbers are multiplied and added to the working accu. The number *No* and the pointers *LB* and *RB* to the boundaries of the living domain of the working accu are stored in the different fields of the accu management register. If the product of the two floating-point numbers has been computed and adjusted to the segment boundaries the leading part of the resulting exponent determines the least significant segment of the working accu that has to be treated. This segment number is additionally stored in a dedicated *segment number register*. The accu number and the content of this segment number register build the address field for cache and segment memory accesses. If the actual accu segment has been updated, the segment number register is incremented and the computation of the next segment is started.

This process is continued until the addition of the mantissa product to the working accu has been completed. Finally, *LB*, *RB*, and the accu flags *N* and *Z* are corrected.

The main control functions of the ARITHMOS processor can be implemented by a microprogrammed *control unit*. The instructions of the ARITHMOS processor are mapped into appropriate entries of the microprogram. The micro operations in turn control the execution of the segment and accu management operations. There are some additional control functions which control the cache memory, the corresponding accesses to the segment memory, and in addition the interface to the host system.

To reduce design effort the conventional floating-point operations (from the first instruction group) are executed by a standard *floating-point unit*, which is added on the board. Therefore the ARITHMOS chip additionally contains dedicated control functions which initialize and synchronize the operating of this external floating-point unit.

4. Architecture Evaluation

The architect and designer of a complex hardware system has to find a trade-off between functionality, universality, flexibility, and performance on the one hand and costs (i.e. design effort, ultimate production costs) on the other. This can be achieved by consulting design experience and by evaluating the various design alternatives.

The primary design decisions are in most cases based on experience. In that way we have found adequate solutions for the system integration, the system partitioning, and the definition of the programming model. Moreover the choice of our memory model results from empirical knowledge.

More precise determination of the expected performance, detailed design decisions, and proper architecture tuning however require powerful methods for the evaluation of architectures. The usual approach which we have adopted too, is based on modeling and simulation of different solutions.

Our estimation of the expected performance is based on a high-level simulation model of the ARITHMOS processor we have developed in the programming language Modula-2. The model is structured conforming to the top level block architecture. It is however independent of implementation details such as the number representation system or the choice of adder and multiplier modules.

To study the consequences of more detailed design decisions (e.g. implementation of the multiplier) and for advanced architecture optimization (e.g. pipelining or increased parallelism) refined models of the architecture would be necessary.

Our main focus lies on the optimization of the memory system. The characterizing parameters are the *living domain* of the accus, the number of segments transferred between the execution unit and the on-chip cache and off-chip segment memory respectively (*segment transfer rate*), and the *cache hit rate*. Our simulation model of the ARITHMOS processor allows the extraction of these parameters. In consequence of this evaluation the architectural parameters of the memory system (cache strategy, cache size, and cache granularity) have been adjusted to achieve maximum performance.

Various benchmarks have been executed to extract relevant values of the parameters of the memory system. As we have not succeeded in getting the profile of an actual ARITHMOS user we have constructed parametrizeable benchmarks based on hypothetical user profiles. Our benchmarks mainly contain the accurate summation

$$S := x_1 + x_2 + \ldots + x_n$$

and the accurate scalar product

$$S := x \times y = x_1 * y_1 + x_2 * y_2 + \ldots + x_n * y_n$$

for various $n \in \mathbb{N}$ and different sets of randomly chosen floating-point numbers x_i and y_i. The numbers x_i and y_i are equally, normally, or logarithmically distributed on the intervals $[10^{-37}, 10^{+38}]$ (spectrum of short IEEE floating-point numbers) or $[10^{-307}, 10^{+308}]$ (spectrum of long IEEE floating-point numbers). The benchmarks make use of the instructions AFADD, AFMAC, MOV, AFRND, and CLAC in a realistic mix depending on $n \in \mathbb{N}$. All other instructions are considered to be of secondary importance regarding performance evaluation.

The first two rows of table 1 contain the results for the summation of short ($a := a + s$) and long numbers ($a := a + d$), logarithmically distributed on the interval $[10^{-37}, 10^{+38}]$. The last two rows show the corresponding parameters for the scalar product with short and long floating-point numbers. We can see that even with rather small caches (e.g. 8 segments of 64 bit length) excellent cache hit rates are obtained.

Accu Operation	Average Transfer Rate	Maximum Number of Living Segments	Cache Miss Rate for different cache sizes		
			4	8	12
$a := a + s$	4.7	2	2%	<1%	<1%
$a := a + d$	4.9	5	12%	<1%	<1%
$a := a + s_1 \times s_2$	5.9	7	17%	4%	<1%
$a := a + d_1 \times d_2$	7.4	8	38%	6%	<1%

Table 1: Simulated Cache Behavior
($n = 1000$ floating-point numbers randomly distributed on the interval $[10^{-37}, 10^{+38}]$; s : single format floating-point number; d : double format floating-point number)

Table 2 summarizes the resulting average performance based on reasonable assumptions on the execution times of the different modules involved. With an area efficient but rather slow Booth multiplier (see [3], [4]) requiring in the average $k/2$ cycles for the multiplication of two k-bit numbers an elementary scalar product instruction can be computed in 33 cycles for short floating-point numbers and 52 cycles for long numbers respectively.

Accu Operation	Cycles	Execution Time @ 15 MHz [μs]
$a: = a + s$	13	0.9
$a: = a + d$	14	0.9
$a: = a + s_1 \times s_2$	33	2.2
$a: = a + d_1 \times d_2$	52	3.5

Table 2: Performance of the Accu Operations
(s : single format floating-point number; d : double format floating-point number)

A comparison with standard floating-point coprocessors executing ordinary IEEE arithmetic (see [6], [13]) on the basis of these performance estimations is contained in table 3. The resulting values include the loading of two long floating-point numbers, their multiplication, and the addition of the product to an accumulating variable. The performance of the ARITHMOS processor is comparable with the one for standard FPUs. Note however that in contrast to standard floating-point units our processor computes the scalar product with maximum accuracy.

Processor	Cycles			Clock [MHz]	Time for one Scalar Product Operation [μs]
	Add.	Mult.	Scalp.		
MC 68882 ①	75	95	139	25	5.6
Intel 80387 ①	29–37	32–57	80–114	25	3.2–4.6
ARITHMOS Processor ②	14	52	52	15	3.5

Table 3: Performance Comparison with Monolithic Floating-Point Units
(Values include loading of operands; all floating-point operands are in double format; ① ordinary floating-point operation; ② maximum accurate accu operation)

The simulation results and some additional realization studies show the possibility of realizing our concept with affordable effort. Exploratory design studies using our in-house standard cell design system and a 1.5 µm CMOS-technology furthermore anticipate a chip core area of 60 mm^2 and a clock rate of at least 15 MHz. The predicted clock rate results in an overall performance of 450 000 maximum accurate scalar product operations per second with single floating-point operands and of 300 000 operations per second with double format operands. Or stated in conventional terms, we could achieve 0.6 MFlops double precision and 0.9 MFlops single precision performance.

5. Conclusions

In this paper we describe a specialized hardware executing a maximum accurate arithmetic and designated to be integrated into standard workstations. Two main restrictions exist for such a system: It should be highly flexible to be easily incorporated into various computers and its design and production costs should be as low as possible.

On the basis of these (and some additional) goals an appropriate architecture has been developed. Its main characteristics are the usage of standard components and one application-specific VLSI chip.

The architecture has been developed and evaluated by simulations based on hypothetical user profiles. The results obtained clearly demonstrate the efficiency of the cache concept. Due to the usage of a well tailored special chip the data throughput for the accu operations is comparable to the one for conventional floating point operations on ordinary monolithic floating point units. Even better performance should be possible if the architecture is further tuned by introducing increased parallelism, especially pipelining.

Acknowledgement

This work has been carried out by M. Deppermann, S. Wallstab, and the author at the Siemens R&D Labs. It has been supported in part by the ESPRIT research program (project number 1072-DIAMOND). However the author alone is responsible for the content.

References

[1] *ARITHMOS (BS 2000) – Benutzerhandbuch*, Siemens AG, München 1986.

[2] T. Brightman: *Advancing the standard in floating-point performance*, High Performance Systems, Nov. 1989, pp. 59–64.

[3] J. Cavanagh: *Digital Computer Arithmetic – Design and Implementation*, McGraw-Hill, New York 1985.

[4] K. Hwang: *Computer Arithmetic – Principles, Architecture, and Design*, Wiley, New York 1979.

[5] *IEEE Standard for Binary Floating-Point Arithmetic, ANSI/IEEE Std 754-1985*, IEEE, New York 1985.

[6] *Intel Microprocessor and Peripheral Handbook – Volume I*, Intel Corporation, Santa Clara 1988.

[7] T. Kohonen: *Content-Addressable Memories*,Springer, Berlin 1980.

[8] U. Kulisch: *Schaltungsanordnung und Verfahren zur Bildung von Skalarprodukten und Summen von Gleitkommazahlen mit maximaler Genauigkeit*, Offenlegungsschrift DE 3144015 A1, Deutsches Patentamt, 5.11.1981.

[9] U. Kulisch, R. Kirchner: *Schaltungsanordnung zur Bildung von Summen, insbes. von Skalarprodukten*, Patentanmeldung P 36 26 353.2, Deutsches Patentamt, 1.8.1986.

[10] U. Kulisch, R. Kirchner: *Schaltungsanordnung zur Bildung von Summen,* Patentanmeldung P 36 30 625.5, Deutsches Patentamt, 8.9.1986.

[11] U. Kulisch, R. Kirchner: *Schaltungsanordnung zur Bildung von Summen, insbes. von Skalarprodukten,* Patentanmeldung P 37 03 440.5, Deutsches Patentamt, 4.2.1987.

[12] U. Kulisch, W.L. Miranker: *Computer Arithmetic in Theory and Practice,* Acad. Press, New York 1981.

[13] *MC 68881/MC68882 Floating-Point Coprocessor – User's Manual,* Prentice Hall, Englewood Cliffs, N.J. 1987.

[14] T. Teufel: *Ein optimaler Gleitkommaprozessor,*Dissertation an der Universität Karlsruhe, Karlsruhe 1984.

Enclosure Methods

G. Alefeld
Institut für Angewandte Mathematik
Universität Karlsruhe
Federal Republic of Germany

Abstract: We present an overview on existing methods for including the range of functions by interval arithmetic tools.

1. Introduction

In this paper we do not try to give a precise definition of what we mean by an enclosure method. Instead we first recall that the four basic interval operations allow to include the range of values of rational functions. Using more appropriate tools also the range of more general functions can be included. Since all enclosure methods for the solution of equations which are based on interval arithmetic tools are finally enclosure methods for the range of some function we concentrate ourselves on methods for the inclusion of the range of functions. We limit our discussion to the case of functions of one real variable. Most of the material in the present paper is well known. See [1], Chapter 3 and the books [9], [10] by Ratschek and Rokne, for example. However, there are also some new results. See Theorem 2, for example.

Computer Arithmetic and Self-Validating
Numerical Methods

ISBN 0-12-708245-X

2. Notation

The notation in this paper is essentially the same as in [1]. We repeat the most important ones in order that the non-specialist can also read this paper. Real intervals $[a_1;a_2]$, $[b_1;b_2]$,... are denoted by $[a]$, $[b]$,... . The four arithmetic operations for intervals are defined by

$$[a]*[b] = \{a*b \mid a \in [a] , b \in [b]\} , \quad * \in \{+,-,\times,/\} .$$

The result is an interval whose bounds can be computed from the bounds of $[a]$ and $[b]$. Similarly, vectors with intervals as components, so-called interval vectors, are denoted by $[a] = ([a]_i)$, $[b] = ([b]_i)$,... . Analogously, $[A] = ([a]_{ij})$ denotes an interval matrix. $m[a]$ is the center of $[a]$, $d[a] = a_2-a_1$ is the diameter of $[a]$, $q([a],[b]) = \max\{|a_1-b_1| , |a_2-b_2|\}$ denotes the distance of $[a]$ and $[b]$ and $q([a],0) = |[a]|$ is called the absolute value of $[a]$. These concepts are defined for interval vectors and interval matrices via the components. For additional details we refer to [1].

3. Interval arithmetic evaluation

The four basic operations for intervals are inclusion monotone:

If $[a] \subseteq [c]$, $[b] \subseteq [d]$ then $[a]*[b] \subseteq [c]*[d]$, $* \in \{+,-,\times,/\}$.

From this it follows that for rational functions (and more generally for all functions f which have an interval arithmetic evaluation (see [1], Chapter 3)) the range $R(f;[x])$ of f over the interval $[x]$ is contained in the interval arithmetic evaluation $f([x])$:

(1) $R(f;[x]) \subseteq f([x])$.

Example 1. Let

$$f(x) = \frac{x}{1-x} , \quad x \neq 1$$

and $[x] = [2\,;\,3]$. Then

$$R(f;[x]) = [-2\,;\,-\tfrac{3}{2}] ,$$

$$f([x]) = \frac{[x]}{1-[x]} = \frac{[2;3]}{1-[2;3]} = [-3\,;\,-1] ,$$

and therefore $R(f;[x]) \subset f([x])$ holds.

For $x \neq 0$ we can rewrite $f(x)$ as

$$f(x) = \frac{x}{1-x} = \frac{1}{\frac{1}{x} - 1} , \quad x \neq 0 .$$

For the interval arithmetic evaluation of this function over $[2;3]$ we get

$$\tilde{f}([x]) = \frac{1}{\frac{1}{[2;3]} - 1} = [-2\,;\,-\tfrac{3}{2}] = R(f;[x]) .$$

□

The preceding example shows that the overestimation of $R(f;[x])$ by $f([x])$ is strongly dependent on the arithmetic expression which is used for the interval arithmetic evaluation of the given function.

Moore [6] has shown that under reasonable assumptions the following inequality holds for the distance between R(f;[x]) and f([x]):

(2) $$q(R(f;[x])\ ,\ f([x])) \le \gamma\ d[x]\ ,\ [x] \subseteq [x]^{0}\ ,\ \gamma \ge 0\ .$$

This means that the overestimation of R(f;[x]) by f([x]) goes linearly to zero with d[x] . We illustrate this using the following example.

Example 2. Let

$$f(x) = x - x^2\ , \quad x \in [x]^{0} = [0;1]\ .$$

Set $$[x] = [\tfrac{1}{2} - r\ ;\ \tfrac{1}{2} + r]\ ,\quad 0 \le r \le \tfrac{1}{2}\ .$$

A simple discussion gives

$$R(f;[x]) = [\tfrac{1}{4} - r^2\ ;\ \tfrac{1}{4}]\ .$$

For f([x]) we get

$$\begin{aligned} f([x]) &= [\tfrac{1}{2} - r\ ;\ \tfrac{1}{2} + r] - [\tfrac{1}{2} - r\ ;\ \tfrac{1}{2} + r][\tfrac{1}{2} - r\ ;\ \tfrac{1}{2} + r] \\ &= [\tfrac{1}{4} - 2r - r^2\ ;\ \tfrac{1}{4} + 2r - r^2]\ . \end{aligned}$$

From this we get

$$q(R(f;[x])\ ,\ f([x])) = \max\ \{|\tfrac{1}{4} - 2r - r^2 - \tfrac{1}{4} + r^2|\ ,\ |\tfrac{1}{4} + 2r - r^2 - \tfrac{1}{4}|\}$$

$$= \max \{2r \ , \ 2r - r^2\}$$

$$= 2r = \gamma \ d[x] \ , \ \gamma = 1 \ ,$$

as predicted by Moore's result (2). □

The second part of Example 1 rises the question whether it is possible to rearrange the variables of the given function in such a manner that the interval arithmetic evaluation gives higher than linear convergence to the range of values. The answer is "yes". Before we state the general result we consider again an example.

Example 3. The function $f(x) = x - x^2$, $x \in [0;1]$, from the preceding example can be rewritten as

$$f(x) = x - x^2 = \frac{1}{4} - (x - \frac{1}{2})(x - \frac{1}{2}) \ , \quad x \in [0;1] \ .$$

Plugging in intervals we get for the interval arithmetic evaluation

$$\tilde{f}([x]) = \frac{1}{4} - ([\frac{1}{2} - r \ ; \frac{1}{2} + r] - \frac{1}{2}) \ ([\frac{1}{2} - r \ ; \frac{1}{2} + r] - \frac{1}{2})$$

$$= \frac{1}{4} - [- r \ ; \ r] \ [- r \ ; \ r] = \frac{1}{4} + [- r^2 \ ; \ r^2]$$

$$= [\frac{1}{4} - r^2 \ ; \frac{1}{4} + r^2] \ .$$

Hence we get

$$q(R(f;[x]) \ , \ \tilde{f}([x])) = \max \{|\frac{1}{4} - r^2 - (\frac{1}{4} - r^2)| \ , \ |\frac{1}{4} + r^2 - \frac{1}{4}|\}$$

$$= r^2 = \frac{1}{4}(d[x])^2$$

which means that the distance goes quadratically to zero with $d[x]$.

The general result is as follows:

Theorem 1. (The centered form). *Let the (rational) function* $f : \mathbb{R} \to \mathbb{R}$ *be represented in the form*

(3) $$f(x) = f(z) + (x-z) \cdot h(x)$$

for some $z \in [x]$. *If we define*

(4) $$f([x]) := f(z) + ([x] - z)\, h([x])$$

then (under weak conditions on the interval arithmetic evaluation $h([x])$, *see Theorem* 2) it holds that

a) $$R(f;[x]) \subseteq f([x])$$

and

(5) b) $$q(R(f;[x])\ ,\ f([x])) \leq \gamma\, (d[x])^2 \ . \qquad \square$$

Inequality (5) is called "Quadratic approximation property" of the centered form. (3) was introduced by Moore in [6], where he conjectured that (5) holds. (5) was first proved by E. Hansen [5].

How can one find the centered form?

Consider first the case that $f(x)$ is a polynomial

$$f(x) = a_o + a_1 x + \ldots + a_n x^n .$$

Using the Taylorpolynomial (which can be computed by applying the complete Horner-scheme) we get

$$f(x) = f(z) + \frac{f'(z)}{1!}(x-z) + \ldots + \frac{f^{(n)}(z)}{n!}(x-z)^n$$

$$= f(z) + (x-z)\, h(x)$$

where

$$h(x) = f'(z) + \frac{x-z}{2!} f''(z) + \ldots + \frac{f^{(n)}(z)}{n!}(x-z)^{n-1} .$$

If $f(x)$ is a general rational function then (see Ratschek [7], [8]) for $s(x) \neq 0$ we can write

$$f(x) = \frac{r(x)}{s(x)} = f(z) + \frac{r(x) - f(z)\, s(x)}{s(x)}$$

$$= f(z) + (x-z)\, \frac{r(x) - f(z)\, s(x)}{(x-z)\, s(x)}$$

$$= f(z) + (x-z)\, h(x)$$

where

$$h(x) = \frac{r(x) - f(z)\ s(x)}{(x-z)\ s(x)} \ .$$

Since $r(z) - f(z)\ s(z) = 0$ and $s(z) \neq 0$ the term $x-z$ appears both in the nominator and in the denominator of $h(x)$ and therefore can be cancelled out.

<u>Example</u> 4. Let

$$f(x) = \frac{r(x)}{s(x)} = \frac{x-x^2}{x-3}, \quad x \neq 3 \ .$$

For $z = \frac{1}{2}$ we have $f(z) = -\frac{1}{10}$ and therefore

$$f(x) = -\frac{1}{10} + (x - \tfrac{1}{2}) \frac{r(x) - (-\frac{1}{10})\ s(x)}{(x - \frac{1}{2})\ s(x)}$$

$$= -\frac{1}{10} + (x - \tfrac{1}{2}) \frac{\frac{1}{4} - (x - \frac{1}{2})\ (x - \frac{1}{2}) - (-\frac{1}{10})\ ((x - \frac{1}{2}) - \frac{5}{2})}{(x - \frac{1}{2})\ ((x - \frac{1}{2}) - \frac{5}{2})}$$

$$= -\frac{1}{10} + (x - \tfrac{1}{2}) \frac{\frac{1}{10} - (x - \frac{1}{2})}{-\frac{5}{2} + (x - \frac{1}{2})}$$

$$= -\frac{1}{10} + (x - \tfrac{1}{2})\ h(x)$$

where

$$h(x) = \frac{\frac{1}{10} - (x - \frac{1}{2})}{-\frac{5}{2} + (x - \frac{1}{2})} \ . \qquad \square$$

The question whether there exists a representation of f such that for the interval arithmetic evaluation of this representation it holds that

$$q(R(f;[x]), f([x])) \leq \gamma\, (d[x])^m$$

where $m \geq 3$ is an open question. However, in special cases this can be achieved.

Theorem 2. (Generalized centered forms). Let the (rational) function $f : \mathbb{R} \to \mathbb{R}$ be represented in the form

$$(6) \qquad f(x) = \varphi_0 + \ell(x) \cdot h(x) \quad , \quad x \in [x] \, ,$$

where $\varphi_0 \in \mathbb{R}$. Assume that there exist intervals $\ell([x])$ and $h([x])$ such that

$$(7) \qquad \ell(x) \in \ell([x]) \, , \quad x \in ([x]) \, ,$$

$$(8) \qquad h(x) \in h([x]) \, , \quad x \in [x] \, ,$$

$$(9) \qquad |\ell([x])| \leq \tau (d[x])^n \, ,$$

$$(10) \qquad d(h([x])) \leq \sigma\, d[x] \, .$$

If we define

(11) $f([x]) := \varphi_0 + R(\ell;[x]) \cdot h([x])$

then

(12) $R(f;[x]) \subseteq f([x])$,

(13) $q(R(f;[x]) , f([x])) \leq \kappa(d[x])^{n+1}$. □

A proof of Theorem 2 has been performed in [1].

Example 5. a) Assume that

(14) $\ell(x) = (x-c)^n , \quad c \in [x]$.

Then

$$|\ell([x])| = |([x] - c)^n| \leq (d[x])^n$$

and therefore (9) holds.

For $n = 1$ in (14) we have the classical centered form (see Theorem 1). For $n > 1$ in (14) the result of Theorem 2 was already proved in [2].

b) Assume that

(15) $\ell(x) = (x-x_1) \cdot \ldots \cdot (x-x_n) , \quad x_i \in [x] , \quad i = 1(1)n$.

Then again

$$|\ell([x])| \le (d[x])^n$$

and therefore (9) holds.

Whereas $R(\ell;[x])$ is easy to compute in case a) this is in general not true in case b). On the other hand it might be much easier to find a representation of $f(x)$ of the form (6) with $\ell(x)$ defined by (15) compared with finding such a representation using (14). □

We illustrate the preceding Theorem 2 by a simple example.

Example 6. Consider the real polynomial

$$f(x) = x^3 - 6x^2 + (12 - \epsilon^2)x - (8 - 2\epsilon^2), \quad \epsilon \ge 0,$$

which has the zeroes

$$x_1 = 2 - \epsilon, \quad x_2 = 2, \quad x_3 = 2 + \epsilon.$$

For $[x] = [2-\delta; 2+\delta]$, $\epsilon \le \delta \le 2$, the zeroes are contained in $[x]$.

a) If $\delta^3 - \epsilon^2\delta \ge \frac{2}{9}\sqrt{3}\,\epsilon^3$ then

$$R(f;[x]) = [-\delta^3 + \epsilon^2\delta; \delta^3 - \epsilon^2\delta].$$

b) If $\delta^3 - \epsilon^2\delta < \frac{2}{9}\sqrt{3}\,\epsilon^3$ then

$$R(f;[x]) = \frac{2}{9}\sqrt{3}\,\epsilon^3\,[-1\,;\,1]\,.$$

We consider three different cases for the inclusion of $R(f;[x])$ by the evaluation of interval expressions.

A) $f([x]) = f([2-\delta\,;\,2+\delta])$

$$=[-\delta^3 + \delta\epsilon^2 - 48\delta\,;\,\delta^3 - \epsilon^2\delta + 48\delta]$$

from which it follows that

$$q(R(f;[x])\,,\,f([x])) \leq \gamma\,d[x]\,.$$

This agrees with Moore's result (2).

B) $f(x)$ can be written as

$$f(x) = \varphi_o + \ell(x)\cdot h(x)$$

where $\varphi_o = 0$, $\ell(x) = x - 2$, $h(x) = x^2 - 4x + 4 - \epsilon^2$.

From this we get

$$f([x]) := \varphi_o + R(\ell;[x])\cdot h([x])$$

$$= [-\ \delta\ ;\ \delta]\ ([2-\delta\ ;\ 2+\delta][2-\delta\ ;\ 2+\delta] - 4[2-\delta\ ;\ 2+\delta] + 4 - \epsilon^2)$$

$$= [-\ \delta^3 + \delta\epsilon^2 - 8\delta^2\ ;\ \delta^3 - \delta\epsilon^2 + 8\delta^2]$$

and therefore

$$q(R(f;[x])\ ,\ f([x])) \leq \gamma\ (d[x])^2$$

which agrees with the statement (5) of Theorem 1.

C) If we write $f(x)$ as

$$f(x) = \varphi_o + \ell(x) \cdot h(x)$$

where

$$\varphi_o = 0\ ,\ \ \ell(x) = (x-2)\ (x - (2+\epsilon))\ ,\ h(x) = x - (2-\epsilon)$$

then

(16) $$f([x]) = \varphi_o + R(\ell;[x]) \cdot h([x])$$

$$= [\min\ \{-\ \frac{\epsilon^2}{4}\ (\delta + \epsilon)\ ,\ \ -\ (\delta - \epsilon)(\delta^2 + \epsilon\delta)\}\ ;$$

$$\max\ \{\ \frac{\epsilon^2}{4}\ (\delta - \epsilon)\ ,\ \ (\delta^2 + \epsilon\delta)(\delta + \epsilon)\}]$$

and therefore

(17) $q(R(f;[x]) , f([x])) \le \gamma (d[x])^3$

which agrees with the statement (13) for $n = 2$ of Theorem 2. See Example 5, case a). It is important to note that (17) is no longer true if we replace $R(\ell;[x])$ by $\ell([x])$ (the interval arithmetic evaluation) in (16). □

Cornelius and Lohner [4] had the idea to consider so-called remainder forms of f for including the range of f by higher order methods.

Theorem 3. (Remainder form). Let $f : D \subseteq \mathbb{R} \to \mathbb{R}$ have a representation of the form

$$f(x) = g(x) + s(x) \quad , \quad x \in D \ .$$

Assume that

$$R(s;[x]) \subseteq s([x]) \quad , \quad [x] \subseteq D \ .$$

Define

(18) $f([x]) := R(g;[x]) + s([x])$.

Then

a) $R(f;[x]) \subseteq f([x])$

b) $$q(R(f;[x]), f([x])) \leq d(s([x])) \leq 2|s([x])| . \qquad \square$$

How can one find a remainder form of f ?

Suppose that f has derivatives of sufficiently high order. Let $p_\sigma(x)$ be the unique polynomial of degree $\sigma \geq 0$ solving the Hermite interpolation problem:

$$p_\sigma^{(j)}(x_i) = f^{(j)}(x_i) \;; \quad j = 0(1)m_i - 1 , \quad m_i \in \mathbb{N} , \quad i = 0(1)n , \quad n \geq 0 , \tag{19}$$

where $x_o, x_1, \ldots, x_n \in [x]$ are pairwise distinct and

$$\sigma + 1 = \sum_{i=o}^{n} m_i .$$

Then it is well known that

$$f(x) = p_\sigma(x) + \frac{f^{(\sigma+1)}(\xi(x))}{(\sigma+1)!} \prod_{i=o}^{n} (x-x_i)^{m_i} \tag{20}$$

$$= g(x) + s(x) , \quad x \in [x]$$

where we have set $g(x) = p_\sigma(x)$ and $s(x)$ is the remainder term. Assume now that the derivative $f^{(\sigma+1)}$ has an interval arithmetic evaluation over $[x]$. Then, since $\xi(x) \in [x]$, we can set

$$s([x]) = \frac{f^{(\sigma+1)}([x])}{(\sigma+1)!} \prod_{i=o}^{n} ([x]-x_i)^{m_i} .$$

Using this $s([x])$ in (18) Lohner and Cornelius [4] have proved that

$$q(R(f;[x]), f([x])) \le \gamma(d[x])^{\sigma+1} . \tag{21}$$

Of course it must be stressed that practically only small values of σ are possible for finding $R(g;[x])$ in (18).

<u>Example</u> 7. Take

$$n = 0 , \quad m_0 = 3 , \quad \sigma = 2 .$$

Then we have given

$$p_2^{(j)}(x_0) = f^{(j)}(x_0) , \quad j = 0(1)2 , \quad x_0 \in [x] ,$$

and (20) reads

$$f(x) = f(x_0) + \frac{f'(x_0)}{1!}(x-x_0) + \frac{f''(x_0)}{2!}(x-x_0)^2 + \frac{1}{3!}f'''(\xi(x))(x-x_0)^3$$

$$= g(x) + s(x)$$

where

$$g(x) = f(x_0) + \frac{f'(x_0)}{1!}(x-x_0) + \frac{f''(x_0)}{2!}(x-x_0)^2 ,$$

$$s(x) = \frac{1}{3!}f'''(\xi(x))(x-x_0)^3 .$$

$R(g;[x])$ is easy to compute in this case since $g(x)$ is a quadratic polynomial. For

$$f([x]) := R(g;[x]) + \frac{1}{3!} f'''([x])([x]-x_0)^3$$

we have by (21)

$$q(R(f;[x]) , f([x])) \leq \gamma(d[x])^3 . \qquad \square$$

4. Outlook

The discussion in the preceding chapter shows that although it is easy to include the range of functions using interval arithmetic tools it is in general not obvious how to find very good inclusions with a reasonable amount of work. Therefore this problem needs very careful further investigations.

We have not considered functions of several variables. From a practical point of view including the range of such a function is even of much greater importance. See [10], for example, where optimization algorithms, based on interval arithmetic tools, are discussed. In principle all results of the present paper hold for the multidimensional case. However, getting good inclusions is in general much more laborious than for the one dimensional case.

References

[1] Alefeld, G.: On the approximation of the range of values by interval expressions. Submitted for publication.

[2] Alefeld, G., Lohner, R.: On higher order centered forms. Computing 35, 177-184 (1985).

[3] Alefeld, G., Herzberger, J.: Introduction to Interval Computations. New York: Academic Press 1983.

[4] Cornelius, H., Lohner, R.: Computing the range of values with accuracy higher than second order. Computing 33, 331-347 (1984).

[5] Hansen, E.R.: The centered form. In Topics in Interval Analysis, ed. E. Hansen. Oxford 1969, pp. 102-105.

[6] Moore, R.E.: Interval Analysis. Prentice Hall, Englewood Cliffs, N. J., 1966.

[7] Ratschek, H.: Zentrische Formen. Z. Angew. Math. Mech. 58 (1978), T 434- T 436.

[8] Ratschek, H.: Centered forms. SIAM Journal on Numerical Analysis, 17, 656-662, 1980.

[9] Ratschek, H., Rokne, J.: Computer Methods for the Range of Functions. Ellis Horwood, Chichester (1984).

[10] Ratschek, H., Rokne, J.: New Computer Methods for Global Optimization. Ellis Horwood, Chichester (1988).

Differentiation Arithmetics

L.B. Rall
Department of Mathematics
University of Wisconsin-Madison
Madison 53 706
USA

Abstract. Automatic methods for computation of Taylor coefficients and partial derivatives of functions without resort to approximations or symbolic differentiation have been in use for some time. A unified approach to these methods based on formal power series arithmetic is given, in which the computations are viewed as use of suitably defined computer arithmetics on data of appropriate types. It is shown that roundoff error in implementation of these arithmetics can be minimized, and details of serial and parallel computation of Taylor coefficients and derivatives are discussed. Some important applications are indicated.

1. Evaluation Arithmetics. Differentiation arithmetics can be considered to be special cases of the more general concept of an evaluation arithmetic, which includes also ordinary real, complex, and interval arithmetics, as well as vector and matrix arithmetics. An *evaluation arithmetic* $\mathcal{A}$ consists of a set of elements E, a set U of *unary operations* with domains and ranges in E, and a set B of *binary operations* with domains in $E \times E$ and ranges in E.

For example, the *real* evaluation arithmetic $\mathcal{R}$ will be defined by taking $E = \mathbf{R}$, the set of real numbers, the binary operations to be the arithmetic operations $\{+, -, \cdot, /\}$, and the set of unary operations to consist of unary $+, -$ and a set Φ of real *standard functions*, say

$$\Phi = \{\text{abs}, \text{sqr}, \text{sqrt}, \exp, \ln, \cos, \sin, \arctan\}, \tag{1.1}$$

as in Pascal and Pascal-SC [2], [17]. Of course, this set could easily be expanded to include other useful standard functions, but it is adequate for the present purpose.

$\mathcal{R}$ will be extended to differentiation evaluation arithmetics $\mathcal{D}$ which contain $\mathcal{R}$ as a special case. The method is the same as the extension of $\mathcal{R}$ to complex evaluation arithmetic $\mathcal{C}$ ($E = \mathbf{C}$) or interval evaluation arithmetic $\mathcal{I}$ ($E = \mathbf{I}$).

ISBN 0-12-708245-X

A consistent extension $\mathcal{A}$ of $\mathcal{R}$ has the properties that there is a subset $R \subset E$ which can be identified with the set $\mathbf{R}$ of real numbers, and the arithmetic operations and standard functions of $\mathcal{A}$ acting on R give the same results as in $\mathcal{R}$. The differentiation arithmetics considered below will be extensions of the real evaluation arithmetic $\mathcal{R}$ in this sense. Evaluation arithmetics are special cases of what are called functoids by Kaucher and Miranker [7], and inclusion algebras by Neumaier [13].

2. Code List Representation of Functions. Functions in an evaluation arithmetic $\mathcal{A}$ can be defined by code lists [14], [15]. Given a set $I = \{i_1, i_2, \ldots, i_\delta\} \subset E$ of *inputs*, a *code list* $t = \{t_1, t_2, \ldots, t_n\}$ is a finite sequence of *terms* $t_k \in E$, each of which is of the form

$$t_k = u(a), \quad u \in U, \quad a \in I \cup \{t_1, t_2, \ldots, t_{k-1}\}, \tag{2.1}$$

or

$$t_k = a_1 \circ a_2, \quad \circ \in B, \quad a_1, a_2 \in I \cup \{t_1, t_2 \ldots, t_{k-1}\}. \tag{2.2}$$

In other words, each argument of the operation which defines t_k is required to be an input or a previous term $t_1, t_2, \ldots, t_{k-1}$ of the code list.

If all terms of the code list t are defined, then the final term t_n is said to be the *value* of the function $f : I \to E$ represented by the code list t, that is,

$$t_n = f(i_1, i_2, \ldots, i_\delta). \tag{2.3}$$

This value will also be called the *output* of the code list.

In practice, the set I of inputs is usually considered to consist of *variables* $V = \{x_1, x_2, \ldots, x_\alpha\}$, *constants* $C = \{c_1, c_2, \ldots, c_\beta\}$, and *parameters* $P = \{p_1, p_2, \ldots, p_\gamma\}$ with $\alpha + \beta + \gamma = \delta$. In this context, the dependence of the function f on constants and parameters is usually suppressed, and one writes $t_n = f(x_1, x_2, \ldots, x_\alpha)$.

For example, the real function f represented by the formula

$$f(x, y) = (xy + \sin x + 4)(3y^2 + 6) \tag{2.4}$$

is also represented in the real evaluation arithmetic $\mathcal{R}$ by the code list

$$(2.5)\qquad \begin{aligned} t_1 &= x, \\ t_2 &= y, \\ t_3 &= t_1 \cdot t_2, \\ t_4 &= \sin(t_1), \\ t_5 &= t_3 + t_4, \\ t_6 &= t_5 + 4, \\ t_7 &= \mathrm{sqr}(t_2), \\ t_8 &= 3 \cdot t_7, \\ t_9 &= t_8 + 6, \\ t_{10} &= t_6 \cdot t_9. \end{aligned}$$

Neither representation is unique. Algorithms for the conversion of formulas into code lists [15] or equivalent push-down stacks of operators and operands are a standard feature of compilers for translation of computer languages into machine code.

This code list also provides a way to compute values of a complex or interval *extension* of the function f to the evaluation arithmetics $\mathcal{C}$ or $\mathcal{I}$, respectively. All that is required is the redefinition of the arithmetic operations and standard functions for the new set of elements E upon which the evaluation arithmetic is based. The interval extension of f provides an *inclusion* of the values of the real function f for real values ξ, η of its variables belonging respectively to the input intervals x, y. Code list representations of functions have been exploited by Rall [14] and Kedem [8] for automatic differentiation and by Neumaier [13] for definition of functions in inclusion algebras.

3. Formal Power Series Arithmetic. The fundamental differentiation arithmetic considered here is the arithmetic $\mathcal{F}$ of formal power series ([5], Chap. 1). Elements of $\mathcal{F}$ are the infinite real sequences

$$(3.1)\qquad a = (a_0, a_1, \ldots, a_k, \ldots).$$

The components of a can be taken to be coefficients of power series, giving the identification

$$(3.2)\qquad a \Leftrightarrow a(\xi) = \sum_{k=0}^{\infty} a_k \xi^k.$$

Arithmetic operations and standard functions are defined accordingly, but without regard to convergence. Hence, the elements (3.1) are referred to as formal power series. Formal manipulation of power series has a long history in applied mathematics, and the rules for arithmetic operations are well understood.

Addition and Subtraction $c = a \pm b$:

$$c_k = a_k \pm b_k, \qquad k = 0, 1, 2, \ldots. \tag{3.3}$$

Multiplication $c = a \cdot b$:

$$c_k = \sum_{j=0}^{k} a_j \cdot b_{k-j}, \qquad k = 0, 1, 2, \ldots. \tag{3.4}$$

Division $c = a/b$ is defined by $b \cdot c = a$, which gives a lower triangular system of equations for $c_0, c_1, \ldots, c_k, \ldots$ with solutions

$$c_k = \frac{1}{b_0}\left(a_k - \sum_{j=1}^{k} b_j \cdot c_{k-j}\right), \quad b_0 \neq 0, \qquad k = 0, 1, 2, \ldots. \tag{3.5}$$

$\mathcal{F}$ is an integral domain with respect to arithmetic operations, which means that it is a commutative ring with identity without divisors of zero [5]. Elements of the form $v = (0, v_1, v_2, \ldots, v_k, \ldots)$ do not have multiplicative inverses, and are called *nonunits*. Although division by nonunits is undefined in general, u/v can be computed for nonunits of the form

$$u = (0, \ldots, 0, u_m, \ldots), \qquad v = (0, \ldots, 0, v_m, \ldots), \quad v_m \neq 0. \tag{3.6}$$

The lower-triangular system of equations $vw = u$ then has solutions

$$w_0 = u_m / v_m, \tag{3.7}$$

which is L'Hôspital's rule of elementary calculus, and in general,

$$w_k = \frac{1}{v_m}\left(u_{m+k} - \sum_{j=0}^{k-1} v_{m+k-j} w_j\right), \qquad k = 0, 1, 2, \ldots. \tag{3.8}$$

Definitions of standard functions $\phi(a)$ can of course be obtained by substitution of the power series for $a(\xi)$ into the power series for $\phi(\xi)$, but it is generally easier to use the differential equation satisfied by the given standard function [11]. The derivative a' of a formal power series a is defined to be the formal power series

$$a' = (a_1, 2a_2, \ldots, (k+1)a_{k+1}, \ldots). \tag{3.9}$$

Standard functions ϕ in $\mathcal{R}$ can thus be extended to $\mathcal{F}$ by solving the initial value problem

$$c' = \phi'(a) \cdot a', \qquad c_0 = \phi(a_0), \tag{3.10}$$

for $c = \phi(a)$. For example, $c = \exp(a)$ is defined by $c' = c \cdot a'$, $c_0 = \exp(a_0)$, so $c_k = c_0 b_k$, where

$$b_0 = 1, \qquad b_k = \frac{1}{k}\sum_{j=1}^{k}(ja_j) \cdot b_{k-j}, \quad k = 1, 2, 3, \ldots, \tag{3.11}$$

[13] The remaining standard functions (1.1) for $\mathcal{R}$ can be extended to $\mathcal{F}$ in a similar fashion. (The function abs can be defined directly; for sqr, one can use (3.4) with $a = b$ and take advantage of symmetry [15].)

A formal power series of the form $(r, 0, 0, \ldots, 0, \ldots)$ is called a *constant*, and is identified with the real number r. Formal power series arithmetic on constants thus reduces to real arithmetic, so $\mathcal{F}$ is an extension of $\mathcal{R}$ in this sense.

4. Automatic Differentiation. It follows from the definition of the evaluation arithmetic $\mathcal{F}$ that if the inputs to a code list containing arithmetic operations and the standard functions (1.1) are formal power series, then the output of the code list, if defined, is the formal power series for the function defined by the code list. Thus, if the code list defines a real function f in the evaluation arithmetic $\mathcal{R}$, then it defines the formal power series for the same function f in $\mathcal{F}$ with formal power series as inputs. In essence, one has the result of substitution of the power series expansions of the inputs into the power series expansion of the function, but without regard to convergence. Of course, if the inputs and output all corresponding to convergent power series, then the output represents the Taylor series expansion of the real function f in the classical sense.

Hence, evaluation of the function $f(x, y)$ given by the code list (2.5) in the formal power series arithmetic $\mathcal{F}$ using the "variables"

$$x = (x_0, h, 0, \ldots, 0, \ldots), \quad y = (y_0, k, 0, \ldots, 0, \ldots), \tag{4.1}$$

gives the formal power series f for which

$$f_m = \frac{1}{m!}\sum_{j=0}^{m}\binom{m}{m-j}\frac{\partial^m f(x_0,y_0)}{\partial^{m-j}x\partial^j y}h^{m-j}k^j, \qquad m=0,1,2,\ldots. \tag{4.2}$$

This mth *Taylor coefficient* of $f(x,y)$ is the same as would be obtained by symbolic differentiation of $f(x,y)$ m times with respect to x and y, followed by evaluation of the resulting partial derivatives and Taylor coefficients in the real arithmetic $\mathcal{R}$. Evaluation of functions in a differentiation arithmetic such as $\mathcal{F}$ is often called *automatic differentiation*, [21], [15], although the terminology *algorithmic differentiation* [1] is perhaps more descriptive, because the algorithm used to compute the function (the code list in this case) also computes its derivatives if a differentiation arithmetic is used. The point which should be clear is that *exact* values of Taylor coefficients are obtained without symbolic differentiation of the function.

Formal power series arithmetic can be used to obtain power series solution of initial-value problems for systems of ordinary differential equations [11], [12]. In this application, if initial conditions and coefficients are specified as intervals, then interval formal power series arithmetic will give bounds for solutions.

5. Taylor Arithmetics. The formal power series arithmetic $\mathcal{F}$ can be extended to the corresponding complex or interval evaluation arithmetics by taking the set of elements E to consist of complex or interval sequences, and using the appropriate definitions of arithmetic operations and standard functions. Interval power series arithmetic will in particular provide interval inclusions of real Taylor coefficients, and thus gives a method to obtain guaranteed bounds for truncation errors [11], [12].

For computational purposes, these formal power series arithmetics are restricted to the corresponding *Taylor arithmetics* $\mathcal{T}_N$ of finite sequences

$$a = (a_0, a_1, \ldots, a_N), \tag{5.1}$$

which correspond to the the Taylor polynomials

$$a(\xi) = \sum_{k=0}^{N} a_k\xi^k = \sum_{k=0}^{N}\frac{1}{n!}a^{(k)}(0)\xi^k \tag{5.2}$$

of degree N in the formal variable ξ. The definition of arithmetic operations and standard functions for $\mathcal{T}_N$ is the same as for $\mathcal{F}$, with the range of indices restricted to $k = 0, 1, \ldots, N$. The truncation of a power series to a Taylor polynomial of degree N is called *Taylor rounding* by Kaucher and Miranker [7], and is analogous to the approximation of a decimal number $a_0.a_1a_2a_3\ldots$ by its first $N+1$ digits ($\xi = 10^{-1}$).

In the real case, one has $\mathcal{T}_0 = \mathcal{R}$. For finite $N \geq 1$, the arithmetics $\mathcal{T}_N$ are commutative rings with identity [16]. Constants $r = (r, 0, 0, \ldots, 0)$ and "variables" $x = (x_0, h, 0, \ldots, 0)$ in $\mathcal{T}_N$ are identified with the corresponding Taylor polynomials. Complex and interval extensions of $\mathcal{T}_N$ are defined as for $\mathcal{F}$.

If $f(x_1, x_2, \ldots, x_\alpha)$ is a function defined by a code list, and the output $f = t_n = (f_0, f_1, \ldots, f_N)$ of the code list is defined for $x_i = (x_{0_i}, h_i, 0, \ldots, 0)$, $i = 1, 2, \ldots, \alpha$, then it follows that f is differentiable at least N times at the point $x_0 = (x_{0_1}, x_{0_2}, \ldots, x_{0_\alpha})$, and the f_k are its corresponding Taylor coefficients for $k = 0, 1, \ldots, N$. Suppose that the code list for f can also be evaluated in the interval Taylor arithmetic $\mathcal{IT}_N$ for variables $X_i = (X_{0_i}, H_i, 0, \ldots, 0)$, giving the result $F = T_n = (F_0, F_1, \ldots, F_N)$. In this case, a stronger assertion can be made: The function f is differentiable at least N times for all $x_{0_i} \in X_{0_i}$, $i = 1, 2, \ldots, \alpha$. Furthermore, for all $h_i \in H_i$, one has $f_k \in F_k$, $k = 0, 1, \ldots, N$, for the Taylor coefficients of f. Thus, by Taylor's theorem, one has

$$f(\xi_1, \xi_2, \ldots, \xi_\alpha) \in f_0 + f_1 + \cdots + f_{N-1} + F_N, \tag{5.3}$$

for all $\xi_i = x_{0_i} + h_i$, $x_{0_i} \in X_i$, $h_i \in H_i$, $i = 1, 2, \ldots, \alpha$. This gives rigorous bounds for the approximation $f(\xi) = f(\xi_1, \xi_2, \ldots, \xi_\alpha)$ by its Taylor polynomial of degree $N-1$ expanded at x_0. Thus, estimates of truncation error in many numerical processes based on polynomial approximation (such as numerical integration [3] or the solution of initial-value problems for systems of ordinary differential equations [11], [12]) can be obtained by the use of interval Taylor arithmetic. Ordinarily, one takes

$$X_i = x_{0_i} + H_i, \qquad i = 1, 2, \ldots, \alpha, \tag{5.4}$$

and $H_i = [0, \eta_i]$ or $H_i = [-\eta_i, \eta_i]$.

6. Rounded Taylor Arithmetic. Real and interval formal power series and Taylor arithmetics are valuable theoretical tools, and the algorithms defining arithmetic operations and standard functions in these arithmetics are simple to program for digital computers. This is particularly true in languages which permit overloading of operators and functions, such as Pascal-SC [2], see [10], pp. 291–309. The results are no longer exact, of course, because of roundoff error. This is the only source of error, however, because the basic formulas are exact, and not approximate. One would also have roundoff error (likely more) if symbolic differentiation were used to obtain formulas for the Taylor coefficients which would then be evaluated using floating-point arithmetic. By suitable implementation, the level of roundoff error for the arithmetic operations and standard functions of the Taylor arithmetic $\mathcal{T}_N$ can be reduced to the same level as for the real evaluation arithmetic $\mathcal{R}$.

An evaluation arithmetic will be called a *computer arithmetic* if the sets E, U, B are finite. For evaluation arithmetics defined in terms of real numbers, the corresponding computer arithmetics are usually obtained by restriction of E to a set of floating-point numbers. This process of mapping from the reals to floating-point approximations is commonly called "rounding." If $\mathbf{F}$ denotes the available set of floating-point numbers, and $\Box$ denotes a rounding operator $\Box : \mathbf{R} \to \mathbf{F}$ which is a semimorphism as defined by Kulisch and Miranker [9], then $\Box\,\mathcal{R}$ will denote the computer arithmetic corresponding to the evaluation arithmetic $\mathcal{R}$. Arithmetic operations and standard functions in $\Box\,\mathcal{R}$ are defined respectively by

$$(6.1) \qquad a \boxdot b = \Box\,(a \circ b), \qquad (\Box\,\phi)(a) = \Box\,(\phi(a))$$

for all $a, b \in \mathbf{F}$. Such roundings produce results of *maximum quality*, which means there are no floating-point numbers between the true and rounded results.

The computer differentiation arithmetic $\Box\,\mathcal{T}_N$ corresponding to $\mathcal{T}_N$ is defined by componenentwise rounding,

$$(6.2) \qquad \Box\, a = (\Box\, a_0, \Box\, a_1, \ldots, \Box\, a_N).$$

Implementation of the arithmetic operations $\pm$ and the standard function abs are straightforward in $\Box\,\mathcal{T}_N$. Multiplication and the standard function sqr can be based on the maximum quality scalar product of vectors $\mathbf{u}, \mathbf{v} \in \mathbf{F}^{m+1}$ for $m = 1, 2, \ldots, N$,

$$(6.3) \qquad \Box\,(\mathbf{u} \cdot \mathbf{v}) = \Box \left(\sum_{j=0}^{m} u_i \cdot v_i \right),$$

which is also useful for other purposes. Division requires the maximum quality solution of the nonsingular lower-triangular system of linear equations corresponding to (3.5). This is known to be possible using iterative defect correction and interval arithmetic [19]. In order to compute the defects with sufficient accuracy, the maximum quality scalar product (6.3) is used.

The standard functions sqrt, exp, ln, sin, cos, arctan can also be obtained from simple systems, provided the corresponding real functions are available with sufficient precision. For example, the standard function exp can be computed by solving the lower-triangular *linear* system corresponding to (3.10),

$$\begin{aligned} b_0 &= 1, \\ -a_1 b_0 + b_1 &= 0, \\ -2a_2 b_0 - a_1 b_1 + 2b_2 &= 0, \\ &\cdots \\ -\sum_{j=0}^{N-1} (N-j) a_{N-j} \cdot b_j + N b_N &= 0, \end{aligned} \tag{6.4}$$

for $b_0, b_1, \ldots, b_N$ [13], and computing $c_0 = \exp(a_0)$ with sufficient precision so that maximum quality floating-point approximations to the products $c_k = c_0 b_k$ are obtained. (The coefficients $(N-j)a_{N-j}$ may also have to be computed with additional precision to guarantee maximum quality of the solution of the linear system.)

Similarly, for $c = \ln(a)$, one has $c_0 = \ln(a_0)$, and the remaining coefficients are determined by solving the linear system $a \cdot c' = a'$. Likewise, for $c = \arctan(a)$, one has $c_0 = \arctan(a_0)$ and the linear system $(1 + sqr(a)) \cdot c' = a'$ for $c_1, c_2, \ldots, c_N$. The trigonometric functions $s = \sin(a)$ and $c = \cos(a)$ satisfy the linear system $s' = c \cdot a'$, $c' = -s \cdot a'$, with the initial conditions $s_0 = \sin(a_0)$, $c_0 = cos(a_0)$. One has

$$s_i = s_0 \cdot d_i + c_0 \cdot t_i, \qquad c_i = c_0 \cdot d_i - s_0 \cdot t_i, \tag{6.5}$$

$i = 0, 1, \ldots, N$, where $d_0 = 1$, $t_0 = 0$, and the remaining coefficients of d, t satisfy the system of $2N$ linear equations

$$\begin{aligned} \sum_{j=0}^{N-1} (N-j) a_{N-j} t_j + i d_i &= 0, \\ -\sum_{j=0}^{N-1} (N-j) a_{N-j} d_j + i t_i &= 0, \end{aligned} \tag{6.6}$$

$i = 1, 2, \ldots, N$ [13]. If the solutions of this linear system and the values of $\sin(a_0), \cos(a_0)$ are computed with sufficient precision, then maximum quality values of the coefficients (6.3) are obtained as in the case of exp.

The function sqrt requires the solution of the lower-triangular quadratic system $sqr(c) = a$ or $c \cdot c' = a'/2$ with $c_0 = \mathrm{sqrt}(a_0)$. This can be done with maximum quality by using Newton's methods for iterative refinement, given a sufficiently high precision evaluation of $\mathrm{sqrt}(a_0)$, since the residual can then be computed with as high precision as necessary by use of the maximum quality scalar product (6.3).

It follows from the above that maximum quality floating-point approximations to the results arithmetic operations and standard functions of the Taylor arithmetic $\mathcal{T}_N$ can be computed, provided the maximum quality scalar product and sufficiently high precision versions of the real standard functions (1.1) are available. Thus, the level of roundoff error in Taylor arithmetic can be made the same as in the real floating-point arithmetic $\mathcal{R}$.

Of course, while a semimorphism produces maximum quality floating-point arithmetic, the converse is not necessarily true. However, it is possible to ensure that the rounding to maximum quality approximations satisfy the axioms for a semimorphism (preservation of the floating-point screen, monotonicity, and antisymmetry [9]).

It is known that the composition of semimorphisms is a semimorphism, and since the rounding $\square_N$ such that $\square_N\ \mathcal{F} = \mathcal{T}_N$ is a semimorphism, the rounded Taylor arithmetic $\square\ \mathcal{T}_N$ is a semimorphic image of the formal power series arithmetic $\mathcal{F}$.

In the case of interval Taylor arithmetic, maximum quality interval inclusions of the solutions of the linear and nonlinear equations defining division and standard functions may not be obtainable, although methods which give high-quality inclusions are known. However, the interval Taylor coefficients in any case include the corresponding real Taylor coefficients for real inputs belonging to the interval inputs, and thus give valid bounds for them.

7. Partial Derivatives. Differentiation arithmetics can also be constructed to compute partial derivatives [15]. These arithmetics are based on the evaluation arithmetic $\mathcal{S}_m$ of sequences

$$a = (a^0, a', a'', \ldots, a^{(k)}, \ldots), \tag{7.1}$$

where a^0 is a scalar, a' a vector in $\mathbf{R}^m$, a'' an $m \times m$ matrix, and so on. (More precisely, $a^{(k)}$ is a representation of a k-linear operator from $\mathbf{R}^m$ into $\mathbf{R}$ [14], so a' is a row vector, a'' an array with one row and m^2 columns, etc.) Arithmetic operations and standard functions are again defined according to the corresponding rules for differentiation. Given variables $x_i = (x_i^0, e_i, 0, \ldots, 0, \ldots)$, $i = 1, \ldots, m$, where the e_i are the unit vectors in $\mathbf{R}^m$, evaluation of a code list for f in this arithmetic gives $f^{(k)}$ as an array of all m^k of the kth partial derivatives of f,

$$f^{(k)} = \left[\frac{\partial^k f(x)}{\partial^{i_1} x_1 \cdots \partial^{i_m} x_m} \right], \qquad i_1 + i_2 + \cdots + i_m = k. \tag{7.2}$$

For $h = (h_1, h_2, \ldots, h_m)$, the corresponding Taylor coefficients of f are given by the multilinear forms

$$f_k = \frac{1}{k!} f^{(k)} h^k, \qquad k = 0, 1, 2, \ldots. \tag{7.3}$$

Computer arithmetics are obtained from $\mathcal{S}_m$ as before by truncation to finite sequences followed by rounding to floating-point approximations of elements. The evaluation arithmetics $\mathcal{G}_m$ of pairs (f^0, f') and $\mathcal{H}_m$ of triples (f^0, f', f'') (called gradient and Hessian arithmetic, respectively) are the most important for applications, because f' can be identified with the gradient (column) vector ∇f of f, and f'' with the Hessian matrix Hf of second partial derivatives of f.

8. Gradient and Hessian Arithmetic. The rules for gradient and Hessian arithmetic are simple enough to be written explicitly [15]. Let $u = (u^0, u', u'') = (u^0, \nabla u, Hu)$ and $v = (v^0, \nabla v, Hv)$ denote generic Hessian elements. Arithmetic operations are defined as follows:

Addition and Subtraction, $w = u \pm v$:

$$w^0 = u^0 \pm v^0, \qquad \nabla w = \nabla u \pm \nabla v, \qquad Hw = Hu \pm Hv. \tag{8.1}$$

Multiplication, $w = u \cdot v$:

$$\begin{aligned} w^0 &= u^0 \cdot v^0, \qquad \nabla w = u^0 \cdot \nabla v + v^0 \cdot \nabla u, \\ Hw &= u^0 \cdot Hv + \nabla u \nabla v^T + \nabla v \nabla u^T + v^0 \cdot Hu. \end{aligned} \tag{8.2}$$

Division, $w = u/v$:

$$\begin{aligned} w^0 &= u^0/v^0, \qquad \nabla w = (\nabla u - w^0 \cdot \nabla v)/v^0, \\ Hw &= (Hu - \nabla w \nabla v^T - \nabla v \nabla w^T - w^0 \cdot Hv)/v^0, \qquad v^0 \neq 0. \end{aligned} \tag{8.3}$$

Standard Functions, $w = \phi(u)$, $\phi \in \Phi$:

$$w^0 = \phi(u^0), \qquad \nabla w = \phi'(u^0) \cdot \nabla u, \qquad Hw = \phi''(u^0) \cdot \nabla u \nabla u^T + \phi'(u^0) \cdot Hu, \tag{8.4}$$

provided the real standard function ϕ is twice differentiable at u^0.

Computation of only w^0 and ∇w by the above formulas defines gradient arithmetic. The extension of gradient and Hessian arithmetics to the corresponding interval arithmetics and rounding of these arithmetics to floating-point computer arithmetics is carried out in the same way as for Taylor arithmetics. As before, the rounding of real gradient and Hessian arithmetics to the corresponding floating-point computer arithmetics can be done with maximum quality.

Gradient and Hessian arithmetics are useful in computational solution of nonlinear optimization problems. The interval gradient and Hessian arithmetics are suitable for validation, as usual, and also for search procedures. For example, if $0 \notin \nabla F(X)$, then it follows that there are no critical points of the function f in the region X. In this way, the search for extrema of f can be confined to promising regions which are generally much smaller than given initial regions.

9. Serial Computation of Gradients and Hessians. Programming the algorithms for gradient and Hessian arithmetic operations and standard functions is straightforward and simple. One can thus proceed through the code list (or equivalent stack), computing function values, gradients, and Hessians at each step until the final results are obtained. This is the so-called "forward" method of automatic differentiation, which is easy to understand. However, storage of arrays and componentwise modification of them at each step of the evaluation are required. Of course, the storage allocated to terms which are no longer needed in the evaluation of the code list can be released, but updating all components of arrays at each step generally requires additional time proportional to m for gradient arithmetic and m^2 for Hessian arithmetic in $\mathbf{R}^m$ as compared to scalar evaluation in $\mathcal{R}$ [15].

One also can note that the ith component of the gradient vector ∇f is $\partial f/\partial x_i$, and since $f = t_n$, one has

$$\frac{\partial f}{\partial x_i} = \frac{\partial t_n}{\partial x_i} = \sum_{j=1}^{n-1} \frac{\partial t_n}{\partial t_j}\frac{\partial t_j}{\partial x_i}, \tag{9.1}$$

where the summation contains at most two nonzero terms because of the structure of the code list. In general, in the code list in $\mathcal{R}$ can be considered to be a sequence of transformations of $t_1, \ldots, t_{k-1}$ into t_k for $k = 1, \ldots, n$, with

$$\frac{\partial t_k}{\partial x_i} = \sum_{j=1}^{k-1} \frac{\partial t_k}{\partial t_j}\frac{\partial t_j}{\partial x_i}, \tag{9.2}$$

again with at most two nonzero terms on the right. If the code list is begun with $t_1 = x_1, \ldots, t_m = x_m$, then

$$\frac{\partial t_k}{\partial x_i} = \delta_{ik} = \begin{cases} 1, & i = k, \\ 0, & i \neq k, \end{cases} \tag{9.3}$$

for $k = 1, \ldots, m$. Formula (9.2) then gives a simple way to compute $\partial t_k/\partial x_i$ for $k = m+1, \ldots, n$. It has the advantage for serial computation of dealing with only scalar quantities, and terms in the code list which are independent of x_i ($\partial t_k/\partial x_i = 0$) can be skipped. However, the code list has to be processed m times, at least in part, to compute the entire gradient vector.

This method requires the storage of all terms $\{t_1, t_2, \ldots, t_n\}$ of the code list evaluated in the scalar arithmetic $\mathcal{R}$ (actually $\square\mathcal{R}$ for floating-point evaluation). In addition, the Jacobian matrix of the code list can be computed and stored at the same time the code list is evaluated to provide the partial derivatives $\partial t_i/\partial t_j$ entering into formula (9.2). The Jacobian matrix

$$J(t) = \left[\frac{\partial f_i}{\partial t_j}\right] \tag{9.4}$$

of a code list is lower-triangular and sparse, since $t_i = f_i(t_1, t_2, \ldots, t_{i-1})$ and by (2.1)–(2.2), no row has more than two nonzero elements. For example, the Jacobian matrix of the example code list (2.5) is

$$J(t) = \begin{bmatrix} 0 & . & . & . & . & . & . & . & . & 0 \\ 0 & 0 & . & . & . & . & . & . & . & . \\ t_2 & t_1 & 0 & . & . & . & . & . & . & . \\ \cos(t_1) & 0 & 0 & 0 & . & . & . & . & . & . \\ 0 & 0 & 1 & 1 & 0 & . & . & . & . & . \\ 0 & 0 & 0 & 0 & 1 & 0 & . & . & . & . \\ 0 & 2t_2 & 0 & 0 & 0 & 0 & 0 & . & . & . \\ 0 & 0 & 0 & 0 & 0 & 0 & 3 & 0 & . & . \\ 0 & 0 & 0 & 0 & 0 & 0 & 0 & 1 & 0 & . \\ 0 & 0 & 0 & 0 & 0 & t_9 & 0 & 0 & t_6 & 0 \end{bmatrix} \tag{9.5}$$

If the term t_i of the code list depends explicity on t_j ($j < i$), then t_j and $\partial t_i/\partial t_j$ can be obtained by evaluation of the term in the simple scalar differentiation arithmetic $\mathcal{G}_1$, with the variable $(t_j, 1)$, otherwise, $\partial t_i/\partial t_j = 0$. Because of the sparseness of $J(t)$, its storgage requires fewer than $2n$ locations. The elements of $J(t)$ can then be used with formula (9.2) for the evaluation of gradient vectors.

The structure of $J(t)$ also reveals dependence of terms on variables. The nonzero elements of $J(t)$ define the incidence matrix and thus the computational (or Kantorovich) graph of the function f defined by the code list t [15]. Nonzero entries in the jth column of $J(t)$ show which terms t_i, $i > j$, depend directly on t_j. Thus, starting with the first column, all terms depending directly or indirectly on the variable $x_1 = t_1$ can be found, and so on. For the above example, it is evident that only $t_3, t_4, t_5, t_6, t_{10}$ depend on x, so

$$\frac{\partial t_2}{\partial x} = \frac{\partial t_7}{\partial x} = \frac{\partial t_8}{\partial x} = \frac{\partial t_9}{\partial x} = 0. \tag{9.6}$$

It has been noted recently [20] (see also [6], or [4] for a particularly clear explanation), that components of the gradient vector can be computed efficiently using only scalar quantities obtained by evaluating the code list t and its Jacobian matrix $J(t)$. It has been shown that the time required by this so-called "backward" method to compute the gradient vector depends on the length n of the code list, not the dimensionality m of the space of variables, and is bounded by five times what it takes to evaluate the function value in $\mathcal{R}$, where the actual ratio may well be smaller [4]. Since what are wanted are the values of $\partial t_n/\partial t_i$, $i = 1, 2, \ldots, m$, one could think of generating the sequence of partial derivatives

$$\frac{\partial t_n}{\partial t_n} = 1, \frac{\partial t_n}{\partial t_{n-1}}, \ldots, \frac{\partial t_n}{\partial t_{i+1}}, \frac{\partial t_n}{\partial t_i}, \ldots, \frac{\partial t_n}{\partial t_2}, \frac{\partial t_n}{\partial t_1}, \tag{9.7}$$

in reverse order.

In the code list t, each term t_j will depend explicitly on terms t_k with indicies $k \in K_j \subset \{1, 2, \ldots, j-1\}$. A subset of the subsequent terms $t_{j+1}, \ldots, t_n$ will depend explicitly on t_j, namely the terms t_i with indicies $i \in I_j = \{i \mid j \in K_i\}$. By the chain rule [4],

$$\frac{\partial t_n}{\partial t_j} = \sum_{i \in I_j} \frac{\partial t_n}{\partial t_i} \frac{\partial t_i}{\partial t_j}. \tag{9.8}$$

Since $I_j \subset \{j+1, j+2, \ldots, n\}$, it follows from (9.8) that $\partial t_n/\partial t_j$ can be computed from the entries of the Jacobian matrix (9.4) and the previously computed derivatives

$$\frac{\partial t_n}{\partial t_n} = 1, \frac{\partial t_n}{\partial t_{n-1}}, \ldots, \frac{\partial t_n}{\partial t_{j+1}}. \tag{9.9}$$

The total computation of the gradient vector thus amounts to a *single* backward sweep through the code list, after forward computation of t and $J(t)$, which have to be stored. The sets I_j and K_i are obtained easily from the indices of the nonzero elements of $J(t)$ [4].

The same technique can be used with interval arithmetic to obtain interval inclusions of gradient vectors of real functions defined by code lists. The Jacobian matrix $J(t)$ is also useful for improving the accuracy of evaluation of functions by iterative defect correction based on Newton's method [18]. The code list $\{t_1, t_2, \ldots, t_n\}$ can be written as the lower-triangular nonlinear system

$$t_i - f_i(t_1, t_2, \ldots, t_{i-1}) = 0, \qquad i = 1, 2, \ldots, n, \tag{9.10}$$

with invertible Jacobian matrix $I - J(t)$. Of course, as in the case of linear systems, the defect has to be computed with sufficient accuracy to obtain improvement by iteration. This means essentially that the arithmetic operations and standard functions must be available with higher precision, if needed.

By differentiation of (9.2) with respect to x_j, the resulting formula can be used to compute second partial derivatives $\partial^2 t_i / \partial x_i \partial x_j$ and thus the Hessian matrix Hf of f by a "backward" method similar to the one described above for first partial derivatives and gradient vectors. Note that the Hessian operator of a code list,

$$H(t) = \left[\frac{\partial^2 f_i}{\partial t_j \partial t_k} \right] \tag{9.11}$$

is also sparse and is symmetric in j, k, and can likewise be evaluated along with the real or interval evaluation of f and $J(t)$. Here, if t_i depends explicitly on t_j and t_k, then the simple differentiation arithmetic $\mathcal{H}_2$ with two variables $(t_j, 1, 0)$ and $(t_k, 1, 0)$ is used to compute the desired result. Second derivatives of t_i with respect to all other terms of the code list with indicies less than i will be zero. Of the arithmetic operations and standard functions, only division can result in three nonzero terms in a row of $H(t)$, multiplication can result in at most two, and all other arithmetic operations and standard functions in at most one. Hence, the storage of $H(t)$ requires less than three times the space allotted to the terms of the code list t.

10. Parallel Implementation of Differentiation Arithmetics. Many applications lead to large scale optimization problems, which ordinarily require massive computation for their numerical solution. Because of this need, highly parallel or vectorized "supercomputers" have come into use to perform calculations beyond the effective reach of even the fastest serial machines. Gradient and Hessian arithmetics are well suited for vector and parallel processors. Since all components of gradient vectors and Hessian matrices are computed according to the same rules, even single instruction multiple data (SIMD) processors can be used effectively. For example, given the gradient variables

$$u = (u^0, u'_1, \ldots, u'_m), \qquad v = (v^0, v'_1, \ldots, v'_m), \tag{10.1}$$

and $m + 1$ processors $0, 1, \ldots, m$, then multiplication by scalars,

$$v^0 \cdot u = (u^0 \cdot v^0, v^0 \cdot u'_1, \ldots, v^0 \cdot u'_m), \qquad u^0 \cdot v = (u^0 \cdot v^0, u^0 \cdot v'_1, \ldots, u^0 \cdot v'_m), \tag{10.2}$$

followed by addition (with processor 0 disabled) gives the product

$$u \cdot v = (u^0 \cdot v^0, u^0 \cdot v'_1 + v^0 \cdot u'_1, \ldots, u^0 \cdot v'_m + v^0 \cdot u'_m), \tag{10.3}$$

and so on.

It follows that if enough parallelism is available, one forward pass through the code list (or processing of the equivalent stack) will yield the function value, gradient vector, and Hessian matrix in time which is bounded by a small constant multiple of the time to compute the function value alone, and with fewer idle processors. Because of the symmetry of the Hessian matrix H, only $n(n+1)/2$ elements H_{ij}, $i = 1, \ldots, m$, $j = 1, \ldots, i$ need be computed and stored. The ratio of times will be bounded by the slowest of the componentwise gradient or Hessian arithmetic operations and standard functions as compared to the scalar case. It appears that backward evaluation, which is inherently serial, would have little or no advantage in this case. However, if the number of processors is much smaller than the number of variables (which is certainly true in the serial case), a combined parallel-serial mode using the backward method could be formulated. Furthermore, results of maximum quality can be obtained, especially if the arithmetic units of the processors are properly designed.

A number of physical systems, such as helicopters and power plants, can be described by large systems of coupled ordinary differential equations which are not particularly stiff. Thus, solution by Taylor series methods may be appropriate, but require the computation of many Taylor coefficients for each of the large number of variables. However, each Taylor coefficient of a given order is computed in exactly the same way for different data, so here again vectorization or parallelization of the computation is appropriate, even on computers with SIMD architecture. Thus, the method of prolongation of Taylor series with possible interval inclusion of solutions for bounding global error [11], [12] could be carried out for systems larger than feasible on present serial machines. This could be of critical importance in so-called "real time" applications, which require prediction of the state of a rapidly changing system in advance. Once a set of Taylor coefficients has been computed, the corresponding function can be evaluated over a range of variables before the coefficients have to be recomputed. In particular, interval bounds need only show limits in which the Taylor expansion is valid to the desired accuracy, and thus may not have to be recomputed very often.

References

[1] P. Arbenz and W. Gander. Solving nonlinear eigenvalue problems by algorithmic differentiation, Computing **36** (1986), 205–215.

[2] G. Bohlender, C. Ullrich, J. Wolff von Gudenberg, L. B. Rall. Pascal-SC: A Computer Language for Scientific Computation, Academic Press, 1987.

[3] G. F. Corliss and L. B. Rall. Adaptive, self-validating numerical quadrature, SIAM J. Scientific and Statistical Computing **8** (1987), 831–847.

[4] A. Griewank. On automatic differentiation, in Mathematical Programming: Recent Developments and Applications, ed. by M. Iri and K. Tanabe, KTK Scientific/Kluwer Academic Publishers, 1989.

[5] P. Henrici. Applied and Computational Complex Analysis, Vol. 1, Wiley, 1974.

[6] M. Iri. Simultaneous computations of functions, partial derivatives and estimates of rounding errors—complexity and practicality, Japan Journal of Applied Mathematics **1** (1984), 223–252.

[7] E. W. Kaucher and W. L. Miranker. Self-Validating Numerics for Function Space Problems, Academic Press, 1984.

[8] G. Kedem. Automatic differentiation of computer programs, ACM Trans. Math. Software **6** (1980), 150–165.

[9] U. W. Kulisch and W. L. Miranker. Computer Arithmetic in Theory and Practice, Academic Press, 1981.

[10] U. W. Kulisch and W. L. Miranker (Eds.) A New Approach to Scientific Computation, Academic Press, 1983.

[11] R. E. Moore. Interval Analysis, Prentice-Hall, 1966.

[12] R. E. Moore. Methods and Applications of Interval Analysis, SIAM, 1979.

[13] A. Neumaier. Interval Methods for Systems of Equations, Cambridge, 1989.

[14] L. B. Rall. Computational Solution of Nonlinear Operator Equations, Wiley, 1969. (Reprinted by Krieger, 1979.)

[15] L. B. Rall. Automatic Differentiation: Techniques and Applications, Lecture Notes in Computer Science No. 120, Springer, 1981.

[16] L. B. Rall. The arithmetic of differentiation, Math. Mag. **59** (1986), 275–282.

[17] L. B. Rall. Pascal and Pascal-SC, Encyclopedia of Physical Science and Technology, Vol. 10, pp. 183–209, Academic Press, 1987.

[18] L. B. Rall. Tools for mathematical computation, Proceedings of the Conference on Computer Assisted Proofs in Analysis, University of Cincinnati, 1989.

[19] S. Rump. Solving algebraic problems with high accuracy, [10], pp. 51–120 (1983).

[20] B. Speelpennig. Compiling Fast Partial Derivatives of Functions Given by Algorithms, Ph.D. Dissertation, Department of Computer Science, University of Illinois at Urbana-Champaign, 1980.

[21] R. E. Wengert. A simple automatic derivative evaluation program, Communications ACM **7** (1964), 463–464.

Industrial Applications of Interval Techniques

G.F. Corliss
Marquette University
Milwaukee, WI 53233
USA

Abstract. Tools for achieving high accuracy and guaranteed inclusions in scientific computations have been promoted by the research community for many years. Commercial products have made those tools widely available. We apply the tools of ACRITH to a least squares problem, a problem involving matrix products and inverses, the conjugate gradient algorithm, and a partial differential equation with a moving boundary. We discuss the lessons learned from these experiences about how interval techniques and high accuracy computations can be used cost-effectively for industrial problems.

1. Introduction

Since Moore's *Interval Analysis* [13] appeared in 1966, the academic research community has been promoting the use of interval techniques to achieve guaranteed bounds for the results of scientific computations. Conventional algorithms, or *point* algorithms, compute an estimate for an answer and perhaps an error estimate. The user cannot tell how accurate the estimated answer may be without extensive, expensive error analysis. The quality of scientific software packages has improved, but there are still problems for which catastrophically incorrect answers are returned to the user with no warning.

Interval techniques compute an interval in which the correct answer is guaranteed to lie. If interval techniques compute an answer $[1.23456, 1.23457]$, then 6 decimal places are *known* to be correct. If an algorithm yields the interval $[-10^{30},\ +10^{30}]$, then you *know* that you do not know the answer. Hence, interval results carry with them an assurance of their quality.

In the early days, it was hard for a scientist or engineer to apply interval techniques because suitable software was not widely available. Now there are at least four com-

Computer Arithmetic and Self-Validating
Numerical Methods

ISBN 0-12-708245-X

mercially available systems which support interval arithmetic and the Kulisch-Miranker accurate scalar product [8]:

Pascal-SC - Variant of Pascal for scientific computation. Runs on popular microcomputers. Available from Wiley-Teubner. [10]

ACRITH - Fortran callable subroutine library. Runs on IBM 370 architecture. Available from IBM. [7] Fortran-SC [12] makes ACRITH subroutine calls easier to program.

ARITHMOS - Fortran callable subroutine library. Runs on Siemens BS 2000. Available from Siemens. [21]

NAG - Ada programming language library. Requires VAX/VMS Ada compiler. Available from Numerical Algorithms Group. [4]

Hence, a scientist or engineer now has ready access to these tools on a variety of commonly used computers.

This paper addresses ways in which tools for interval methods can be applied to industrial problems drawn from the author's consulting experience. We illustrate with a least squares problem, a problem involving matrix products and inverses, the conjugate gradient algorithm, and a partial differential equation with a moving boundary. We consider when interval tools should be used and how to use them.

2. High Accuracy

Interval techniques compute an interval in which the correct answer is validated to lie. On the other hand, techniques for high accuracy compute an answer which is accurate to as many places as possible. Interval and high accuracy techniques are related, because we want the widths of interval inclusions to be as narrow as possible. However, techniques for high accuracy computation are very useful in their own right, with no reference to intervals. This section describes the support provided by ACRITH for high accuracy computations and how those tools have are applied to some actual problems.

2.1. Accurate scalar product

The fundamental tool provided by ACRITH for high accuracy computation is the accurate scalar product. Its use often prevents the error from growing with problem size. Let $\mathbf{w} = (w_1, w_2, \ldots, w_n)$ and $\mathbf{v} = (v_1, v_2, \ldots, v_n)$ be vectors in $\mathbf{R^n}$. Then ACRITH's scalar product computes

$$\mathbf{w} \bullet \mathbf{v} = \sum_{i=0}^{n} w_i \cdot v_i \tag{1}$$

with only *one rounding error*. The ACRITH scalar product is more accurate than the corresponding Fortran DO loop. On some hardware, it is also about 15

A Fortran DO loop for evaluating Equation (1) commits $2n$ rounding errors, while the ACRITH program commits only one rounding error *independent* of the size of n. This is important since industrial problems tend to be large. Since the ACRITH program commits only one rounding error at the conclusion of the calculation, it is not subject to catastrophic cancellation errors or to intermediate overflow.

Conventional wisdom in numerical analysis suggests that it is often useful to accumulate scalar products in a higher precision. If a Fortran DO loop accumulates the sum in double precision, the $2n$ rounding errors are of the order of double precision machine roundoff. The result is more accurate one computed by the single precision program, but it is not as accurate as the results from ACRITH. Further, the double precision accumulation of scalar products is still sensitive to catastrophic cancellation and overflow of intermediate results.

2.2. How does it work?

A detailed discussion of the algorithms used by ACRITH is beyond the scope of this paper (see [7], pp. 57–61 for details). ACRITH uses a long accumulator with a capacity for 328 hexadecimal digits. The accumulation of the scalar product is done in fixed-point arithmetic with partial products being added into the accumulator in the appropriate digit positions. When the summation is completed, the result is rounded once for storing back into a single or double precision memory location.

2.3. An example – Least squares

Let A be an $m \times n$ matrix, with $m > n$. We wish to solve

$$Ax = \mathbf{b} \tag{2}$$

in the least squares sense. This problem arises often in many industrial applications, and it is often very large. The least squares solution to Equation (2) is the same as the solution to the square system

$$A^T A\mathbf{x} = A^T \mathbf{b}, \tag{3}$$

which are called the normal equations. It is well known that the normal equations are very poorly conditioned. The conventional wisdom in numerical analysis is that one can almost never solve a least squares problem using the normal equations. Methods such as Gram-Schmidt orthonormalization, Q-R factorization, or singular valued decomposition were developed to overcome the numerical difficulties of ill conditioned systems.

The accurate scalar product calls this conventional wisdom into question. The reason the ill conditioning of the normal equations is a problem is that with conventional arithmetic, roundoff errors accumulate so fast that the computed solution is worthless. With ACRITH's scalar product, the accumulation of roundoff errors is much less severe, and it is independent of the size of the problem.

The program in Listing 1 solves a modest least squares problem. This program uses ACRITH routines to form $A^T A$, $A^T b$, and to solve the normal equations.

```
      PARAMETER (M = 228, N = 178)
      DOUBLE PRECISION A (M, N), AT(N, M), APROD(N, N), B(M),
     +                 BT(N), XL(N), XU(N)

C                                Get matrix A
C                                AT = A transpose
C                                APROD = A transpose * A
      CALL DMAMN (N, M, N, AT, A, APROD, IER)
C                                BT = A transpose * B
```

```
      CALL DMAMN (N, M, 1, AT, B, BT, IER)
C                                     Solve A trans A x = A trans b
      CALL DLIN (N, APROD, BT, BT, XL, XU, IER)
```

Listing 1. Solve the normal equations.

For the data we were given, ACRITH was able to solve the normal equations with an absolute residual error of 4.02E-14. For comparison, Listing 2 is code using IMSL routines to form and solve the normal equations.

```
C                                     APROD = A transpose * A
      CALL DMXTXF (M, N, A, M, N, APROD, N)
C                                     BT = A transpose * B
      CALL DMXTYF (M, N, A, M, M, 1, B, M, N, 1, BT, N)
C                                     Solve A trans A x = A trans b
      CALL DL2ARG (N, APROD, N, BT, 1, XL, FAC, IPVT, WK)
```

Listing 2. Solve the normal equations using IMSL routines.

Since the difficulty of using tools for high accuracy is an issue, we remark that changing from the IMSL subroutines to the ACRITH subroutines requires only the replacing of one set of subroutine calls by another. The two programs are of the same complexity.

The IMSL library provides a routine L2BRR for solving linear least squares problems using a Q-R decomposition with optimal column pivoting. This routine was unable to solve this problem; it fails because of an underflow. In Section 4, we will see that L2BRR fails because the problem is very close to a singular problem.

If the ACRITH linear equation solving routine DLIN is unable to solve the normal equations because of excess width arising from the explicit formation of $A^T A$ and $A^T b$, then the normal equations are recast as an $(2m+n) \times (2m+n)$ block system with no matrix products:

$$\begin{pmatrix} A & -I & \\ & A^T & \\ & & I \end{pmatrix} \begin{pmatrix} x \\ y \\ z \end{pmatrix} = \begin{pmatrix} 0 \\ 0 \\ b \end{pmatrix}. \tag{4}$$

Then DLIN is called to solve Equation (4).

2.4. An example - Conjugate gradient

A second approach to solving Problem (2) is the conjugate gradient algorithm [5], which involves many scalar products. We suspected that an accumulation of roundoff errors was forcing a conventional program to take several times as many iterations than should be necessary. We replaced the existing scalar products with ACRITH calls to improve the accuracies. We present an outline of the analysis and our experiences as an example of how to apply the tools of ACRITH.

Suppose that A is an $m \times n$ matrix. The conjugate gradient algorithm for solving the least squares problem

$$A\mathbf{x} = \mathbf{b}$$

is:

$$\begin{aligned}
&\text{Choose } \mathbf{x}_0; \\
&\text{Put } \mathbf{s}_0 = \mathbf{b} - A\mathbf{x}_0; \\
&\qquad \mathbf{r}_0 = \mathbf{p}_0 = A^T(\mathbf{b} - A\mathbf{x}_0); \\
&\qquad \mathbf{q}_0 = A\mathbf{p}_0; \\
&\text{For } k = 0, 1, \ldots \\
&\qquad \alpha_{k+1} = (\mathbf{r}_k, \mathbf{r}_k)/(\mathbf{q}_k, \mathbf{q}_k); \\
&\qquad \mathbf{x}_{k+1} = \mathbf{x}_k + \alpha_{k+1}\mathbf{p}_k; \\
&\qquad \mathbf{s}_{k+1} = \mathbf{s}_k - \alpha_{k+1}\mathbf{q}_k; \\
&\qquad \mathbf{r}_{k+1} = A^T\mathbf{s}_{k+1}; \\
&\qquad \beta_{k+1} = (\mathbf{r}_{k+1}, \mathbf{r}_{k+1})/(\mathbf{r}_k, \mathbf{r}_k); \\
&\qquad \mathbf{p}_{k+1} = \mathbf{r}_{k+1} + \beta_{k+1}\mathbf{p}_k; \\
&\qquad \mathbf{q}_{k+1} = A\mathbf{p}_{k+1};
\end{aligned}$$

If arithmetic is done with infinite precision, the conjugate gradient algorithm converges in n iterations. In practice, the accumulation of roundoff errors requires extra iterations. We observed $4 - 5 \times n$ iterations being required for satisfactory convergence.

Initially, we replaced each of the scalar products $(\cdot,\cdot)$ by a call to the scalar product from ACRITH. Calculations such as $\mathbf{b} - A\mathbf{x}$ or $\mathbf{x} + \alpha\mathbf{p}$ are potentially subject to catastrophic cancellation, so each was reformulated to use the ACRITH scalar product. It is critical to evaluate the expression

$$\mathbf{x}_{k+1} = \mathbf{x}_k + \alpha_{k+1}\mathbf{p}_k = \mathbf{x}_0 + \sum_{j=0}^{k} \alpha_{j+1}\mathbf{p}_j. \tag{5}$$

accurately. By storing $\mathbf{p}_0, \ldots, \mathbf{p}_k$, each component of $\mathbf{x}_{k+1}$ can be computed with a single rounding error. Similarly, each component of $\mathbf{s}_{k+1}$ can be computed with a single rounding error. This requires storing all previous iterates and negates one of the advantages of the conjugate gradient algorithm. We developed a compromise algorithm which retained about 10 previous iterates.

It is common to design algorithms to achieve high accuracy by reformulating a problem as a residual correction. Suppose we wish to compute the double precision value $y = \sum_{j=1}^{n} u_j v_j$ as for each component of $\mathbf{x}_{k+1}$ in Equation (5). The value of y is computed with only one rounding error, but suppose that still introduces too much error into succeeding calculations. If so, we compute $y_1 = \sum_{j=1}^{n} u_j v_j$ and then $y_2 = -y_1 + \sum_{j=1}^{n} u_j v_j$ as a single scalar product. The effect is that $y_1 + y_2$ is a quadruple precision value of y. Later computations use $y_1 + y_2$ in place of y. We applied this trick to the computation of $\mathbf{x}_{k+1}$ and $\mathbf{s}_{k+1}$.

We had eliminated several different sources of accumulation of rounding error from the original program, so we ran the modified program with great anticipation. The resulting vectors $\mathbf{x}_k$ agreed to several figures with those calculated previously. That is, there was no significant improvement.

Close inspection of the intermediate results showed that while the vectors $\mathbf{x}_k$ agreed with those calculated previously, the search directions $\mathbf{p}_k$ and the residual vectors $\mathbf{r}_k$ were completely different after the first few iterations. When we checked the orthogonalities

$$(\mathbf{r}_j, A\mathbf{r}_{k+1}) = 0 \text{ for } j = 0, \ldots, k$$

predicted by the theory, the "accurate" residuals were only slightly more nearly orthogonal than the original residuals. This suggests that to improve the accuracy of the algorithm, we must do a partial reorthonormalization.

What did we learn from this attempt to apply the ACRITH scalar product to the conjugate gradient algorithm? First, we were able to replace calls to conventional scalar product routines by calls to the ACRITH scalar product routine. Second, we learned about the behavior of the algorithm. We now understand that the source of the inaccuracies is a gradual deterioration in the orthonormality of the residual vectors. Finally, we are reminded that hard problems are hard. The scalar product from ACRITH is a useful tool to achieve increased accuracies, but it is not a magic wand. To use the tool effectively, we must understand the algorithm to which it is being applied. A mechanical replacement of one scalar product by another is not necessarily enough.

Since the human time required to use these methods is an issue, we remark that these results have required about 3 man-days of an expert on ACRITH who was unfamiliar with the conjugate gradient algorithm.

2.5. An example - Hat matrices

Another example using ACRITH's high accuracy arises in computing Allen's PRESS criterion for selecting variables in regression [1]. Let A be an $m \times n$ matrix of derivatives of a model function with respect to n model parameters at m values of the independent variable. Problems of interest are about 100×5. The PRESS criterion requires the diagonal elements of the $m \times m$ "hat" matrix

$$H = A(A^T A)^{-1} A^T. \tag{6}$$

The sum of the diagonal elements of H is n, and H is idempotent ($H \cdot H = H$). These two properties can be used to check the computed values.

Hoaglin and Welsch [6] suggest using Householder transformations or singular value decompositions to minimize roundoff errors. If possible, we preferred a simpler implementation based on the definition for ease of coding and maintenance. Neither single nor double precision routines from the Harwell or IMSL libraries were sufficiently accurate, even for systems as small as 10×4. When the calls to Harwell or IMSL library routines are replaced by calls to double precision ACRITH routines, the sum of the diagonal elements is n, and H is idempotent to full machine accuracy.

The human and machine time required is an issue. The engineer who did this work remarks, "In my application, the extra time required by ACRITH routines is insignificant in comparison to the time it would have required to program more complex methods and to maintain the program as changes in FORTRAN, etc. occur. It certainly is less than the clock and CPU time I would have wasted by now in trying to make sense of less accurate computed results."

If the accumulation of errors is still too great using ACRITH routines to evaluate H, Equation (6) can be recast as a block system in $m \times m + n \times m$ unknowns

$$\begin{pmatrix} I & -A \\ A^T & \end{pmatrix} \begin{pmatrix} H \\ V \end{pmatrix} = \begin{pmatrix} 0 \\ A^T \end{pmatrix}. \tag{7}$$

Then DLIN is called to solve Equation (7).

2.6. How should you use techniques for high accuracy?

Scientists and engineers can begin at once to apply tools for high accuracy computation to their industrial problems.

Use accurate scalar product in new or existing code.

The accurate scalar product can be profitably used whenever a scalar or dot product is required. Both authors of new development codes and those maintaining existing codes can look for code segments which sum vectors or compute scalar products. Replacing conventional code with a call to the appropriate ACRITH scalar product routine improves the accuracy. Depending on the host hardware, the execution speed may also be improved.

This strategy is similar to that of seeking to recognize scalar products in algorithms being implemented on vector or parallel machines. In both cases, the strategy works because the scalar product is treated as an elementary operation by the machine.

Use high accuracy general-purpose problem solving routines.

ACRITH provides several powerful, general-purpose problem solving routines which can serve as major building blocks. Examples include solving linear equations, eigenvalue/eigenvector analysis, polynomial rootfinding, and solving nonlinear equations. These routines can be used modularly in the same manner as routines from the NAG or IMSL libraries. Hence, a scientist or engineer can include the ACRITH library among the libraries searched for existing software to help solve applications problems.

Use simpler methods before more complicated ones.

The accurate scalar product from ACRITH can be used to make algorithms which have historically been discarded as numerically unstable behave satisfactorily. Examples include solving a least squares approximation problem using the normal equations or by Gram-Schmidt orthonormalization. Such methods are poorly conditioned using conventional arithmetic. However, they may be well behaved when carefully programmed using the accurate scalar product. The simpler methods may work using accurate arithmetic, and they are much easier to program. Of course, some bad methods are still bad methods, even with the accurate scalar product.

2.7. What are the costs?

The primary cost of using the ACRITH tools for high accuracy is the human cost of learning to use them. Those costs are highly variable and difficult to quantify. Learning costs may be minimal, since one form of a scalar product is being replaced with another. Learning costs may be modest, since problem solving routines from the ACRITH library replace similar routines from other libraries. Learning costs also may be large, since new programming tools are replacing familiar ones. Further, a fundamental analysis of the algorithm being used is sometimes required to achieve the potential benefits of the accurate scalar product.

3. When should interval techniques be considered?

Now we turn our attention to our main focus: interval techniques for computing validated inclusions. We address a scientist or engineer who is acquainted with conventional point numerical algorithms and who is interested in appropriate uses of interval algorithms for their problems.

Interval techniques should be considered in two circumstances:

1. Accuracy of the computed answer is in doubt.

2. Input data or parameters in the problem are uncertain.

We consider each of these circumstances separately.

3.1. Accuracy of the computed answer is in doubt.

Interval techniques should be considered whenever to provide some assurance about how accurately the answer is computed. An interval answer $[1.23456, 1.23457]$ provides assurance that the answer is known correctly to 6 decimal places, while a point answer 1.23456 may have the wrong sign. The accuracy of many scientific and engineering calculations is in doubt. Sections 4 and 5 give two examples.

3.2. Uncertain data or parameters.

The second circumstance in which interval techniques should be considered are problems whose input data or parameters are uncertain. Often, data are known to be accurate to only 3 or 4 decimal places. How does the uncertainty in the data contribute to uncertainty in the answer? This question is especially important when the problem is known to be ill conditioned. The question may be ignored, or it may be addressed by hard analysis, by simulations, or by appealing to experience. The analysis is often difficult or impossible; simulations are expensive, especially in many dimensions; and experience may be lacking or misleading.

In contrast, interval techniques provide guarantees, and they may be applied almost automatically. If we start with intervals in which the input data or parameters must lie, then computations using interval arithmetic give intervals in which the answers must lie.

4. An example - Least squares

Interval techniques should be considered when the accuracy of the computed solution is in doubt. For example, we asserted in Section 2.4 that the program in Listing 1 solved a least squares problem accurately. We gave as evidence a residual of 4.02E-14. However, a small residual does not assure a small relative error in the solution, and the conventional wisdom suggests that it is not possible to solve the normal equations as accurately as we claim to have done. Hence, the accuracy of our answer is in doubt.

The ACRITH routine DLIN in Listing 1 computes an interval [XL(J), XU(J)] containing each component of the solution:

```
  J                      [XL(J), XU(J)]
  1  (  0.1044068767878713D+02,  0.1044068767878714D+02 )
  2  (  0.1182954682872744D+02,  0.1182954682872745D+02 )
        .
        .
        .
177  ( -0.1808366597477822D+02, -0.1808366597477821D+02 )
178  (  0.6148360648647853D-01,  0.6148360648647854D-01 )
```

Listing 3. Inclusions of least squares solution.

For each of these intervals, XL(J) differs from XU(J) by one unit in the last hexadecimal place. These interval answers verify that we have computed the answers with full double precision accuracy. This example illustrates the power of interval techniques for automated error analysis. An interval answer carries with it a guarantee of its uncertainty. By contrast, a point value carries no measure of its uncertainty. Even a sound error analysis can only estimate the error.

Interval techniques should also be considered when input data or parameters in the problem are uncertain. For example, if we replace the point valued coefficients of A in Equation (2) by intervals whose endpoints differ in the 4th decimal place, we get intervals containing the solution. The width of the solution intervals is a measure of the uncertainty in the answer as a result of uncertainty in the data. This is a type of sensitivity analysis which can replace costly, repeated simulation runs.

However, if the interval coefficients of A in our problem are only one double precision unit wide, ACRITH's linear equation solver finds that A contains a singular matrix. That is, the data differs from a singular problem by less than double precision accuracy. Since the data is certainly not known with double precision accuracy, the tight inclusions shown in Listing 3 are meaningless. In the terminology of [11], we have solved the specified least squares problem with high accuracy, but the specified problem has no physical meaning. We would prefer to be assured that everything is fine, but if we are attempting to solve a meaningless problem, we welcome a warning of that fact.

This example illustrates that

1. It is possible to solve poorly conditioned linear systems with high accuracy.

2. It is possible to be guaranteed of the accuracy attained.

3. ACRITH can tell us when our problems are too poorly conditioned relative to their input data to render any computed solution physically meaningless.

5. An example - Nonlinear systems

Another example of the application of interval techniques is the Nordgren crack model [19]. The underlying mathematical problem is a partial differential equation with a moving boundary. Let $w(x,t)$ be the width of the crack at position x and time t. Let $L(t)$ be the length of the crack at time t, and $\tau(x)$ be the time required for the crack to reach length x. Then $w(x,t)$ satisfies the PDE:

$$\begin{aligned}
(w^4)_{xx} &= w_t + \frac{1}{\sqrt{t-\tau(x)}}, \quad 0 \le x \le L(t), 0 \le t, \\
-(w^4)_x(0,t) &= 1 \quad \text{(flow from wellbore is constant)} \\
w^4(x,0) &= 0 \quad \text{(crack starts opening at } t=0\text{)} \\
w^4(L(t),t) &= 0 \quad \text{(no flow at the crack tip)} \\
-(w^4)_x(L(t),t) &= \frac{dL}{dt} \cdot w(L(t),t)
\end{aligned} \tag{8}$$

(fluid velocity equals tip propagation speed)

or

$$-(w^4)_x(L(t),t) \quad = \quad 0 \quad \text{(no volume of fluid flow at the tip)}.$$

A standard algorithm uses a discretization in x and t to form finite difference approximations to the partial derivatives appearing in Equation (8). At each time step t_j, one must solve a nonlinear system of equations for $w_{i,j} = w(x_i, t_j)$ using a Newton iteration. A conventional implementation of this algorithm gave results which agreed with asymptotic results and with results computed by Nordgren. However, after about 100 time steps, the solution failed when the program attempted to take the square root of a negative number.

Consider the approximations made in modeling the crack:

1. The physical behavior of cracking rock is approximated by Equation (8).

2. Equation (8) has exact solution $w(x, t)$.

3. Equation (8) is approximated by finite difference equations.

4. The finite difference equations have exact solution $w_{i,j}$ which approximates $w(x, t)$.

5. At time step t_j, $w_{i,j}$ is approximated by a sequence of Newton iterates based on approximations to $w_{i,j-1}$.

No interval algorithms for Problem (8) are known, but we can use interval techniques to explore the behavior of the point algorithm. We do not attempt to compute an inclusion for $w(x, t)$. Instead, we take as our specified problem the finite difference equations which define $w_{i,j}$.

Suppose we assume that $w_{i,j-1}$ at time step t_j are known exactly. Then the values $w_{i,j}$ satisfy a system of $j + 1$ nonlinear equations. The interval Newton algorithm [14] computes 1 ULP inclusions for $w_{i,j}$. This gives inclusions for $w_{i,j}$, not for $w(x, t)$, and not for the width of the physical crack. Further, the algorithm of assumes that $w_{i,j-1}$ are exact, which they are not. Hence, we have an inclusion of the answer which the point algorithm in Step 5 above should have computed.

The interval program described in the preceding paragraph showed that as time passes, the Newton iteration becomes extremely sensitive to the initial guess for $w_{i,j}$ used to start the iteration. The problem is not with the convergence of the Newton iteration, but that a very carefully constructed initial guess is required to evaluate the first Newton iteration. The interval algorithm gives us a better understanding of the behavior of the point algorithm, so we can improve its performance.

Another approach is to view the entire set of $w_{i,j}$ for $0 < j \leq N$ as determined by a single system of $(N+1)(N+2)/2$ nonlinear equations. Solving as a single system rather than solving one time step at a time allows us to compute inclusions for $w_{i,j}$ without assuming that $w_{i,j-1}$ are exact. Further, the widths at earlier times are automatically computed with sufficient accuracy to allow the widths at later times to be computed with 1 ULP inclusions. This is a slight improvement over the previous algorithm because it gives inclusions for all the $w_{i,j}$ without assuming that some are known exactly.

The result is *not* an inclusion for $w(x,t)$ because we have taken the finite difference equations as our specified problem. To compute an inclusion for $w(x,t)$ requires an interval algorithm, not just an interval version of a point algorithm. Interval algorithms for PDE's are under development. One could capture the truncation error in the finite difference approximations to the derivatives, but a better interval algorithm will probably result from an application of a maximum principle or from reformulating the PDE as an integral equation and applying a contractive mapping.

The nonlinear equation solver in ACRITH was overly restrictive for this application, so we had to write our own. We estimate that it took approximately 60 – 100 times as long to write our own interval Newton algorithm following [14] than would be required to simply use an existing point nonlinear equation solver from the IMSL or NAG libraries.

6. What are some limitations of interval techniques?

We do not pretend that interval techniques are the solution to all scientific computation problems. Among the difficulties involved in using interval techniques are:

1. What do we get an inclusion of?

2. How do we develop interval algorithms?

3. How do we develop interval models?

4. Is it worth the cost?

We address each of these in turn.

6.1. What do we get an inclusion of?

Interval problems should be interpreted in the sense of sets. For example, consider the interval valued system of linear equations

$$A\mathbf{x} = \mathbf{b}, \tag{9}$$

where the elements of the matrix A and of the vector $\mathbf{b}$ are intervals. Let $\tilde{A}$ be any point valued matrix such that $\tilde{A}_{i,j} \in A_{i,j}$, and let $\tilde{\mathbf{b}}$ be any point valued vector with $\tilde{b}_i \in b_i$. Let $\tilde{\mathbf{x}}$ be the solution to the point valued problem $\tilde{A}\tilde{\mathbf{x}} = \tilde{\mathbf{b}}$. Then, the meaning of having an interval valued vector $\mathbf{x}$ which solves Equation (9) is that it must hold that $\tilde{x}_i \in x_i$. That is, the solution to the interval problem contains the solution to every possible point problem.

Having an interval answer does not guarantee that it includes anything of interest, as we saw in the discussion above of the Nordgren crack model. To achieve a meaningful inclusion requires careful mathematical reasoning at all stages of algorithm development and implementation.

For example, suppose that we have a code which computes a numerical solution to a system of ordinary differential equations using a Runge-Kutta algorithm. Suppose that we are concerned about the accuracy of the computed solution, so we convert the point arithmetic in the code to interval arithmetic using calls to ACRITH routines. The resulting interval answers are *useless* for two reasons:

1. They enclose the Runge-Kutta approximation to the solution, not the solution itself. In order to enclose the solution, we must extend the algorithm to include an inclusion of the truncation error term.

2. They are hopelessly wide. In order to achieve the desired *tight* inclusions, algorithms must be interval algorithms, not naive interval versions of point algorithms.

Both of these reasons suggest that it is often necessary to develop algorithms which are designed from the beginning as interval algorithms.

As a practical matter, it may not be necessary or cost-effective to develop an interval algorithm for an entire problem. Instead, interval techniques can often be profitably applied to small subproblems of a larger problem. For example, the original algorithm may require the solution of a linear system, a nonlinear system, or an optimization problem. If the accuracy or reliability of one of these subproblems is suspected, an interval solution of the subproblem is suggested. The result will not be an inclusion for the solution of the entire problem, but the result may provide increased confidence or insight.

The example given above of the nonlinear system from a Nordgren crack model illustrates these points. There, we applied interval techniques the subproblem of solving the nonlinear system, either at one time step at a time, or for the entire time domain of interest. We were very careful to identify what we were computing an inclusion of. The use of interval techniques was valuable because we gained insights which allowed us to improve performance of our calculations.

6.2. How do we develop interval algorithms?

Good point algorithms rarely make good interval algorithms. Hence, when developing interval algorithms, it is necessary to discard much of the conventional wisdom of numerical analysis and perform a fresh analysis of each problem. In what follows, we give some useful hints.

The development of good interval algorithms begins with a search of computer and written libraries for existing tools or algorithms which can be applied. The ACRITH library contains many useful problem solving routines.

Interval algorithms often involve computing a point approximation, followed by computing either an interval inclusion near the approximation or an interval inclusion of the error of the approximation.

When developing an algorithm using such a mix of point and interval valued objects, special care must be taken to keep track of which objects are points and which are intervals. It is usually helpful to view an interval as a primitive data object on which operations are performed using calls to ACRITH routines. In computer science jargon, intervals form an abstract data type. Thinking of the two endpoints as separate objects often leads to confusion. Fortran-SC [12] supports interval as a primitive data type at the language level.

Interval algorithms are developed to enclose mathematical results. For example, if $a \in A = [1,2]$, then $a - a = 0$, while $A - A = [-1,1]$. Similarly, if $B = [-1,2]$, then $B * B = [-2,4]$, while $B^2 = [0,4]$. Hence, equivalent mathematical expressions are not necessarily equivalent interval expressions.

Fundamental tools for achieving a guaranteed inclusion use interval arithmetic and must capture of all sources of errors including modeling errors, approximation errors, truncation errors, discretization errors, and roundoff errors. Differentiation arithmetic [20] is often helpful to capture derivatives evaluated at unknown points which appear in error terms for many algorithms. Differentiation arithmetic uses recurrence relations for the fast, accurate generation of the terms of the Taylor series.

Achieving an inclusion is often a separate issue from achieving a *tight* inclusion. Mathematical "tricks" which are commonly used in good interval algorithms to achieve inclusions which are tight include

- interval arithmetic,
- formula + remainder,
- subinterval adaptation,
- monotonicity,
- use of point values,
- high order methods,

- differentiation arithmetic,
- intersection of multiple inclusions, and
- contractive iteration.

In [3], Corliss illustrates these techniques using several algorithms for computing bounds for the range of a function on an interval.

6.3. How do we develop interval models?

Pursuit of an interval strategy often affects the way we build the original mathematical models. In general, we want to solve a problem as close as possible to the original one. If we formulate problem P_1, do some approximation to get problem P_2, do some further approximation to get problem P_3, and finally solve several subproblems, we have many sources of error which must be bounded in order to get a guaranteed inclusion of the solution to the original problem. It is usually better to solve problem P_1, if we can.

We cannot always solve P_1. Even when we can, we may not be able to compute an inclusion. As a compromise, it may be useful to find guaranteed inclusions of the solution to some subproblems or of some approximate solution to the original problem. In this case, be careful to understand clearly what the answer is an inclusion of.

Pursuit of an interval strategy often encourages one to develop models in terms of lower and upper bounds. This is especially useful when details of the physical problem are uncertain but can be shown to have little affect on the outcome.

6.4. Is it worth the cost?

Interval algorithms incur costs for learning an unfamiliar approach, for programmer time, and for CPU time.

Interval techniques are not usually as well known among scientists and engineers as point techniques, so there is a cost associated with learning new approaches and new tools.

There are fewer ready-made interval programs in either computer software libraries or in the literature. Further, an interval algorithm is usually more difficult to write than a point algorithm for the same problem. For these reasons, the design and implementation of a complete integral algorithm may require 4 – 8 times as much programmer time as the design and implementation of a point algorithm. If the choice is between designing and writing your own interval algorithm and using an existing high quality point routine, the cost may be a factor of hundreds. Hopefully, this will be less frequent as more high quality interval routines are developed.

This cost should be balanced against the value of increased insight provided by verified bounds for the computed answers.

An interval algorithm carries lower and upper bounds throughout its calculations. Hence, it nearly always requires twice as much CPU time as a corresponding point algorithm. In practice, a good interval algorithm typically takes about 3 – 5 times as much CPU time as a good point algorithm. It is possible for an interval algorithm to be faster than a point algorithm, especially when a tolerance is requested, because the interval program can determine exactly when the requested tolerance is met. It is also possible for an interval algorithm to take much longer.

The increased CPU cost should be balanced against the cost of repeated runs for simulation, with perturbed data, or in higher precisions.

7. What should you DO?

We recommend:

- Determine whether the ACRITH accurate scalar product routines are faster or slower than the corresponding DO loop on your machine. If they are faster, use them whenever possible.

- Use ACRITH accurate scalar product routines for vector summations and scalar products.

- Include the ACRITH library among the libraries you customarily search when you are looking for existing software to help solve your problem.

- Consider older, simpler algorithms whose bad reputation is due to an accumulation of roundoff errors which can be minimized by an accurate scalar product.

- Determine whether the potential benefits of ACRITH's high accuracy tools justify your effort to learn how to use them by applying them first to small problems to gain experience in your own setting.
- Learn to recognize problems for which interval techniques should be considered because you would like to know the accuracy of the answer you have computed.
- Learn to recognize problems for which interval techniques should be considered because the input data or the parameters in the problem are uncertain.
- Consider interval techniques as an alternative to repeated simulation runs for sensitivity analysis.
- Apply interval techniques to small subproblems of larger, more complicated problems.

The following references are good places to start.

R. E. Moore, *Methods and Applications of Interval Analysis,* [14] and G. Alefeld and J. Herzberger, *Introduction to Interval Computations,* [2] are good introductory surveys of interval methods.

K. Nickel, ed., *Interval Mathematics, 1975,* [16], *Interval Mathematics, 1980,* [17], and *Interval Mathematics, 1985,* [18] are each collections of papers presented at a prestigious conference held in Freiburg.

R. E. Moore, ed., *Reliability in Computing: The Role of Interval Methods in Scientific Computing,* [15] is a collection of 22 research articles in the field.

The ACRITH User's Guide [7] and the Pascal-SC User's Manual [10] are excellent descriptions of two commercially available products which make interval tools widely accessible.

Acknowledgments

The author wishes to thank a client whose wish to remain anonymous prevents giving the appropriate credit for work done by several of its scientists. Professor Edgar

Kaucher suggested the matrix reformulations in Section 2. The author also wishes to thank Professors L. B. Rall and Gary S. Krenz for many helpful discussions.

References

[1] Allen, D. M. The relationship between variable selection and data augmentation and a method for prediction, *Techonometrics,* **16** (1974), 125–127.

[2] Alefeld, G.; and Herzberger, J. *Introduction to Interval Computations,* Academic Press, New York 1983.

[3] Corliss, George F. How can you apply interval techniques in an industrial setting? Technical Report No. 301, Department of Mathematics, Statistics and Computer Science, Marquette University, Milwaukee, WI 1988.

[4] Delves, L. M.; Ford, B.; and Hodgson, G. S. The MAP 750 Ada Library Project, *NAG Newsletter,* **2** (1987), 29–38.

[5] Hestenes, M.; and Stiefel, E. Methods of conjugate gradients for solving linear systems, *Nat. Bur. Standards J. Res.,* **49** (1952), 409–436.

[6] Hoaglin, David C.; and Welsch, Roy E. The hat matrix in regression and ANOVA, *The American Statistician,* **32** (1978) 1, 17–22.

[7] IBM. *High-Accuracy Arithmetic Subroutine Library (ACRITH),* Program Description and User's Guide, SC 33–6164–2, 3rd Edition 1986.

[8] Kulisch, Ulrich W.; and Miranker, Willard L. *Computer Arithmetic in Theory and Practice,* Academic Press, New York 1981.

[9] Kulisch, Ulrich W.; and Miranker, Willard L. *A New Approach to Scientific Computation,* Academic Press, New York 1983.

[10] Kulisch, Ulrich W. (ed.). *Pascal-SC: A PASCAL Extension for Scientific Computation,* Information Manual and Floppy Disks, Teubner, Stuttgart, and Wiley, New York 1987.

[11] Kulisch, Ulrich W.; and Stetter, Hans J. Automatic result verification. In: Kulisch, Ulrich W.; and Stetter, Hans J. (eds.): *Scientific Computation with Automatic Result Verification*, Springer, New York 1988, pp. 1 – 8.

[12] Metzger, M. FORTRAN-SC, A FORTRAN extension for engineering/scientific computing with access to ACRITH. In: Moore, Ramon E. (ed.): *Reliability in Computing: The Role of Interval Methods in Scientific Computing*, Academic Press, New York 1988, pp. 63–80.

[13] Moore, Ramon E. *Interval Analysis*, Prentice-Hall, Englewood Cliffs, NJ 1966.

[14] Moore, Ramon E. *Methods and Applications of Interval Analysis*, SIAM, Philadelphia, PA 1979.

[15] Moore, Ramon E. (ed.). *Reliability in Computing: The Role of Interval Methods in Scientific Computing*, Academic Press, New York 1988.

[16] Nickel, Karl L. E. (ed.). *Interval Mathematics, 1975*, Springer, New York 1975.

[17] Nickel, Karl L. E. (ed.). *Interval Mathematics, 1980*, Academic Press, New York 1980.

[18] Nickel, Karl L. E. (ed.). *Interval Mathematics, 1985*, Springer, New York 1985.

[19] Nordgren, R. P. Propagation of vertical hydraulic fracture, *Society of Petroleum Engineers Journal* **12** (1972) 4, 306–314.

[20] Rall, Louis B. *Automatic Differentiation: Techniques and Applications*, Lecture Notes in Computer Science No. 120, Springer, Berlin 1981.

[21] Siemens. *Arithmos, (BS 2000) Benutzerhandbuch*, U2900–J–Z87–1, Siemens, GmbH 1986.

Programming Languages for Enclosure Methods

C. Ullrich
Institut für Informatik
Universität Basel
CH-4056 Basel
Switzerland

Abstract: In spite of the development and implementation of high level programming languages in the past forty years, the ease and handling of numerical computation is not much improved.

This paper, with regard to this oversight, initially considers desirable concepts for programming languages from a numerical view point. It becomes apparent that existing computer arithmetic falls far short of our needs and that new innovations will be required to overcome the difficulties facing technical scientific computation. In response, the newly developed languages FORTRAN-SC, PASCAL-SC and MODULA-SC are presented. These languages provide a large variety of features especially adapted for numerics.

In particular, they contain a new data type which allows the exact calculation of scalar products of arbitrarily long vectors as well as a special kind of expression for highly precise computation of complete vector and matrix expressions. Finally we finish our discussion with the introduction of program blocks which effect calculation of all variables in a program area with high precision and shall open a window to the near future.

1 Introduction

The design of new languages, modified and improved versions of old standbys such as FORTRAN, PASCAL and MODULA must be carefully balanced against the need for them. Examining the current situation of computer aided numerical calculation,

Computer Arithmetic and Self-Validating
Numerical Methods

ISBN 0-12-708245-X

we discover a large gap between existing programming tools and those needed by engineers, physicists, mathematicians, and other scientists.

More than thirty years ago the first higher programming languages were developed, FORTRAN in 1954 ([5]) and ALGOL 60 in 1958 ([35]). This was an enormous step in the use of digital computers for technical scientific applications. With this advent, the average user (who was so inclined) could accomplish with little effort what was previously possible only for specialists trained in assembler and/or machine languages.

FORTRAN and ALGOL 60, "formula languages," made mathematical algorithm programming possible. These algorithmic expressions could only be formulated for scalar data types - **integer**, **real**, and **boolean** in ALGOL 60 and **integer**, **real**, **double precision**, **complex**, and **logical** in FORTRAN. Both languages provided the user with the possibility to structure data by aggregating scalar values into arrays. Programming of subalgorithms was already possible due to function and subroutine subprograms.

In the meantime, enormous developments took place in the class of imperative programming languages. New more strongly developed languages were added to this class and existing languages were improved to meet new needs. FORTRAN is an excellent example of this kind of development. It was improved in different company versions until in 1966 and then in 1977 it was standardized by the American National Standards Institute (ANSI) and gained worldwide acceptance as a language standard. Since 1977, experts have been working on a new standard FORTRAN-8x which will likely be released this year and will be supported by both ANSI and ISO (International Standardization Organization) ([40]).

ALGOL 60's development was skew to that of FORTRAN's. It was the first higher programming language to be defined abstractly by scientists. However, its successor ALGOL 68 ([41]) was too complex and abstract, and both are no longer in use.

At present, the scientific computing languages PASCAL, MODULA-2 and ADA, which can't deny their relation to ALGOL 60, play important roles ([7, 20, 21, 22, 25, 42]). PASCAL and MODULA-2 have taken prime positions in education, while ADA is used as an all purpose language for the development of complex programs for distributed systems. Despite the support ADA gets from the Department of Defense (DOD) of the United States, it is inestimable whether or not this language will gain much in popularity with respect to FORTRAN.

A rather different development took the programming language C ([17]) which is today the primary language of UNIX systems but also available in several other environments. The design of C is more influenced by BASIC and Fortran than by ALGOL-like languages. Its generality and an absence of restrictions make it very convenient and effective for many tasks than supposedly more powerful languages. C has been used for a wide variety of programs, including the UNIX operating system, the C compiler itself and essentially all UNIX applications software. Although sufficiently expressive the language is rather low-level and cannot compete with FORTAN-8x, PASCAL, MODULA-2 or ADA.

We would briefly like to mention here the concepts that were of importance to the evolution of all these programming languages:

- New data structures such as "records," "sets," and "files," were introduced and "pointers" allowed the building of dynamic data structures.
- The operator concept allowed the user to declare operators and thus employ mathematical notation of nonstandard formulas.
- The development of larger software packages was simplified by the programming of modules or packages. (Modules can be implemented and translated separately.)
- Parallel process programming or concurrency was developed to meet modern computer architecture.

Interestingly, no further development took place in the field of computer arithmetic.

2 The Role of Arithmetic

In higher programming languages, a finite set of real numbers called machine numbers is provided for numerical calculations. These are usually implemented as normalized floating point numbers, $x = m * b^e$ with mantissa $m = \pm 0.d_l, d_{l-1} \ldots d_1$, $d_i \in \{0, 1, 2, \ldots, b-1\}$ for $i = 1(1)l, d \neq 0$ and an exponent e with $e_{min} \leq e \leq e_{max}$. The base b is a natural number greater than one. The set of machine numbers, or the floating point system, is labelled $\Re(b, l, e_{min}, e_{max})$.

It is normal that higher programming language floating point numbers can only be written in the decimal system though nearly all digital computers use binary, octal or hexadecimal internal representations. Thus, all input data must be translated to the internal representation. For a large portion of values, this is only achievable by approximation. As a tacit conclusion, we have to accept that the program does not execute with the original data, but with perturbed data; only a relatively similar problem to the original is being solved. The error that occurs during the mapping is known as the conversion error.

For work with floating point numbers, a floating point arithmetic with the operations $+, -, *, /$ was provided. Floating point systems are not closed under these operations. Therefore, descriptions of these operations are merely approximations of the mathematical operations. No other programming language but ADA, with the model number concept ([26, 27]), contains a kind of quality statement for floating point arithmetic. Example: To which machine number is the intermediate result $x \circ y$ for the operations $+, -, *, /$ rounded? (See Figure 1)

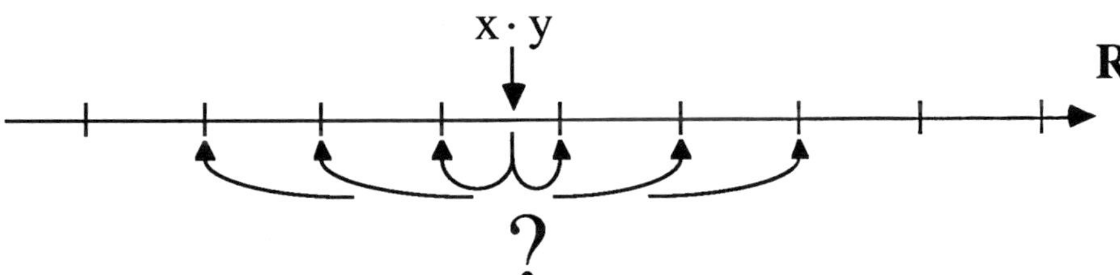

Figure 1: Result of a Floating Point Operation

The most useful results would come from using the two closest floating point numbers. Although this method is postulated in an internationally accepted standard and is made possible now by a variety of microprocessors, it is yet to be introduced in higher programming language definitions other than in ADA, PASCAL-SC, and FORTRAN-SC. The nature of the problem is obvious if we look at the following:

Problem: Compute the lower bound for the sum of two floating point numbers x and y.

For clarity's sake, let us denote real addition by $+$ and machine addition by $\boxplus$. Then we first ask, is $x \boxplus y \leq x + y$? We cannot give an answer in general. Now begins the search through computer manuals. Three situations typically occur:

a. There is no statement regarding arithmetic.

b. The nearest machine number is always calculated. Then the lower bound can be calculated by either subtraction or addition of 1 to the last digit of the mantissa of $x \boxplus y$: subtraction when $x \boxplus y > 0$, addition when $x \boxplus y < 0$. If $x \boxplus y = 0$, the value $-0.1 * b^{e_{min}}$ can immediately be employed. We see that we need a lot of information about the internal number representation to get our answer. Only a sophisticated user will be able to do so.

c. Rounding can be controlled in a way which is implementation dependent, e.g. by referencing a function "setround" before addition.

```
uses sane;
........
setround (Downward);
.........
u := x+y;
```

or by executing the directed operation by a function call itself:

```
u := nextdown (x+y);
```

The above mentioned procedures should, however, only be regarded as emergency solutions since they interfere with the portability of software.

The problems with usability are immediately obvious when evaluating larger expressions. Below, we consider a simple, but elementary problem of computer geometry.

> *Problem: Does a point (x, y) lie above or below a line that is defined by the points (x_1, y_1) and (x_2, y_2)?*

Using the equation [37]

$$y(x) = \frac{x - x_1}{x_2 - x_1} * y_2 + \frac{x_2 - x}{x_2 - x_1} * y_1$$

the test can be performed in the following manner:

If $x_1 = x_2$: Compare y_0 and y_1

If $x_1 > x_2$: Interchange (x_1, y_1) and (x_2, y_2), and continue in the following case.

If $x_1 < x_2$: Compare y_0 with $y(x_0)$, i.e.

$y_0 * (x_2 - x_1)$ with $(x_0 - x_1) * y_2 + (x_2 - x_0) * y_1$ or

$y_0 * x_2 - y_0 * x_1 - x_0 * y_2 + x_1 * y_2 - x_2 * y_1 + x_0 * y_1$ with 0.

The evaluation of the expression in the last line above using machine operations

$$w := y_0 \boxast x_2 \boxminus y_0 \boxast x_1 \boxminus x_0 \boxast y_2 \boxplus x_1 \boxast y_2 \boxminus x_2 \boxast y_1 \boxplus x_0 \boxast y_1$$

cannot give us the certainty that in the cases $w < 0$, $w = 0$ or $w > 0$ the exact result fulfills the same property. The difficulty with this problem has become famous in computer geometry as the phenomenon of "dangling lines," lines that seem to clasp each other. Almost parallel lines, each defined by two points, lying very close, can lead to contrary statements regarding their mutual positions.

The point-line test can be applied to cases for which w is a lower bound for $w > 0$ or an upper bound for $w < 0$ to obtain the exact result. This could be possible by the earlier mentioned rounding control. Due to changing signs, a closer analysis of the expression now becomes necessary. Notice that application of

```
setround (Downward);
```

$$w := y_0 * x_2 - y_0 * x_1 - x_0 * y_2 + x_1 * y_2 - x_2 * y_1 + x_0 * y_1$$

in general evolves no lower bound. However, the optimal solution is the exact computer evaluation or at least the computation of the closest machine number, if rounding towards zero is not used. For later use, we wish to mention that the scalar product of two vectors $y = (y_0, -y_0, -y_2, y_2, -y_1, y_1)$ and $x = (x_2, x_1, x_0, x_1, x_2, x_0)$ leads to the same expression.

In practical applications, data are often not precise, but rather the machine equivalent solution of previous calculations. For this reason, we must assume that a tolerance accompanies the data. For example, in our point-line test instead of the points (x_1, y_1) and (x_2, y_2), we have $(x_1 \pm \varepsilon_1, y_1 \pm \delta_1)$ and $(x_2 \pm \varepsilon_2, y_2 \pm \delta_2)$ and maybe $(x_0 \pm \varepsilon_0, y_0 \pm \delta_0)$.

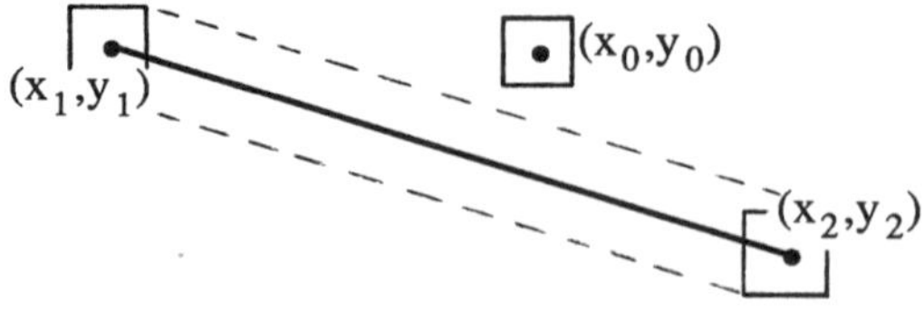

Figure 2: Point Line Test with Tolerance for the Data

It naturally follows to employ interval arithmetic, a simple tool for calculations involving tolerances (see [8]). Real and/or complex vector matrices appear in most numerical applications, yet correspondingly, there has been no arithmetic developed to deal with these in higher programming languages. Thus the operations must be implemented by software, inducing the problem of evaluation of compound expressions. Complex multiplication is normally programmed in the form:

$$z \boxast w = ((z_1 \boxast w_1) \boxminus (z_2 \boxast w_2), (z_1 \boxast w_2) \boxplus (z_2 \boxast w_1))$$

for complex machine numbers, $z = (z_1, z_2)$ and $w = (w_1, w_2)$, while the desired values for the real part and the imaginary parts, $(z_1 * w_1 - z_2 * w_2)$ and $(z_1 * w_2 + z_2 * w_2)$, respectively, are the machine numbers closest to the exact values.
Accordingly, the product of two matrices $a = (a_{ij})$ and $b = (b_{ij})$, for $i, j = 1(1)n$ is given as $a \boxast b = \square \sum_{k=1}^{n} (a_{ik} \boxast b_{kj})$ although, in each component the value $\square(\sum_{k=1}^{n} a_{ik} * b_{kj})$ would be appropriate.
Vector and parallel computers, a separate aspect of supercomputer technology, are calculators which characteristically provide the user with vector operations that cannot be used by or are only available for higher programming languages in a restricted manner. The procedure in Figure 3

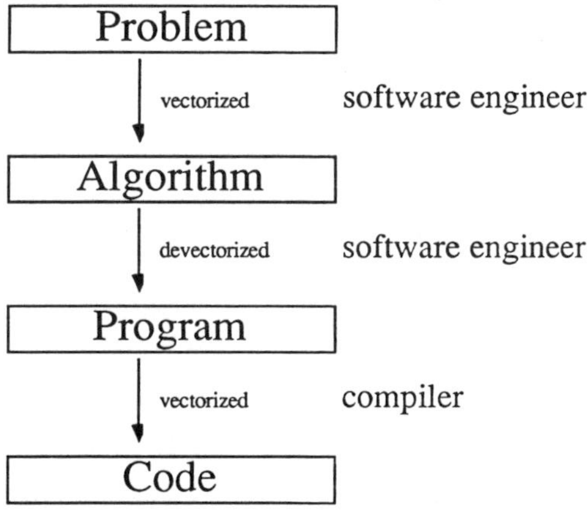

Figure 3: Vectorization for Supercomputers

works on a typical problem.
The vector algorithm indicated by the mathematical model of the original problem must be changed to a scalar algorithm by the programmer. Then to run the program, vectorizing supercomputer compilers create a code that utilizes the vector operations. Much of this effort could be saved if machine independent vector operations existed for the programming language. These operations could then be translated directly in corresponding machine dependent code.

3 New Developments

Some high level programming languages developed in the past ten years have overcome many of the aforementioned difficulties. Since 1978, PASCAL-SC has been defined as an enhancement of PASCAL and has been implemented on processors with various missions ([12, 13, 14, 15, 36]). PASCAL-SC systems run on IBM PCs und ATARI STs ([30, 29]). FORTRAN-SC, started in 1983 ([10, 9]), an enhancement of FORTRAN 77 ([5]), evolved in another direction. This language originally was designed to simplify the use of the ACRITH Subroutine Library which offered the arithmetical facilities of PASCAL-SC as subprograms. A precompiler from FORTRAN-SC to FORTRAN-77 was developed which makes possible the use of the language supported by software on IBM PCs, AT/370s, and IBM/370s, and by hardware on the IBM 4361s and 9377s ([2, 3]).
The bulk of the development work was done at the Institut für Angewandte Mathematik (Prof. U. Kulisch) of the Universität Karlsruhe, West Germany, with the temporary cooperation of the Institut für Informatik of the Universität Kaiserslautern, Nixdorf AG (Paderborn), IBM Deutschland (Böblingen), Siemens AG (München), CWI (Amsterdam) and NAG (Oxford).
With the collaboration of the latter three partners from 1986 to 1989, the ESPRIT project DIAMOND (Development and Integration of Accurate Mathematical Operations in Numerical Data Processing) was succesfully prepared.
In the last year we designed MODULA-SC, an extension of MODULA-2 for scientific computation at our Institute. The language was partly implemented by my collaborators C. Falco Korn, S. König and S. Gutzwiller as a precompiler to MODULA-2. Since even the basic arithmetic is written in software, the result is a highly portable

tool which can be transferred with small effort to each computer system providing MODULA-2 ([28]).

Let us briefly summarize the new possibilities of FORTRAN-SC, PASCAL-SC and MODULA-SC:

- Real Operations with accuracy guarantee (according to IEEE standard)

  $+, -, *, /$ (with rounding to the next machine number)
  $+<, -<, *<, /<$ (with downwardly directed rounding)
  $+>, ->, *>, />$ (with upwardly directed rounding)

- Complex arithmetic with accuracy guarantee
- Vector and matrix arithmetics (**real** and **complex**) with accuracy guarantee
- Interval arithmetic for all numeric data types
- Precise evaluation of vector and matrix expressions by so called scalar product expressions
- A new data type, **dotprecision**, that is used to calculate and store the result of arbitrarily long scalar products.
- A great number of numeric standard functions for which the precision of one ulp (unit last place) is guaranteed.
- Dynamic arrays which allow numeric software to be programmed and translated without set dimension for vectors and matrices.
- Functions with structured result types
- Declaration of operators, thus preserving the mathematical notation in programs.
- Modular programming (only in PASCAL-SC and MODULA-SC)

With exception of the module concept, all events are implemented equivalently in both languages. In consideration of paper length, we will only take a closer look at the new data type **dotprescision** and the scalar product expressions, and then

finally examine the so-called accurate blocks, program parts using highly precise evaluation.
Further details on these as well as on all remaining language elements are to be found in the corresponding literature [28, 3, 13, 36].

4 New Datatype Dotprecision

In applications, it is often necessary to calculate scalar products of vectors. This need can be hidden, as in the point-line test, or it can be very obvious as in complex arithmetic or vector matrix arithmetic. In any case, the new data type **dotprecision** computes these scalar products exactly and stores the values. Variables of this simple datatype structure are declared in the usual form, as for example in PASCAL-SC by **var** d : dotprecision.
The assignment of values to variables of type **dotprecision** is only possible with dotprecision expressions. Figure 4 shows the syntax of such expressions using the usual syntax diagrams. The structure of these diagrams and their use will not be further explained.

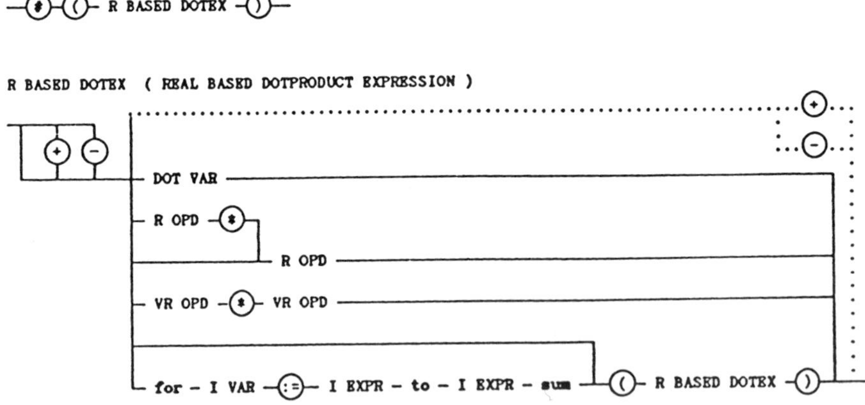

Figure 4: Syntax of the Dotprecision Expressions

Dotprecision expression can consist of a series of dotprecision variables, products of real operands, and scalar products of real vectors, all of which may only be added and/or subtracted. The last line of the second diagram allows the building of expressions dynamically by a kind of loop. The following example shows the various possibilites of calulation of the scalar product of two vectors, x and y:

```
{Possibility 1 : }
d := # (x * y);
{Possibility 2 : }
d := # (for i := 1 to n sum (x [i] * y [i]));
{ Possibility 3 : }
d := # (0);
for 1 := 1 to n do d :=# (d + x [i] * y [i]);
```

A special feature of the dotprecision variable is the possibilty to accumulate scalar prodcuts, i.e. to use exact values for further calculation of scalar products. Given x, y, z, and w as real vectors, the exact evaluation of the expression $x * y + z * w$ is possible as follows:

```
d:= # (x * y + z * w);
```

but also,

```
d := # (x * y);
.......
d := # (d + z * w);
```

In the second and third statements above, the values of x and y, and z and w are not needed at the same time.

The special properties of the type **dotprecision** are achieved by its implementation as a fixed point number with the necessary number of digit positions (for detailed description see [11]). In comparison to the representation of real numbers, the memory used for storage of dotprecision values is much larger. It is therefore suggested to use these values rather rarely and only when writing algorithms for which high precision is needed.

The conversion to other data types is provided by usage of the symbols $\# <$, $\# >$, $\#*$, or $\#\#$ which convert to the next smaller, larger, or nearest floating point number, or the smallest enclosing real interval respectively.

<u>Example:</u>

```
var  i : interval;
     d : dotprecision;
     v : rvector
.................
     i.inf := #< (d);
     i.sup := #> (d);
alternatively:
     i := ## (d);
.........
     for j := 1 to k do {readout digits for d>0}
     begin
          v[i] := #< (d);
          d := # (d - v[i])
     end;
```

Execution of the loop assigns the first $l * k$ significant digits of d as values with corresponding scaling factors to the k components $v[1], v[2], ...v[k]$ of the real vector v. This allows us to store a certain number of significant digits with less effort, especially when the number of these digits is much smaller than the total number of digits for dotprecision values.

Finally, we would like to comment that the difficulties facing the point-line test are now relieved by use of the exact value: $\#(y_0*x_2-y_0*x_1-x_0*y_2+x_1*y_2-x_2*y_1+x_0*y_1)$

5 Scalar Product Expressions

Scalar product expressions make it possible to enhance the principle of highly precise or rather exact evaluation of expressions for all data types, which had previously been possible only for real expressions with real operands by the dotprecision expression. Now, operands of the types **complex**, **interval** or **complex interval** can be terms in these scalar product expressions, as well as vector and matrix operands with these component types, as far as this makes sense from a mathematical point of view. Variables of component type **dotprecision** are not allowed; each result of a scalarproduct expression is converted to a **real**, **complex** or **interval** type. The rounding symbols #*, # <, # > and ## are provided for this reason. Figure 5 shows the syntax of scalar product expression for real vectors. Real vectors and products of real vectors and real values or matrices are allowed as addition terms.

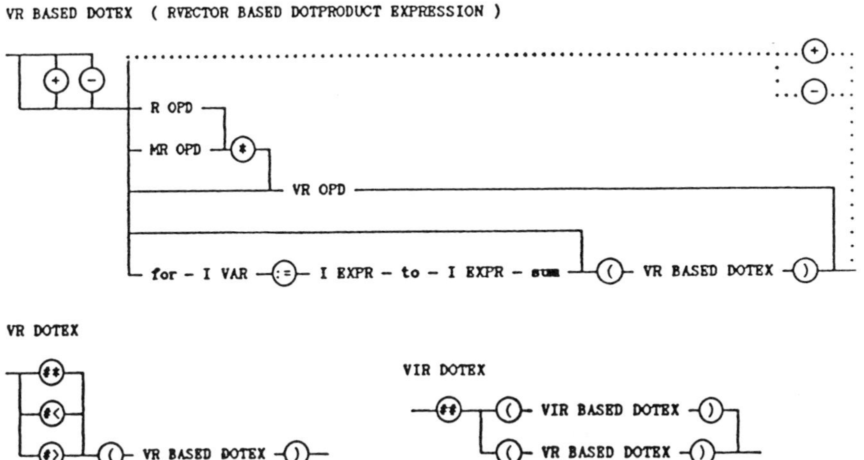

Figure 5: Syntax of Scalar Product Expression for Real Vectors

Again, the rounding symbols #*, # <, and # > provide the conversion to the respective real vectors, and the ## converts values to the smallest enclosing interval vectors.

Scalar product expressions can only appear on the righthand side of assignment statements.

Example:

```
v, w, x, y : rvector
..........................
A : rmatrix
y := #* (v + w + A * x + r * v)
```

During compilation the assignment statement is transferred into a program section which is equivalent to the following source code:

```
dim := length(y);
if (length(v) = dim) and (length(w) = dim)
   and (length(A,1) = dim) and (length(x) = dim)
   then
```

```
for i := 0 to dim do
        y[lbound(y)+i] := #* (v[lbound(v) + i] +
        w[lbound(w) + i] + A[lbound(A,1) + i] * x +
        r * b[lbound(v) + i]
else
        indexerror(y)
```

It is obvious that scalar product expressions are evaluated componentwise. This allows the use of dotprecision storage only when really needed. In the example above, storage space is only used for one dotprecision value. The automatic check of vector lenghts greatly reduces errors in scalar product expressions.

The following PASCAL-SC procedure for the verified solution of a system of linear equations, $A * x = b$, shows the usage of scalar product expressions at critical points in the algorithm, such as during the defect computations of $b - A * xt$ and $E - R * A$ with an approximated solution, xt, and an approximated inverse, R of A. The expression $z + B * xk$, however, is evaluated using simple interval vector operations during the iteration of the inclusion.

```
procedure solve-lin-system (dim: integer);
  var R, A : rmatrix [1..dim, 1..dim];
        B : imatrix [1..dim, 1..dim];
        xt, b : rvector [1..dim];
        xkl, xk, z : ivector [1..dim];
        k : integer;
  begin
        mrread (A); vrread (b);
        R := inverse (A);
        xt := R * b;
        z := ## (b-A*xt);
        B := ## (mrid(dim) - R*A);
        z := intpt (R) * z; {R considered as interval matrix}
        xkl := z;
        k := 0;
        repeat
```

```
        xk := xk1;
        xk1 := z + B * xk;
        k := k + 1
      until (xk1 < xk) or k = 10;
      if k < 10 then
      begin writeln ('verified inclusion');
               viwrite (intpt(xt) + xk1);
  end
  else writeln ('inclusion failed')
  end;
```

6 Program Parts with Highly Accurate Evaluation of Expressions

Generic arithmetical expressions do not fit into the scheme of scalar product expressions and must therefore be treated separately, if an accurate evaluation is desired. Even the evaluation of a simple polynomial (Horner's Method)

$$p(x) = -8118x^4 + 11482x^3 - x^2 - 5741x + 2030$$

at various points in the interval [1.4142, 1.4143] results in a graph (see figure 6) that makes it impossible to determine whether or not a zero lies in this area.

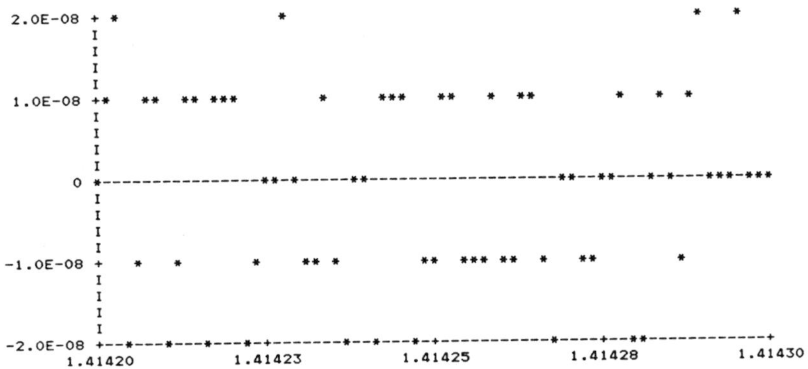

Figure 6: Polynomial Evaluation by Horner's Method

Evaluation by Horner's Method can be described as a system of linear equations with the Horner coefficients as unknown values, if the polynomial is given in the form:

$$p(x) = a_n x^n + a_{n-1} x^{n-1} + ... + a_1 x + a_0$$

$$\begin{pmatrix} -1 & & & \\ -x & \cdot & 0 & \\ & & \cdot & \\ & 0 & & \cdot \\ & & -x & 1 \end{pmatrix} * \begin{pmatrix} u_n \\ \cdot \\ \cdot \\ \cdot \\ \cdot \\ u_0 \end{pmatrix} = \begin{pmatrix} a_n \\ \cdot \\ \cdot \\ \cdot \\ \cdot \\ a_0 \end{pmatrix}$$

Thus the evaluation can be done with satisfactory accuracy using specialized self-validating methods (see [16, 18]). The usage of such methods must be as simple as possible for the user. He would be completely overtaxed if he had to transfer the expression to be evaluated to the corresponding system of equations. For this reason, PASCAL-SC and FORTRAN-SC are currently enhanced in such a way that programming is not altered. Only the keywords **accurate** and **end accurate** indicate that in the included program part, the compilation creates code for the solution of the corresponding system of equations.

```
accurate
        p := a[n];
        for i := n-1 downto 0 do
            p := p * x + a[i]
end accurate;
```

Program parts with accurate evaluation of expressions must be built relatively simply; they should consist of assignment and loop statements. The occuring arithmetical expressions can only contain the standard operators and standard functions. Since there are no restrictions as to the form of expressions, verification of the solution of a system of nonlinear equations is done internally. In principle, the evaluation of standard functions must be able to meet accuracy requirements. As a rule, all

variables that occur on the lefthand side of a assignment statement are computed with an precision of 1/2 ulp (**u**nit **l**ast **p**lace).
The following example is taken from digital circuit theory. It will once more clarify the importance of the **accurate** program parts: For the growth function

$$f(x) = 4 * \lambda * x * (1 - x), 0 < \lambda < 1$$

the series

$$x_{k+1} = f(x_k), k = 0, 1, 2, \ldots\ldots\ldots$$

is to be calculated. Computing the first 80 values with the ad hoc FORTRAN-77 program

```
      PROGRAM R E C U R
      REAL*8 X, LAMBDA
      X=0.3D0
      LAMBDA =0.954D0
      WRITE(*,*) 'STARTING VALUE :',X
      WRITE(*,*)'LAMBDA                :',LAMBDA
      DO 10 I=0, 80
         IF ( MOD(I,10) .EQ. 0 ) THEN
            WRITE(*,*) 'X', I,'         =', X
         ENDIF
      X = 4.D0*LAMBDA*X(1.D0-X)
10    CONTINUE
      END
```

we get the following results:

STARTING VALUE : $0.30000000000000000\underline{3}$

LAMDA : $0.95400000000000000\underline{1}$

X00 $= 0.30000000000000000\underline{3}$

X10 $= 0.8649381492087\underline{71531}$

X20 $= 0.2614885715\underline{30312103}$

X30 $= 0.909597210\underline{049215482}$

X40 $= 0.2840260\underline{99202510110}$

X50 $= 0.59559\underline{5048301181984}$

X60 $= 0.9480\underline{16692296503566}$

X70 $= 0.92\underline{1464704087226846}$

X80 $= 0.\underline{580993658577303035}$

Even though the results develop reasonably until k=70, the inclusion intervals calculated with the ACRITH routine DNLSS indicate that the underlined digits are not correct. Treating the loop statement in the above program as an accurate block would create values x_k from the given interval:

X000 = (0.2999999999999999D+00 , 0.3000000000000001D+00)
X010 = (0.8649381492087607D+00 , 0.8649381492087608D+00)
X020 = (0.2614885715285169D+00 , 0.2614885715285171D+00)
X030 = (0.9095972104041198D+00 , 0.9095972104041199D+00)
X040 = (0.2840260063537763D+00 , 0.2840260063537764D+00)
X050 = (0.5955912948946497D+00 , 0.5955912948946498D+00)
X060 = (0.9480474416038242D+00 , 0.9480474416038243D+00)
X070 = (0.9290968225709420D+00 , 0.9290968225709421D+00)
X080 = (0.9498040714091101D+00 , 0.9498040714091102D+00)

The referenced ACRITH subroutine [4] (ARITHMOS [6] provides a similar subroutine) works in the same manner as the above mentioned principle of solution of equations with the difference that all expressions to be evaluated are to be given in form of strings. With the necessary calculation of a starting approximation and the declaration of all needed variables, a large amount of detailed knowledge is required. This would be an unreasonable method for the user.

7 Final Remarks

The necessity of new programming tools for the numerics can easily be justified, though it is impossible to judge which are the best means. The languages PASCAL-SC and FORTRAN-SC represent an effort to at least satisfy their most urgent needs. Indeed, the evolution of languages has not yet stopped and the efforts of the whole user community are needed to test tools with new applications and to formulate ideas for new developments. Above all, producers of computers, language designers, and system implementators must be made sensitive to the needs of the community; this, however, can only be accomplished with the aid of user feedback.

References

[1] *A Standard for Binary Floating-Point Arithmetic.* American National Standards Institute/Institute of Electrical & Electronic Engineers, New York, 1985. ANSI/IEEE Std. 754–1985.

[2] *FORTRAN-SC General Information Notes and Sample Programs.* IBM Deutschland, Stuttgart, 1989.

[3] *FORTRAN-SC Language Reference and User's Guide.* IBM Deutschland, Stuttgart, second edition, 1989.

[4] *IBM High/Accuracy Arithmetic Subroutine Library (ACRITH): Program Description and User's Guide.* IBM Corporation, 1986.

[5] *Programmiersprache FORTRAN.* Beuth Verlag, Berlin, 1980. DIN 66027.

[6] *Siemens ARITHMOS Benutzerhandbuch.* Siemens AG, München, 1986. Bestellnummer U 2900–J–Z87–1.

[7] *The Programming Language Ada, Reference Manual.* ANSI, Springer, Berlin, 1983. Approved in February 1983.

[8] G. Alefeld. Enclosure Methods. In Chr. Ullrich, editor, *Computer Arithmetic and Self-Validating Numerical Methods*, Academic Press, Boston, 1990.

[9] J.H. Bleher, S.M. Rump, U. Kulisch, M. Metzger, Chr. Ullrich, and W. Walter. FORTRAN-SC: A Study of a FORTRAN Extension for Engineering Scientific

Computation with Access to ACRITH. *Computing Supplementum*, 6:227–244, 1988. Springer, Wien.

[10] J.H. Bleher, S.M. Rump, U. Kulisch, M. Metzger, Chr. Ullrich, and W. Walter. FORTRAN-SC: A Study of a FORTRAN Extension for Engineering Scientific Computation with Access to ACRITH. *Computing*, 39:93–110, 1987.

[11] G. Bohlender. What do we need beyond IEEE Arithmetic? In Chr. Ullrich, editor, *Computer Arithmetic and Self-Validating Numerical Methods*, Academic Press, Boston, 1990.

[12] G. Bohlender, K. Grüner, E. Kaucher, R. Klatte, W. Krämer, U.W. Kulisch, S.M. Rump, Chr. Ullrich, J. Wolff von Gudenberg, and W.L. Miranker. *PASCAL-SC: A PASCAL For Contemporary Scientific Computation.* Research Report RC 9009, IBM Thomas J. Watson Research Center, 1981.

[13] G. Bohlender, L.B. Rall, Chr. Ullrich, and J. Wolff von Gudenberg. PASCAL-SC: A Computer Language for Scientific Computation. *Perspectives in computing*, 17, 1987. Orlando.

[14] G. Bohlender, L.B. Rall, Chr. Ullrich, and J. Wolff von Gudenberg. *PASCAL-SC: Wirkungsvoll programmieren, kontrolliert rechnen.* Bibliographisches Institut, Mannheim, 1986.

[15] G. Bohlender, Chr. Ullrich, and J. Wolff von Gudenberg. New Developments in PASCAL-SC. *SIGPLAN Notices*, 23(8):83–92, 1988.

[16] H. Böhm. *Berechnung von Polynomnullstellen und Auswertung arithmetischer Ausdrücke mit garantierter maximaler Genauigkeit.* PhD thesis, Universität Karlsruhe, 1983.

[17] D.M.Ritchie, S.C.Johnson, M.E. Lesk, and B.W. Kernighan. The C Programming Language. In E. Horowitz, editor, *Programming Languages: A Grand Tour*, pages 458–479, Computer Science Press, Rockville, 1987. Springer, Wien.

[18] H.C. Fischer, G. Schuhmacher, and R. Haggenmüller. Evaluation of Arithmetic Expression with Guaranteed High Accuracy. *Computing, Suppl.*, 6:149–158, 1988. Springer, Wien.

[19] C.A.R. Hoare. Hints on Programming Language Design. In E. Horowitz, editor, *Programming Languages: A Grand Tour*, pages 31–40, Computer Science Press, Rockville, 1987.

[20] K. Jensen and N. Wirth. *PASCAL User Manual and Report. ISO PASCAL Standard.* Springer, Berlin, 3rd edition, 1985.

[21] E. Kaucher, R. Klatte, and Chr. Ullrich. *PASCAL.* Volume 2 of *Programmiersprachen im Griff*, Bibliogr. Institut, Mannheim, 1981.

[22] E. Kaucher, R. Klatte, Chr. Ullrich, and J. Wolff von Gudenberg. *ADA.* Volume 4 of *Programmiersprachen im Griff*, Bibliogr. Institut, Mannheim, 1983.

[23] E. Kaucher, U. Kulisch, and Chr. Ullrich, editors. *Computerarithmetic: Scientific Computation and Programming Languages.* Teubner, Stuttgart, 1987.

[24] E. Kaucher and W.L. Miranker:. *Self-Validating Numerics for Function Space Problems.* Academic Press, New York, 1984.

[25] R. Klatte and Chr. Ullrich. *MODULA-2.* Volume 9 of *Programmiersprachen im Griff*, Bibliographisches Institut, Mannheim, 1988.

[26] R. Klatte, Chr. Ullrich, and J. Wolff von Gudenberg. Arithmetic Specification for Scientific Computation in ADA. *IEEE Transaction on Computers*, C-34:996–1005, 1985.

[27] R. Klatte, Chr. Ullrich, and J. Wolff von Gudenberg. Implementation of Arithmetic for Scientific Computation in ADA. *Journal of PASCAL, ADA & MODULA-2*, 1989. Wiley, Chichester.

[28] C. Falco Korn, S. König, and S. Gutzwiller. MODULA-SC: A Precompiler to MODULA-2. In *Contributions to Computer Arithmetic and Self-Validating Numerical Methods.*

[29] U. Kulisch, editor. *PASCAL-SC: A PASCAL-Extension for Scientific Computation. Inform. Manual and Disks for ATARI SC.* Teubner, Stuttgart, 1987.

[30] U. Kulisch, editor. *PASCAL-SC: A PASCAL-Extension for Scientific Computation. Information Manual and Disks for IBM PC. Wiley-Teubner Series in Computer Science*, Wiley-Teubner, Chichester, 1987.

[31] U. Kulisch and W.L. Miranker, editors. *A New Approach to Scientific Computation.* Academic Press, New York, 1983.

[32] U. Kulisch and W.L. Miranker. *Computer Arithmetic in Theory and Practice.* Academic Press, New York, 1981.

[33] U. Kulisch and H.J. Stetter. Scientific Computation with Automatic Result Verification. *Computing Supplementum*, 6, 1988. Springer, Wien.

[34] U. Kulisch and Chr. Ullrich, editors. *Wissenschaftliches Rechnen und Programmiersprachen.* Volume 10 of *Berichte des German Chapter of the ACM*, Teubner, Stuttgart, 1982.

[35] P. Naur. Report on the algorithmic language ALGOL60. *Numer. Math.*, 2:106–137, 1960.

[36] M. Neaga and J. Wolff von Gudenberg. *The PASCAL-SC level 2 Compiler Architecture and Implementation Description.* Esprit-Project Diamond 03/1-1/1/K2.f, University of Karlsruhe, 1988.

[37] Th. Ottmann, G. Thiemt, and Chr. Ullrich. Numerical Stability of Simple Geometric Algorithms in the plane. *Computation Theory and Logic, Lecture Notes in Computer Science*, 270:277–293, 1987. Springer, Stuttgart.

[38] Th. Ottmann, G. Thiemt, and Chr. Ullrich. On arithmetical problems of geometric algorithms in the plane. *Scientific Computation with Automatic Result Verification, Computing Supplementum*, 6:123–138, 1988. Springer, Wien.

[39] Chr. Ullrich. Neue Programmiersprachen für die Numerik - Theorie und Beispiele. *ZAMM*, 70:T519–T529, 1990.

[40] Chr. Ullrich. Rechnerarithmetik und die Weiterentwicklung von FORTRAN. *Elektronische Rechenanlagen*, 26:71–78, 1984.

[41] A. van Wijngaarden, B.J. Mailloux, J.E.L. Peck, and C.H.A. Kester. Report on the Algorithmic Language ALGOL-68. *Numer. Math.*, 14:79–218, 1969.

[42] N. Wirth. *Programming in MODULA-2.* Springer, Berlin, 1982.

The Determination of Guaranteed Bounds to Eigenvalues with the Use of Variational Methods I

Friedrich Goerisch and Zhiqing He*)
Institut für Mathematik
Technische Universität Clausthal
D-3392 Clausthal-Zellerfeld
Federal Republic of Germany

Abstract

An inclusion theorem by means of which guarenteed bounds to eigenvalues can be calculated is presented. This theorem is especially well suited for the treatment of complicated eigenvalue problems such as those involving systems of partial differential equations. The application of the theorem is discussed and is explained in details for the example of Stokes' eigenvalue problem, which plays an important role in hydrodynamics. The numerical results achieved with the theorem are highly encouraging.

1. Introduction

During the past two decades, interval arithmetic has received a large amount of attention, especially because of numerous interesting results from the Kulisch school. The most spectacular success has hitherto been achieved with linear and nonlinear systems of equations as well as with matrix eigenvalue problems. Many remarkable publications concerning ordinary differential equations have appeared, too. In contrast, the treatment of problems with partial differential equations by means of interval mathematics still appears in the initial stages of development. The purpose of the present publication is to provide a contribution to this field. A procedure which permits the calculation of guaranteed bounds to eigenvalues

*) Permanent address: Department of Mathematics, East China University of Chemical Technology, Shanghai, PR China.

ISBN 0-12-708245-X

for a large class of eigenvalue problems is presented. This class includes the most important eigenvalue problems with partial differential equations encountered in science and engineering.

2. Eigenvalue problems with bilinear forms

In science and engineering, eigenvalue problems of highly diversified form occur, for example, eigenvalue problems with matrices, eigenvalue problems with one (ordinary or partial) differential equation, and eigenvalue problems with systems of differential equations. For treating as large a class as possible of eigenvalue problems in a unified manner, the concept of eigenvalue problem is first formulated — as usual — in an abstract way. For this purpose, the following notations are introduced:

D_N is a real vector space, D_M is a subspace of D_N, $M : D_M \times D_M \to \mathbb{R}$ and $N : D_N \times D_N \to \mathbb{R}$ are bilinear forms.

Eigenvalue problems of the following form are now considered: Find pairs (λ, φ) such that

$$\lambda \in \mathbb{R}, \quad \varphi \in D_M, \quad \varphi \neq 0, \quad M(f,\varphi) = \lambda N(f,\varphi) \quad \text{for all} \quad f \in D_M \tag{1}$$

holds.

If (1) is satisfied, λ is called an *eigenvalue of problem (1),* and φ is called an *eigenelement belonging to the eigenvalue* λ.

If $\tilde{\lambda}$ is an eigenvalue of problem (1),

$$\dim \{\tilde{\varphi} \in D_M : M(f,\tilde{\varphi}) = \tilde{\lambda} N(f,\tilde{\varphi}) \quad \text{for all} \quad f \in D_M\}$$

is designated as the multiplicity of the eigenvalue $\tilde{\lambda}$. Eigenvalues are always counted in accordance with their multiplicity.

Eigenvalue problems with linear operators in a Hilbert space can be regarded as a special case of problem (1):

Let $(H, \langle \cdot \mid \cdot \rangle)$ be a real Hilbert space, let D_M and D_N be dense subsets of H with $D_M \subset D_N$, let $\mathcal{M} : D_M \to H$ and $\mathcal{N} : D_N \to H$ be linear operators. If M and N are defined by

$$M(f,g) := \langle f \mid \mathcal{M}g \rangle \qquad \text{for} \quad f, g \in D_M,$$

$$N(f,g) := \langle f \mid \mathcal{N}g \rangle \qquad \text{for} \quad f, g \in D_N,$$

then the following statements are equivalent:

(i) $\lambda \in \mathbb{R}, \quad \varphi \in D_M, \quad \mathcal{M}\varphi = \lambda \mathcal{N}\varphi,$

(ii) $\lambda \in \mathbb{R}, \quad \varphi \in D_M, \quad M(f,\varphi) = \lambda N(f,\varphi) \qquad \text{for all} \quad f \in D_M.$

All results obtained in this paper can also be applied *mutatis mutandis* to the case in which D_N is not a real, but rather a complex vector space.

3. Determination of guaranteed bounds to eigenvalues by means of matrix eigenvalue problems

In this section, the decisive steps of a procedure for determining guaranteed bounds to the eigenvalues of problem (1) are briefly described. The following assumption is thereby required:

(∗) The bilinear forms, M and N, are symmetric; N is positive definite. Problem (1) possesses at least n eigenvalues $\lambda_1, \ldots, \lambda_n$; for $k = 1, \ldots, n$, λ_k is the kth smallest eigenvalue of problem (1).

Guaranteed bounds to the eigenvalue λ_p (with $p \in \mathbb{N}, p \leq n$) are calculated as follows:

Step 1. In the first step, two matrix eigenvalue problems,

$$A_L x = \mu B_L x \qquad \text{(EVP Low)},$$

$$A_U x = \mu B_U x \qquad \text{(EVP Up)},$$

two rational functions, F_L, F_U, and two natural numbers r, s are determined such that the following conditions are satisfied:

(i) A_L, B_L, A_U, B_U are real symmetric matrices; B_L, B_U are positive definite.

(ii) The elements of the matrices A_L, B_L, A_U, B_U are elementary functions of certain (given or freely chosen) constants.

(iii) $F_L\left(\mu_r^{(L)}\right) \leq \lambda_p \leq F_U\left(\mu_s^{(U)}\right)$ (2)
holds, where $\mu_r^{(L)}$ is the rth smallest eigenvalue of (EVP Low), and $\mu_s^{(U)}$ is the sth smallest eigenvalue of (EVP Up). (Inequality (2) is thereby intended to imply that $F_L\left(\mu_r^{(L)}\right)$ and $F_U\left(\mu_s^{(U)}\right)$ are well-defined.)

(iv) The difference $F_U\left(\mu_s^{(U)}\right) - F_L\left(\mu_r^{(L)}\right)$ is sufficiently small.

Step 2. In the second step, a guaranteed upper bound $\lambda_p^{(U)}$, for $F_U\left(\mu_s^{(U)}\right)$, and a guaranteed lower bound, $\lambda_p^{(L)}$, for $F_L\left(\mu_r^{(L)}\right)$, are calculated with the use of interval arithmetic.

Result. The guaranteed bounds, $\lambda_p^{(L)} \leq \lambda_p \leq \lambda_p^{(U)}$, are obtained.

Remarks. Condition (iv) stated in the first step is understood as follows: If guaranteed bounds which deviate from the exact value, λ_k, by ϵ at the most are being sought, the difference, $F_U\left(\mu_s^{(U)}\right) - F_L\left(\mu_r^{(L)}\right)$, should be definitely smaller than ϵ.

Since F_L and F_U are of very simple form in all applications, the second step consists essentially in the calculation of guaranteed bounds to eigenvalues of the matrix eigenvalue problems, (EVP Up) and (EVP Low).

Assumption ($*$) can be replaced by the following condition:

($**$) The bilinear forms, M and N, are symmetric; M is positive definite. Problem (1) possesses at least n positive eigenvalues $\lambda_1, \ldots, \lambda_n$; for $k = 1, \ldots, n$, λ_k is the kth smallest positive eigenvalue of problem (1).

The procedure just described merely provides a general framework, the decisive question is, of course, how the two steps are to be realized in practice. Two research projects corresponding to the two steps are currently in progress in Clausthal. In the first project, methods are being developed for constructing the quantitites occuring in the first step; condition (iv) is thereby checked only in the sense that $F_L\left(\mu_r^{(L)}\right)$ and $F_U\left(\mu_s^{(U)}\right)$ agree sufficiently well in the case of evaluation with conventional arithmetic, that is, without interval arithmetic. In the second research project, interval algorithms which are well suited for use in the second step of the procedure described are being developed. In the present article, a few results from the first research project are reported; important results achieved by H. Behnke in the course of the second research project can be found in [5], [6], [7].

4. Inclusion theorems and variational methods

If one wishes to carry out the first step of the procedure described in section 3, one must first decide which theorems are to be applied for proving inequality (2). Various theorems are available for this purpose, for example, the important inclusion theorem of N. J. Lehmann [18]. In the following, a new inclusion theorem is stated and proved. This theorem encompasses the theorem of N. J. Lehmann as a special case, and is especially well suited for the treatment of complicated eigenvalue problems.

Theorem 1 (Inclusion Theorem).
Assumptions, notations.

A1 The bilinear forms, M and N, are symmetric; N is positive definite.

A2 $\sigma \in \mathbb{R}$; there exist sequences $(\lambda_i)_{i\in J}$ and $(\varphi_i)_{i\in J}$ such that

$$\begin{aligned} &J = \{i \in \mathbb{N} : i \leq \dim D_M\}, \\ &\lambda_i \in \mathbb{R}, \quad \varphi_i \in D_M, \quad N(\varphi_i, \varphi_k) = \delta_{ik} \qquad \text{for } i, k \in J, \\ &M(f, \varphi_i) = \lambda_i N(f, \varphi_i) \qquad \text{for all } f \in D_M, \; i \in J, \\ &N(f,f) = \sum_{i\in J} \big(N(f, \varphi_i)\big)^2 \qquad \text{for all } f \in D_M, \quad \text{and} \\ &\text{the set } \{i \in J : \lambda_i < \sigma\} \text{ is finite.} \end{aligned}$$

A3 X is a real vector space; $T : D_M \to X$ is a linear mapping; $b : X \times X \to \mathbb{R}$ is a symmetric bilinear form. $b(f,f) \geq 0$ holds for all $f \in X$, and $b(Tf, Tg) = N(f,g)$ holds for all $f, g \in D_M$.

A4 $m \in \mathbb{N}$; $u_i \in D_M$, $w_i \in X$ for $i = 1, \ldots, m$. $b(Tf, w_i) = M(f, u_i)$ for all $f \in D_M$, $i = 1, \ldots, m$.

A5 C is a real m-by-q matrix ($q \in I\!N$). Matrices $\hat{A}, \hat{B}, \tilde{A}, \tilde{B}$ are defined by[1)]

$$\begin{aligned}\hat{A} &:= \left(M(u_i, u_k) - \sigma N(u_i, u_k)\right)_{i,k=1,\dots,m},\\ \hat{B} &:= \left(b(w_i, w_k) - 2\sigma M(u_i, u_k) + \sigma^2 N(u_i, u_k)\right)_{i,k=1,\dots,m},\\ \tilde{A} &:= C'\hat{A}C,\\ \tilde{B} &:= C'\hat{B}C.\end{aligned}$$

$\tilde{B}$ is positive definite. For $i = 1, \dots, q$, the ith smallest eigenvalue of the eigenvalue problem $\tilde{A}x = \mu\tilde{B}x$ is denoted by μ_i.

Assertion. For all $i \in I\!N$ with $i \le q$ and $\mu_i < 0$, the interval $\left[\sigma + \frac{1}{\mu_i}, \sigma\right)$ contains at least i eigenvalues of eigenvalue problem (1).

Two lemmas are provided before beginning the proof of Theorem 1.

Lemma 1. Let the assumptions A1, A2, A3 of Theorem 1 be satisfied. Furthermore, let the following conditions be fulfilled:

(i) $\tilde{\eta} \in I\!R,\quad \tilde{\sigma} \in I\!R,\quad \tilde{\eta} \le \tilde{\sigma},\quad u \in D_M,\quad w \in X,\quad b(w - \tilde{\sigma}Tu, w - \tilde{\eta}Tu) \le 0;$

(ii) $b(Tf, w) = M(f, u)\qquad$ for all $\quad f \in D_M;$

(iii) $N(u, \varphi_i) = 0\qquad$ for all $\quad i \in J\quad$ with $\quad \tilde{\eta} \le \lambda_i \le \tilde{\sigma}.$

Then $u = 0$.

Proof. Since $b(T\varphi_j, T\varphi_k) = \delta_{jk}$ for $j, k \in J$, it follows from Bessel's inequality that

$$b\left(w - \frac{1}{2}(\tilde{\sigma} + \tilde{\eta})Tu, w - \frac{1}{2}(\tilde{\sigma} + \tilde{\eta})Tu\right) \ge \sum_{i \in J}\left(b\left(w - \frac{1}{2}(\tilde{\sigma} + \tilde{\eta})Tu, T\varphi_i\right)\right)^2.$$

Hence, by A2,

$$\begin{aligned}0 &\ge b(w - \tilde{\sigma}Tu, w - \tilde{\eta}Tu)\\ &= b\left(w - \frac{1}{2}(\tilde{\sigma} + \tilde{\eta})Tu, w - \frac{1}{2}(\tilde{\sigma} + \tilde{\eta})Tu\right) - \frac{1}{4}(\tilde{\sigma} - \tilde{\eta})^2 b(Tu, Tu)\\ &\ge \sum_{i \in J}\left(b\left(w - \frac{1}{2}(\tilde{\sigma} + \tilde{\eta})Tu, T\varphi_i\right)\right)^2 - \frac{1}{4}(\tilde{\sigma} - \tilde{\eta})^2 \sum_{i \in J}(N(u, \varphi_i))^2.\end{aligned}$$

1) The transpose of a matrix G is denoted by G'.

Since $b(T\varphi_i, w) = M(\varphi_i, u) = \lambda_i N(\varphi_i, u)$ for all $i \in J$,

$$\begin{aligned}0 &\geq \sum_{i\in J}\left(\lambda_i - \frac{1}{2}(\tilde\sigma + \tilde\eta)\right)^2 (N(u,\varphi_i))^2 - \frac{1}{4}(\tilde\sigma - \tilde\eta)^2 \sum_{i\in J}(N(u,\varphi_i))^2\\ &= \sum_{i\in J}(\lambda_i - \tilde\sigma)(\lambda_i - \tilde\eta)\big(N(u,\varphi_i)\big)^2.\end{aligned}$$

Because of (iii), $(\lambda_i - \tilde\sigma)(\lambda_i - \tilde\eta)\big(N(u,\varphi_i)\big)^2 \geq 0$ holds for all $i \in J$, which implies $N(u, \varphi_i) = 0$ for all $i \in J$. From this it follows by A2 that $N(u,u) = 0$, and hence $u = 0$.

Lemma 2. Let the assumptions A1, A2, A3 of Theorem 1 be satisfied. Furthermore, let the following conditions be fulfilled:

(i) $\quad \eta \in I\!R, \quad \eta < \sigma, \quad u \in D_M, \quad w \in X, \quad b(w - \sigma Tu, w - \eta Tu) \leq 0;$

(ii) $\quad b(Tf, w) = M(f, u) \quad$ for all $\quad f \in D_M;$

(iii) $\quad N(u, \varphi_i) = 0 \quad$ for all $\quad i \in J \quad$ with $\quad \eta \leq \lambda_i < \sigma.$

Then $b(w - \sigma Tu, w - \sigma Tu) = 0$.

Proof (indirect proof). Suppose that $b(w - \sigma Tu, w - \sigma Tu) > 0$. Then

$$\begin{aligned}0 &> b(w - \sigma Tu, w - \eta Tu) - b(w - \sigma Tu, w - \sigma Tu)\\ &= (\sigma - \eta)\big(b(w, Tu) - \sigma b(Tu, Tu)\big),\end{aligned}$$

which implies $0 > b(w, Tu) - \sigma b(Tu, Tu)$. Define

$$h(\xi) := \frac{b(w,w) - \xi b(w,Tu)}{b(w,Tu) - \xi b(Tu,Tu)} \qquad \text{for all} \quad \xi \in I\!R \quad \text{with} \quad b(w,Tu) \neq \xi b(Tu,Tu).$$

Then $h(\sigma) = \dfrac{b(w - \sigma Tu, w - \eta Tu)}{b(w,Tu) - \sigma b(Tu,Tu)} + \eta \geq \eta.$

Because of A2, there is some $\gamma \in I\!R$ with $\gamma < \eta$ such that $\lambda_i \notin [\gamma, \eta)$ for all $i \in J$. Hence, there exists a real number $\tilde\sigma$ with $\tilde\sigma < \sigma$ such that $b(w, Tu) < \tilde\sigma b(Tu, Tu)$, and $h(\tilde\sigma) > \gamma$. Now, let $\tilde\eta := h(\tilde\sigma)$. Then, $b(w - \tilde\sigma Tu, w - \tilde\eta Tu) = 0$.

From $\tilde\eta - \tilde\sigma = \dfrac{b(w - \tilde\sigma Tu, w - \tilde\sigma Tu)}{b(w,Tu) - \tilde\sigma b(Tu,Tu)} \leq 0$, the inequality $\gamma < \tilde\eta \leq \tilde\sigma < \sigma$ is obtained. Hence, $N(u, \varphi_i) = 0$ holds for all $i \in J$ with $\tilde\eta \leq \lambda_i \leq \tilde\sigma$. By Lemma 1, it follows that $u = 0$, which contradicts the inequality $0 > b(w, Tu) - \sigma b(Tu, Tu)$. Thus, $b(w - \sigma Tu, w - \sigma Tu) = 0$ is proved.

Proof of Theorem 1 (indirect proof). Assume that, for some $l \in \mathbb{N}$ with $l \leq q$ and $\mu_l < 0$, the interval $\left[\sigma + \frac{1}{\mu_l}, \sigma\right)$ contains strictly less than l eigenvalues of problem (1).

Since $\tilde{A} = \tilde{A}'$, $\tilde{B} = \tilde{B}'$, and $\tilde{B}$ is positive definite, there exist vectors $x_1, \ldots, x_q$ in $\mathbb{R}^q$ with $x_k = (x_{k,1}, \ldots, x_{k,q})'$ such that $\tilde{A}x_i = \mu_i \tilde{B}x_i$, and $x_i'\tilde{B}x_k = \delta_{ik}$ for $i, k = 1, \ldots, q$.

The elements of C are denoted by c_{ik}, that is $C = (c_{ik})_{i=1,\ldots,m,\ k=1,\ldots,q}$. Define $\tilde{u}_i$ and $\tilde{w}_i$ for $i = 1, \ldots, q$ by

$$\tilde{u}_i := \sum_{k=1}^{q} \sum_{j=1}^{m} x_{i,k} c_{jk} u_j \quad \text{and} \quad \tilde{w}_i := \sum_{k=1}^{q} \sum_{j=1}^{m} x_{i,k} c_{jk} w_j .$$

A simple calculation shows that

$$M(\tilde{u}_i, \tilde{u}_k) - \sigma N(\tilde{u}_i, \tilde{u}_k) = x_i' C' \hat{A} C x_k = \mu_i \delta_{ik} \tag{3}$$

$$b(\tilde{w}_i, \tilde{w}_k) - 2\sigma M(\tilde{u}_i, \tilde{u}_k) + \sigma^2 N(\tilde{u}_i, \tilde{u}_k) = x_i' C' \hat{B} C x_k = \delta_{ik} \tag{4}$$

for $i, k = 1, \ldots, q$.

Now let L be the subspace of D_M spanned by those eigenelements of problem (1) which belong to an eigenvalue contained in the interval $\left[\sigma + \frac{1}{\mu_l}, \sigma\right)$. From the assumption made at the beginning of the proof, it follows that $\dim L < l$. Hence, there exist real numbers, $\beta_1, \ldots, \beta_l$, such that $(\beta_1, \ldots, \beta_l) \neq (0, \ldots, 0)$ and $\sum_{i=1}^{l} \beta_i N(\tilde{u}_i, g) = 0$ for all $g \in L$. Let

$$u := \sum_{i=1}^{l} \beta_i \tilde{u}_i, \quad w := \sum_{i=1}^{l} \beta_i \tilde{w}_i, \quad \eta := \sigma + \frac{1}{\mu_l}.$$

Then $\eta < \sigma$, $u \in D_M$, $w \in X$, $b(Tf, w) = M(f, u)$ for all $f \in D_M$, and $N(u, \varphi_i) = 0$ for all $i \in J$ with $\eta \leq \lambda_i < \sigma$. Furthermore, it can be deduced from (3) and (4) that

$$\begin{aligned}
&b(w - \sigma Tu, w - \eta Tu) \\
&= b(w, w) - 2\sigma M(u, u) + \sigma^2 N(u, u) - \frac{1}{\mu_l}\left(M(u, u) - \sigma N(u, u)\right) \\
&= \sum_{i=1}^{l} \beta_i^2 \left(1 - \frac{\mu_i}{\mu_l}\right) \leq 0
\end{aligned}$$

and

$$b(w - \sigma Tu, w - \sigma Tu) = \sum_{i=1}^{l} \beta_i^2 > 0.$$

This contradicts the equation $b(w - \sigma Tu, w - \sigma Tu) = 0$, which is a consequence of Lemma 2. Thus, Theorem 1 is proved.

The manner in which Theorem 1 can be applied for constructing the quantities, A_L, B_L, F_L, r, which occur in section 3, must now be explained. If

- the assumptions of Theorem 1 are satisfied,
- the sequence $(\lambda_i)_{i \in J}$ is monotonically increasing,
- $p \in \mathbb{N}$, $r \in \mathbb{N}$, $p + r \in J$, $r \leq q$, $\mu_r < 0$, $\sigma \leq \lambda_{p+r}$ holds, and
- $A_L := \tilde{A}$, $B_L := \tilde{B}$, $\mu_r^{(L)} := \mu_r$, $F_L(\xi) := \sigma + \xi^{-1}$ for $\xi \in \mathbb{R}$, $\xi \neq 0$,

the following inequality results,

$$F_L\left(\mu_r^{(L)}\right) \leq \lambda_p,$$

which is the left-hand inequality in (2). This application of Theorem 1 is based on the following idea: If σ is a lower bound to the $(p + r)^{\text{th}}$ eigenvalue, λ_{p+r}, and if the interval, $\left[\sigma + \mu_r^{-1}, \sigma\right)$, contains at least r eigenvalues, then $\sigma + \mu_r^{-1}$ is a lower bound to the pth eigenvalue, λ_p. The decisive feature is the fact that the bound, $\sigma + \mu_r^{-1}$, thereby depends not only on σ, but also on the choice of the trial vectors, u_i and w_i. For an appropriate choice of u_i, w_i, therefore, one can expect that $\sigma + \mu_r^{-1}$ is a very good bound for λ_p, even if σ is only a comparatively rough bound for λ_{p+r}. This expectation has been fully confirmed in the treatment of numerous numerical examples.

The quantities, A_U, B_U, F_U, s, from section 3 can be constructed in a similar manner as A_L, B_L, F_L, r. However, it is frequently more advantageous to apply the theorem on which the Rayleigh-Ritz method is based. This theorem is now expressed in a formulation adapted to problem (1).

Theorem 2 (Rayleigh/Ritz). Let the assumptions A1, A2 of Theorem 1 be fulfilled, let the sequence $(\lambda_i)_{i \in J}$ be monotonically increasing, and let $\alpha \in \mathbb{R}$ be such that $M(f, f) \geq \alpha N(f, f)$ holds for all $f \in D_M$. Furthermore, let $m \in \mathbb{N}$,

$u_i \in D_M$ for $i = 1, \ldots, m$, and let C be a real m-by-q matrix ($q \in \mathbb{N}$). Define matrices A_0, B_0, A_U, B_U by

$$\begin{aligned} A_0 &:= \left(M(u_i, u_k)\right)_{i,k=1,\ldots,m}, & B_0 &:= \left(N(u_i, u_k)\right)_{i,k=1,\ldots,m}, \\ A_U &:= C' A_0 C, & B_U &:= C' B_0 C. \end{aligned}$$

Let B_U be positive definite. For $i = 1, \ldots, q$, denote the ith smallest eigenvalue of the eigenvalue problem $A_U x = \mu B_U x$ by $\mu_i^{(U)}$. Then $\lambda_i \leq \mu_i^{(U)}$ holds for $i = 1, \ldots, q$.

The significance of Theorem 2 for the procedure described in section 3 is obvious. If

- the assumptions of Theorem 2 are fulfilled,
- $p \in \mathbb{N}$, $p \leq q$ holds,
- s and F_U are defined by $s := p$ and $F_U(\xi) := \xi$ for $\xi \in \mathbb{R}$,

then the following inequality results:

$$\lambda_p \leq F_U\left(\mu_s^{(U)}\right).$$

This is the right-hand inequality in (2).

If the assumptions of Theorem 2 are fulfilled, then the eigenvalues λ_i can be characterized by the following variational principle:

$$\lambda_i = \min_{\substack{U \subset D_M \\ U \text{ subspace} \\ \dim U = i}} \max_{\substack{f \in U \\ f \neq 0}} \frac{M(f,f)}{N(f,f)} \qquad \text{for } i \in J.$$

The assertion of Theorem 2 is an immediate consequence of this variational principle. In a similar manner, Theorem 1 can also be derived from a variational principle. It therefore seems justified to classify the numerical procedures based on Theorems 1 and 2 among the variational methods.

It should also be mentioned that theorems which are completely analogous to Theorems 1 and 2 apply to eigenvalue problems of form (1) for which M, rather than N, is positive definite ([7], Theorem 4, and [14]).

5. Bounds to eigenvalues of Stokes' eigenvalue problem

In the following, the first step of the procedure described in section 3 is illustrated by means of an eigenvalue problem involving a system of partial differential equations. A few notations must be introduced for formulating this problem:

$$\Omega := \left\{ \begin{pmatrix} \xi_1 \\ \xi_2 \\ \xi_3 \end{pmatrix} \in I\!R^3 : |\xi_i| < \frac{\pi}{2} \quad \text{for} \quad i = 1,2,3 \right\},$$

$\partial\Omega$ denotes the boundary of Ω,

$$\tilde{D}_M := \left\{ f \in \left(W^{1,2}(\Omega)\right)^3 : \operatorname{div} f = 0 \text{ in } \Omega, \quad f = 0 \text{ on } \partial\Omega \right\},$$

$$\tilde{D}_N := \left(L_2(\Omega)\right)^3,$$

$$\tilde{M}(f,g) := \int_\Omega (\operatorname{curl} f) \cdot (\operatorname{curl} g)\, d\xi_1 d\xi_2 d\xi_3 \quad \text{for } f,g \in \tilde{D}_M,$$

$$\tilde{N}(f,g) := \int_\Omega f \cdot g \; d\xi_1 d\xi_2 d\xi_3 \quad \text{for } f,g \in \tilde{D}_N$$

(for the Sobolev space $W^{1,2}(\Omega)$ see, for example, [20]).

Stokes' eigenvalue problem for the domain Ω is now expressed as follows: Find pairs (λ, φ) such that

$$\lambda \in I\!R, \quad \varphi \in \tilde{D}_M, \quad \varphi \neq 0, \quad \tilde{M}(f,\varphi) = \lambda \tilde{N}(f,\varphi) \quad \text{for all} \quad f \in \tilde{D}_M \tag{5}$$

holds.

This problem is a weak formulation of the following eigenvalue problem: Find

$$\lambda \in I\!R, \quad \varphi \in \left(C^2(\bar{\Omega})\right)^3, \quad \varphi \neq 0, \quad P \in C^1(\Omega) \quad \text{with}$$

$$\operatorname{curl}\operatorname{curl}\varphi + \operatorname{grad} P = \lambda\varphi \quad \text{in } \Omega, \qquad \operatorname{div}\varphi = 0 \quad \text{in } \Omega,$$

$$\varphi = 0 \quad \text{on } \partial\Omega.$$

The smallest eigenvalue of Stokes' eigenvalue problem plays an important role in the theory of hydrodynamic stability ([17], p. 225; [21]).

Stokes' eigenvalue problem for the domain Ω can be subdivided into simpler subproblems. This decomposition is based on the same idea as the decomposition of the matrix eigenvalue problem $Gx = \lambda x$, with a block diagonal matrix,

$$G = \begin{pmatrix} G_{11} & & 0 \\ & \ddots & \\ 0 & & G_{kk} \end{pmatrix},$$

into subproblems, $G_{ii}x = \lambda x$. A mathematically rigorous formulation of the decomposition of Stokes' eigenvalue problem is given in [15]; the decomposition can be justified in analogy with the proofs presented in [2].

In the following, only one of the subproblems resulting from the decomposition of Stokes' eigenvalue problem is discussed. This subproblem yields the smallest eigenvalue of the overall problem. A few notations are required for formulating this subproblem:

$$\begin{aligned} K^{i,j,l} := \{ & f \in L_2(\Omega) : f(\xi_1,\xi_2,\xi_3) = (-1)^i f(-\xi_1,\xi_2,\xi_3) = \\ & = (-1)^j f(\xi_1,-\xi_2,\xi_3) = (-1)^l f(\xi_1,\xi_2,-\xi_3) \quad \text{for} \quad (\xi_1,\xi_2,\xi_3)' \in \Omega \} \\ & \text{for} \quad i,j,l \in \{0,1\}, \end{aligned}$$

$$V^{101} := \left\{ \begin{pmatrix} f_1 \\ f_2 \\ f_3 \end{pmatrix} \in (L_2(\Omega))^3 : f_1 \in K^{0,0,1}, f_2 \in K^{1,1,1}, f_3 \in K^{1,0,0}, \right\},$$

$$(Sf)(\xi_1,\xi_2,\xi_3) := \begin{pmatrix} f_3(\xi_3,\xi_2,\xi_1) \\ f_2(\xi_3,\xi_2,\xi_1) \\ f_1(\xi_3,\xi_2,\xi_1) \end{pmatrix} \quad \text{for} \quad \begin{pmatrix} f_1 \\ f_2 \\ f_3 \end{pmatrix} \in (L_2(\Omega))^3, \quad \begin{pmatrix} \xi_1 \\ \xi_2 \\ \xi_3 \end{pmatrix} \in \Omega,$$

$$\begin{aligned} V^{1011} &:= \{ f \in V^{101} : Sf = -f \}, \\ D_M &:= \tilde{D}_M \cap V^{1011}, \quad M(f,g) := \tilde{M}(f,g) \ \text{for} \ f,g \in D_M, \\ D_N &:= \tilde{D}_N \cap V^{101}, \quad N(f,g) := \tilde{N}(f,g) \ \text{for} \ f,g \in D_N. \end{aligned}$$

It can be shown that a pair (λ, φ) satisfies the following condition,

$$\lambda \in I\!R, \quad \varphi \in D_M, \quad \varphi \neq 0, \quad M(f,\varphi) = \lambda N(f,\varphi) \quad \text{for all} \quad f \in D_M, \tag{6}$$

if and only if this pair fulfills (5) and $\varphi \in V^{1011}$ holds.

Theorem 1 is now applied to problem (6) in the manner described in section 4. Assumptions A1 and A2 of Theorem 1 are satisfied for arbitrary σ (see [15]). A3 is also fulfilled, if

$$X := D_N, \quad Tf := f \ \text{for} \ f \in D_M, \quad b(f,g) := N(f,g) \ \text{for} \ f,g \in X.$$

For constructing the quantities which occur in assumption A4 of Theorem 1, two natural numbers, m_0 and n_0, are first chosen; moreover, the following notations are introduced:

$$\psi_{2i-1}(\xi) := \begin{cases} \sin(i\xi) + \sin\big((i+2)\xi\big) & \text{for } \xi \in \mathbb{R},\ i = 1,3,5,7,\ldots, \\ (i+2)\sin(i\xi) + i\sin\big((i+2)\xi\big) & \text{for } \xi \in \mathbb{R},\ i = 2,4,6,8,\ldots, \end{cases}$$

$$\psi_{2i}(\xi) := \begin{cases} (i+2)\cos(i\xi) + i\cos\big((i+2)\xi\big) & \text{for } \xi \in \mathbb{R},\ i = 1,3,5,7,\ldots, \\ \cos(i\xi) + \cos\big((i+2)\xi\big) & \text{for } \xi \in \mathbb{R},\ i = 0,2,4,6,\ldots, \end{cases}$$

$$\chi_{2i-1}(\xi) := \sin(2i\xi) \qquad \text{for} \quad \xi \in \mathbb{R}, \quad i \in \mathbb{N},$$

$$\chi_{2i}(\xi) := \cos\big((2i-1)\xi\big) \qquad \text{for} \quad \xi \in \mathbb{R}, \quad i \in \mathbb{N},$$

$$v_{1,j,k,l}(\xi_1,\xi_2,\xi_3) := \begin{pmatrix} \chi_{2j-1}(\xi_1)\psi_{2k-1}(\xi_2)\psi_{2l-2}(\xi_3) \\ 0 \\ 0 \end{pmatrix}$$

$$\text{for } j,k,l \in \mathbb{N}, \quad \xi_1,\xi_2,\xi_3 \in \mathbb{R},$$

$$v_{2,j,k,l}(\xi_1,\xi_2,\xi_3) := \begin{pmatrix} 0 \\ \psi_{2j-2}(\xi_1)\chi_{2k}(\xi_2)\psi_{2l-2}(\xi_3) \\ 0 \end{pmatrix}$$

$$\text{for } j,k,l \in \mathbb{N}, \quad \xi_1,\xi_2,\xi_3 \in \mathbb{R},$$

$$v_{3,j,k,l}(\xi_1,\xi_2,\xi_3) := \begin{pmatrix} 0 \\ 0 \\ \psi_{2j-2}(\xi_1)\psi_{2k-1}(\xi_2)\chi_{2l-1}(\xi_3) \end{pmatrix}$$

$$\text{for } j,k,l \in \mathbb{N}, \quad \xi_1,\xi_2,\xi_3 \in \mathbb{R},$$

$$\hat{v}_{j,k,l}(\xi_1,\xi_2,\xi_3) := \sin(i\xi_1)\cos\big((k-1)\xi_2\big)\sin(l\xi_3)$$

$$\text{for } j,k,l \in \mathbb{N}, \quad \xi_1,\xi_2,\xi_3 \in \mathbb{R},$$

$$\Delta f := \begin{pmatrix} \Delta f_1 \\ \Delta f_2 \\ \Delta f_3 \end{pmatrix} \quad \text{for } f = \begin{pmatrix} f_1 \\ f_2 \\ f_3 \end{pmatrix} \in \big(C^2(\Omega)\big)^3.$$

The quantities, m, u_i, w_i, which occur in assumption A4, are now defined as follows:

$$m := 3m_0^3 + n_0^3,$$

$$u_\nu := (I - S)\,\text{curl}\, v_{i,j,k,l}$$

$$\text{for all } \ i,j,k,l,\nu \in \mathbb{N} \ \text{ with } \ i \le 3, \ j \le m_0, \ k \le m_0, \ l \le m_0,$$

$$\nu = i + 3(j-1) + 3(k-1)m_0 + 3(l-1)m_0^2$$

(I denotes the identity operator),

$$w_\nu := -\Delta u_\nu \quad \text{for} \quad \nu = 1, \ldots, 3m_0^3,$$

$$u_\nu := 0 \quad \text{for} \quad \nu = 3m_0^3 + 1, \ldots, 3m_0^3 + n_0^3,$$

$$w_\nu := \text{grad}\, \hat{v}_{j,k,l}$$

$$\text{for all } \ j,k,l,\nu \in \mathbb{N} \ \text{ with } \ j \le n_0, \ k \le n_0, \ l \le n_0,$$

$$\nu = 3m_0^3 + j + (k-1)n_0 + (l-1)n_0^2.$$

With these definitions, A4 is satisfied; for an appropriate choice of C, assumption A5 is also fulfilled.

The eigenvalues of (6) can be arranged into an increasing sequence; for $i \in \mathbb{N}$, let λ_i be the ith smallest eigenvalue of (6). — For the application of Theorem 1, a rough lower bound is still required for one of the eigenvalues, λ_i, with $i \geq 2$. It can be shown that $\lambda_2 \geq 14$ holds. The proof is now briefly sketched (cf. [21]). Let

$$D_0 := \left\{ f \in \left(W^{1,2}(\Omega)\right)^3 \cap V^{1011} : f = 0 \text{ on } \partial\Omega \right\},$$

$$M_0(f,g) := \sum_{i=1}^{3} \int_{\Omega} (\operatorname{grad} f_i) \cdot (\operatorname{grad} g_i)\, d\xi_1 d\xi_2 d\xi_3$$

$$\text{for } \begin{pmatrix} f_1 \\ f_2 \\ f_3 \end{pmatrix} \in D_0, \quad g = \begin{pmatrix} g_1 \\ g_2 \\ g_3 \end{pmatrix} \in D_0,$$

$$N_0(f,g) := \int_{\Omega} f \cdot g\, d\xi_1 d\xi_2 d\xi_3 \quad \text{for } f, g \in D_0.$$

The following eigenvalue problem can then be solved easily in closed form: Find pairs (λ, φ) such that

$$\lambda \in \mathbb{R}, \quad \varphi \in D_0, \quad \varphi \neq 0, \quad M_0(f,\varphi) = \lambda N_0(f,\varphi) \quad \text{for all } f \in D_0. \tag{7}$$

The second smallest eigenvalue of (7) is 14. Since $D_0 \supset D_M$ and $M(f,g) = M_0(f,g)$, $N(f,g) = N_0(f,g)$ holds for all $f, g \in D_M$, it follows from the variational characterization of eigenvalues that $\lambda_2 \geq 14$.

Theorems 1 and 2 can now be applied in the manner described in section 4 (with $r = p = 1$, $\sigma = 14$, u_i in Theorem 2 chosen as in Theorem 1). The bounds for λ_1 given in Tables 1 and 2 are thus obtained. Table 1 illustrates the dependence of the bounds on m_0, and Table 2 shows the dependance on n_0. (The matrices, $\tilde{A}, \tilde{B}, A_U, B_U$, in Theorems 1 and 2 have been calculated by means of conventional arithmetic with quadruple precision. Efforts are now in progress for guaranteeing the results by means of interval arithmetic.)

Table 1: Bounds to the first eigenvalue of Stokes' eigenvalue problem ($n_0 = 5$)

m_0	Lower bounds	Upper bounds
1	6.2016	6.3334
2	6.2471	6.3043
3	6.2871	6.3001
4	6.2972	6.2996
5	6.2988	6.2995

Table 2: Bounds to the first eigenvalue of Stokes' eigenvalue problem ($m_0 = 4$)

Lower bounds	$n_0 = 1$	5.8857
	2	6.2184
	3	6.2862
	4	6.2965
	5	6.2972
	6	6.2973
Upper bound		6.2996

6. Further numerical tests

With the use of Theorem 1 or similar theorems (inclusion theorem of N. J. Lehmann [18] and Theorem 4 in [7]), numerous eigenvalue problems have been treated in Clausthal. The quality of the bounds obtained was thereby at least as high as that of the bounds presented in Table 1. Work has been performed on eigenvalue problems from the following fields of physics and engineering:

1) vibration of circular arches [1],
2) vibration of rhombical membranes [16], [23],
3) vibration of clamped (isotropic or orthotropic) rectangular plates [11],
4) waves in composite materials [8],
5) sloshing of liquids in containers [3], [4],
6) buckling of clamped rectangular plates subject to compression and shear [14],
7) hydrodynamic stability of a liquid in a channel heated from below [13],
8) thermal convection in cylinders [9], [12],
9) energy levels of a two-dimensional bounded harmonic oscillator [1],
10) energy levels of the helium atom [19].

References

[1] **Albrecht, J. and Goerisch, F.:** *Anwendungen des Verfahrens von Lehmann auf Schwingungsprobleme.* In: Albrecht, J.; Collatz, L.; Velte, W. and Wunderlich, W. (Eds.): Numerical Treatment of Eigenvalue Problems, Vol. 4, International Series of Numerical Mathematics (ISNM) 83, Basel/Boston: Birkhäuser Verlag 1987, 1 – 9.

[2] **Bassotti, L.:** *Sottospazi invarianti per l'operatore dell' elasticità in un cubo.* Atti Accad. Naz. Lincei Rend. Cl. Sci. Fis. Mat. Natur. (8) **45** (1968), 485 – 493.

[3] **Behnke, H.:** *Schranken für Schlingerfrequenzen.* Diplomarbeit, Technische Universität Clausthal, 1985.

[4] **Behnke, H.:** *Calculation of bounds for sloshing frequencies with the use of finite elements.* Z. Angew. Math. Mech. **67** (1987), T 347 – T 349.

[5] **Behnke, H.:** *Inclusion of eigenvalues of general eigenvalue problems for matrices.* Computing Suppl. **6** (1988), 69 – 78.

[6] **Behnke, H.:** *Die Bestimmung von Eigenwertschranken mit Hilfe von Variationsmethoden und Intervallarithmetik.* Dissertation, Technische Universität Clausthal, 1989.

[7] **Behnke, H.:** *The determination of guaranteed bounds to eigenvalues with the use of variational methods II.* In this volume.

[8] **Bierwirth, S.:** *Schranken für die Frequenzen von Wellen in Verbundwerkstoffen.* Diplomarbeit, Technische Universität Clausthal, 1986.

[9] **Bürgel, E.:** *Numerische Bestimmung gesicherter Rayleigh-Zahlen und zugehöriger Strömungsformen bei stationärer, rotationssymmetrischer Konvektion in Zylindern.* Dissertation, Technische Universität Clausthal, 1988.

[10] **Collatz, L.:** *Eigenwertaufgaben mit technischen Anwendungen,* 2. Auflage. Leipzig: Akademische Verlagsgesellschaft Geest & Portig K.-G., 1963.

[11] **Frantzen, H.:** *Einschließung von Eigenfrequenzen von Platten.* Diplomarbeit, Technische Universität Clausthal, 1983.

[12] **Girgensohn, R.:** *Bestimmung genauer Eigenwertschranken für eine in der Hydrodynamik auftretende Klasse von Eigenwertaufgaben mit Systemen partieller Differentialgleichungen.* Diplomarbeit, Technische Universität Clausthal, 1989.

[13] **Goerisch, F.:** *Über die Anwendung einer Verallgemeinerung des Lehmann-Maehly-Verfahrens zur Berechnung von Eigenwertschranken.* In: Albrecht, J. and Collatz, L. (Eds.): Numerische Behandlung von Differentialgleichungen, Band 3, International Series of Numerical Mathematics (ISNM) 56, Basel/Boston/Stuttgart: Birkhäuser Verlag 1981, 58 – 72.

[14] **Goerisch, F. and Haunhorst, H.:** *Eigenwertschranken für Eigenwertaufgaben mit partiellen Differentialgleichungen.* Z. Angew. Math. Mech. **65** (1985), 129 – 135.

[15] **Goerisch, F.; He, Z. and Zimmermann, B.:** *Bounds to eigenvalues for eigenvalue problems arising in hydrodynamic stability theory I.* To be published.

[16] **Goerisch, F. and Zimmermann, S.:** *On Trefftz's method and its application to eigenvalue problems.* Z. Angew. Math. Mech. **66** (1986), T 304 – T 306.

[17] **Joseph, D. D.:** *Stability of Fluid Motions I.* Berlin/Heidelberg/New York: Springer-Verlag 1976.

[18] **Lehmann, N. J.:** *Optimale Eigenwerteinschließungen.* Numer. Math. **5** (1963), 246 – 272.

[19] **Schröter, A.:** *Schranken für die Energieniveaus des Heliumatoms.* Diplomarbeit, Technische Universität Clausthal, 1988.

[20] **Temam, R.:** *Navier-Stokes Equations.* Amsterdam/New York/Oxford: North-Holland Publishing Company, 1977.

[21] **Velte, W.:** *Über ein Stabilitätskriterium der Hydrodynamik.* Arch. Rational Mech. Anal. **9** (1962), 9 – 20.

[22] **Velte, W.:** *Eigenwertschranken mit finiten Differenzen beim Stokes'schen Eigenwertproblem.* In: Albrecht, J. and Collatz, L. (Eds.): Numerische Behandlung von Eigenwertaufgaben, Band 2, International Series of Numerical Mathematics (ISNM) 43, Basel/Boston/Stuttgart: Birkhäuser Verlag 1979, 176 – 188.

[23] **Zimmermann, S.:** *Die Methode der Gebietszerlegung und ihre Anwendung zur Berechnung von Eigenwertschranken für das Problem der rhombischen Membran.* Diplomarbeit, Technische Universität Clausthal, 1984.

The Determination of Guaranteed Bounds to Eigenvalues with the Use of Variational Methods II

Henning Behnke
Institut für Mathematik
Technische Universität Clausthal
D-3392 Clausthal-Zellerfeld
Federal Republic of Germany

Abstract: Bounds for the eigenvalues of the matrix eigenvalue problem, $Ax = \lambda Bx$, $A = A^T$, $B = B^T$, B positive definite, are calculated by means of a new procedure which uses variational methods and an interval arithmetic. Furthermore, an eigenvalue problem with an ordinary differential equation is considered; guaranteed bounds for the smallest eigenvalues of this problem are presented.

1 Introduction

The calculation of eigenvalues is an important task in the field of numerical mathematics, as many formulations of questions lead to eigenvalue problems with differential equations. These problems can be solved in closed form only in a few cases; if one needs reliable results, one has to use numerical methods which supply *bounds to eigenvalues*. These numerical methods – for instance, procedures based on *variational methods*, such as the Rayleigh–Ritz procedure for the calculation of upper bounds to eigenvalues and the procedures of Lehmann and Goerisch for the calculation of lower bounds to eigenvalues – again lead to an eigenvalue problem, namely an eigenvalue problem with matrices. Therefore, general eigenvalue problems with matrices are considered in the first part of this paper. A procedure for the calculation of bounds to matrix eigenvalues with the use of *interval arithmetic* is proposed. The second part deals with an eigenvalue problem with an ordinary differential equation.

Computer Arithmetic and Self-Validating
Numerical Methods

ISBN 0-12-708245-X

2 Matrix eigenvalue problems

Consider the general matrix eigenvalue problem,

$$A\,x = \lambda\,B\,x, \quad A = A^T, \quad B = B^T, \quad B \text{ positive definite,} \tag{1}$$

with real $n \times n$ matrices A and B ($n \in \mathbb{N}$). Let the eigenvalues be ordered by magnitude and let them be counted in accordance with their multiplicity: $\lambda_1 \leq \lambda_2 \leq \ldots \leq \lambda_n$.

The calculation of bounds to an eigenvalue λ_j determined by its index j is implemented in three steps:

1. Calculation of approximations for eigenvalues

   $$\tilde{\lambda}_{r-1} < \tilde{\lambda}_r \approx \ldots \approx \tilde{\lambda}_j \approx \ldots \approx \tilde{\lambda}_s < \tilde{\lambda}_{s+1} \tag{2}$$

   ($1 \leq r \leq j \leq s \leq n$) as well as approximations $\tilde{x}_r, \ldots, \tilde{x}_s$ for eigenvectors belonging to $\lambda_r, \ldots, \lambda_s$. In the case $r = 1$ or $s = n$, the approximation $\tilde{\lambda}_{r-1}$ or $\tilde{\lambda}_{s+1}$, respectively, has to be dropped. The notation in (2) expresses the fact that the eigenvalue approximations $\tilde{\lambda}_r, \ldots, \tilde{\lambda}_s$ differ only a little (less than 1 %), whereas $\tilde{\lambda}_{r-1}$ and $\tilde{\lambda}_{s+1}$ differ clearly from these values. Thus, it can be expected that the eigenvalues $\lambda_r, \ldots, \lambda_s$ form a cluster, and that λ_j belongs to this cluster. For a simple (well separated) eigenvalue λ_j, one has, of course, $r = j = s$.

2. Determination of a rough upper bound ρ to λ_{r-1} with $\lambda_{r-1} < \rho < \lambda_r$ (to be dropped, if $r = 1$) and a rough lower bound σ for λ_{s+1} with $\lambda_s < \sigma < \lambda_{s+1}$ (to be dropped, if $s = n$), with the use of the eigenvalue approximations and LDL^T decompositions

3. Calculation of accurate bounds for λ_j with Temple quotients (in the case $r = s$) or a generalization of these quotients by Lehmann (in the case $r < s$) and with the use of $\tilde{x}_r, \ldots, \tilde{x}_s$, ρ and σ

Inclusion theorems

In the following the theorem of Lehmann (cf. [9,10,7]) is given specially for the treatment of matrix eigenvalue problems.

Theorem 1

Let the following assumptions be valid:

1. *A and B are real* $n \times n$ *matrices,* $A = A^T$, $B = B^T$, *B positive definite.*
2. *Let* $m \in \mathbb{N}$; $u_1, u_2, \ldots, u_m$ *are linearly independent elements of* $\mathbb{R}^n$; $v_i := B^{-1} A\, u_i$ *for* $i = 1, \ldots, m$.
3. *Let* $\sigma \in \mathbb{R}$. *The matrices* A_0, A_1, A_2, $\hat{A}$ *und* $\hat{B}$ *are defined by*

$$\begin{aligned} A_0 &:= \left(u_i^T B u_k\right)_{i,k=1,\ldots,m}, \\ A_1 &:= \left(u_i^T A u_k\right)_{i,k=1,\ldots,m}, \\ A_2 &:= \left(v_i^T A u_k\right)_{i,k=1,\ldots,m}, \\ \hat{A} &:= A_1 - \sigma\, A_0 \text{ and } \hat{B} := A_2 - 2\,\sigma\, A_1 + \sigma^2 A_0. \end{aligned}$$

   $\hat{B}$ *is positive definite.* [1]

4. *The eigenvalues* μ_i *of the matrix eigenvalue problem* $\hat{A}\, x = \mu\, \hat{B}\, x$ *are ordered by magnitude:* $\mu_1 \le \mu_2 \le \ldots \le \mu_p < 0 \le \ldots \le \mu_m$.

Assertion: *For* $l = 1, \ldots, p$ *the interval* $[\sigma + \frac{1}{\mu_l}, \sigma)$ *contains at least l eigenvalues of the problem* $A\, x = \lambda\, B\, x$.

Proof: The theorem is a special case of theorem 1 in [7]. Let $D_M := \mathbb{R}^n$, $D_N := \mathbb{R}^n$, the bilinear forms are defined by

$$M(x|y) := x^T A\, y \quad \text{and} \quad N(x|y) := x^T B\, y \quad \text{for } x, y \in \mathbb{R}^n .$$

The eigenvalue problem,

$$M(y|x) = \lambda\, N(y|x) \quad \text{for all } y \in \mathbb{R}^n ,$$

is equivalent to the problem,

$$A\, x = \lambda\, B\, x .$$

A2 is fulfilled. X, $b(.|.)$ and T are defined by $X := \mathbb{R}^n$, $b(x|y) := x^T B\, y$ for $x, y \in \mathbb{R}^n$ and $T : \mathbb{R}^n \to \mathbb{R}^n, T\, x := x$. If $u_1, \ldots, u_m \in \mathbb{R}^n$ are chosen, the definition $v_i := B^{-1} A\, u_i$, $w_i := v_i$ $(i = 1, \ldots, m)$ implies $b(T\, x|w_i) = M(x|u_i)$ for all $x \in \mathbb{R}^n$, $i = 1, \ldots, m$. C is defined as identity matrix.

The object of theorem 1 in this context is the calculation of bounds for eigenvalues of a matrix eigenvalue problem of dimension n by solving matrix eigenvalue problems

[1] If σ is not an eigenvalue of the problem $A\, x = \lambda\, B\, x$, then $\hat{B}$ is always positive definite.

of considerably smaller dimensions m. In the most favourable and common case, m will be equal to 1.

For the algorithmic implementation it is better to use a slight modification of theorem 1:

Theorem 2

Let assumptions 1 and 2 of theorem 1 be valid, and let σ, A_0, A_1 and A_2 be defined as in 3. Let σ not be an eigenvalue of the problem $A\,x = \lambda\,B\,x$. Let the eigenvalues τ_i of the eigenvalue problem,

$$(A_2 - \sigma\,A_1)\,x = \tau\,(A_1 - \sigma\,A_0)\,x\,, \tag{3}$$

be ordered by magnitude: $\tau_1 \le \tau_2 \le \ldots \le \tau_m$. If p is the number of eigenvalues τ_i smaller than σ, then the interval $[\tau_i, \sigma)$ contains at least $p+1-i$ eigenvalues of the problem $A\,x = \lambda\,B\,x$ $(i = 1, \ldots, p)$.

Proof: For a real number μ, $\mu \neq 0$, the following is valid:

μ is an eigenvalue of the problem,

$$(A_1 - \sigma\,A_0)\,x = \tilde{\tau}\,(A_2 - 2\,\sigma\,A_1 + \sigma^2 A_0)\,x \quad \Longleftrightarrow$$

$\frac{1}{\mu}$ is an eigenvalue of the problem,

$$((A_2 - \sigma\,A_1) - (A_1 - \sigma\,A_0))\,x = \hat{\tau}\,(A_1 - \sigma\,A_0)\,x \quad \Longleftrightarrow$$

$\sigma + \frac{1}{\mu}$ is an eigenvalue of the problem,

$$(A_2 - \sigma\,A_1)\,x = \tau\,(A_1 - \sigma\,A_0)\,x\,.$$

Hence, the assertion follows from theorem 1, since $\hat{B}$ is positive definite.

As theorems 1 and 2 result only in intervals containing a certain number of eigenvalues, additional information on the parameter σ is necessary, in order to make a statement on the indices of the enclosed eigenvalues. If σ is a rough lower bound to λ_{s+1} with $\lambda_s < \sigma < \lambda_{s+1}$, and if p is the number of eigenvalues τ_i of problem (3) which are smaller than σ, then τ_i is in general a very precise lower bound to λ_{s+i-p} $(1 \le i \le p)$. This information on σ is calculated in the second step of the procedure and is not derived from a comparison problem, as is done when dealing with differential equations.

Two aspects play a part in the choice of σ: on the one hand, $s - j$ should be as small as possible, in order to be able to select a small m, because the inclusion theorems again result in a matrix eigenvalue problem of order m. On the other hand, experience shows that σ should not belong to the same cluster of eigenvalues to which λ_j possibly belongs. Thus, the following requirement arises: λ_s must be the largest eigenvalue of the cluster to which λ_j belongs. In the case $s = j$ one can achieve $m = 1$; in this special case theorem 2 reduces to Temple's inclusion theorem. If $s = n$ theorem 2 simplifies to the theorem of Rayleigh–Ritz by passage to the limit $\sigma \to \infty$. (The matrix eigenvalue problem $A_1\, x = \tau\, A_0\, x$ has then to be considered.)

In order to get upper bounds, the matrix eigenvalue problem $-A\,x = (-\lambda)\,B\,x$ is considered instead of problem (1). Lower bounds for $(-\lambda)$, that is, upper bounds for λ, are then calculated with the use of theorem 2. This results in

Theorem 3

Let assumptions 1 and 2 of theorem 1 be valid, let $\rho \in \mathbb{R}$ and let A_0, A_1 and A_2 be defined as in 3. Let ρ not be an eigenvalue of the problem $A\,x = \lambda\,B\,x$. Let the eigenvalues τ_i of the eigenvalue problem,

$$(A_2 - \rho\, A_1)\, x = \tau\, (A_1 - \rho\, A_0)\, x\,,$$

be arranged in descending order: $\tau_1 \geq \tau_2 \geq \ldots \geq \tau_m$. If p is the number of eigenvalues τ_i greater than ρ, then the interval $(\rho, \tau_i]$ contains at least $p + 1 - i$ eigenvalues of the problem $A\,x = \lambda\,B\,x$ $(i = 1, \ldots, p)$.

If λ_r denotes the smallest eigenvalue of the cluster, to which λ_j belongs, and if ρ and σ are chosen such that $\rho < \lambda_r \leq \ldots \leq \lambda_j \leq \ldots \leq \lambda_s < \sigma$ holds true, then m is defined by $m := s - r + 1$. Thus, the same elements u_i and hence, the same matrices A_0, A_1 and A_2 can be used in theorems 2 and 3. If $m > 1$, this yields not only bounds for λ_j, but also bounds for $\lambda_r, \ldots, \lambda_s$.

For the application of the theorems, the choice of the elements u_i $(i = 1, \ldots, m)$ has to be explained: the approximate eigenvectors $\tilde{x}_r, \ldots, \tilde{x}_s$ (approximately B–orthonormalized) are used for $u_1, \ldots, u_m$. The generally unknown vectors v_i are replaced by interval vectors $[v_i]$, which result from the solutions of the linear systems $B\,v_i = A\,u_i$ $(i = 1, \ldots, m)$. If one substitutes $[A_2] := ([v_i]^T A\, u_k)_{i,k=1,\ldots,m}$ for A_2, one has $A_2 \in [A_2]$. This choice of the elements u_i offers several advantages: on the one hand, very good bounds can be expected on the basis of convergence theorems [14]; on the other hand, the matrices A_0, A_1 and $[A_2]$ have a nearly diagonal structure. Furthermore, the matrices $-(A_1 - \sigma A_0)$ and $(A_1 - \rho A_0)$ are in general positive definite (for $y \in \mathbb{R}^m$, $y \neq 0$, the Rayleigh quotient $\frac{y^T A_1\, y}{y^T A_0\, y}$ is of the order of λ_j, and $\rho < \lambda_j < \sigma$ holds true). Because of the structure of the matrices, the positive definiteness can easily be proved numerically (for example, with the use of Gerschgorin's theorem).

LDL^T decompositions

The next Lemma [13, p. 255] plays a part in the determination of ρ and σ, as well as (combined with a bisection method) in the calculation of bounds to eigenvalues for matrices of small dimensions (theorems 2 and 3, $m > 1$).

Lemma 1
Let $\tilde{A}$, $\tilde{B}$ be real $q \times q$ matrices, $\tilde{A} = \tilde{A}^T$, $\tilde{B} = \tilde{B}^T$, let $\tilde{B}$ be positive definite, let $c \in I\!R$. The number of eigenvalues $\tilde{\lambda}_i$ of $\tilde{A}\,x = \tilde{\lambda}\,\tilde{B}\,x$ which are smaller than, equal to, or greater than c is equal to the number of negative, zero, or positive eigenvalues of the matrix $(\tilde{A} - c\,\tilde{B})$, respectively.

The number of negative, zero, and positive eigenvalues (the inertia) of a real symmetric matrix $\breve{A}$ can be easily determined with the help of a decomposition $\breve{A} = LDL^T$ [2]. Here, L is a regular lower triangular matrix, and D is a block diagonal matrix with blocks of order 1 or 2. For the implementation of the LDL^T decomposition with an interval arithmetic, the same pivot strategy as in [2] can be used. This results in an interval matrix $[D]$ with a structure corresponding to D. If an interval with zero included should occur during the determination of the inertia for $[D]$ — that is, if zero is included in a block of order 1 or in the determinant of a block of order 2 —, it is impossible to decide whether this interval corresponds to a negative, zero, or positive eigenvalue. The shift parameter c is then altered slightly, and the decomposition is repeated (see [11, p. 46 – 49]).

The algorithm

This results in the following algorithm (see [1]):[2]

Input:	A, B	matrices of problem (1)
	n	dimension of the matrices
	j	index of the eigenvalue to be included

Output:	$r \leq j \leq s$	indices of the included eigenvalues
	$[\lambda_r], \ldots, [\lambda_s]$	inclusion intervals

1. Calculate approximate eigenvalues
   $\tilde{\lambda}_{r-1} < \tilde{\lambda}_r \leq \ldots \leq \tilde{\lambda}_j \leq \ldots \leq \tilde{\lambda}_s < \tilde{\lambda}_{s+1}$

[2]For intervals the notation $[\lambda] = [\underline{\lambda}, \overline{\lambda}]$ is used, if necessary; then $\underline{\lambda}, \overline{\lambda} \in I\!R$ are the lower and the upper boundary points.

and approximate eigenvectors $\tilde{x}_r, \ldots, \tilde{x}_s$;
{ if $r = 1$, $\tilde{\lambda}_{r-1}$ has to be dropped;
if $s = n$, $\tilde{\lambda}_{s+1}$ has to be dropped;
let $\tilde{\lambda}_{r-1} + 0.01|\tilde{\lambda}_{r-1}| < \tilde{\lambda}_r$ and $\tilde{\lambda}_s < \tilde{\lambda}_{s+1} - 0.01|\tilde{\lambda}_{s+1}|$;
let $s - r$ be minimal }

2. If $r \neq 1$, then
   $\rho := \tilde{\lambda}_{r-1} + 0.005|\tilde{\lambda}_{r-1}|$;
   repeat
      calculate an LDL^T decomposition of the matrix $A - \rho B$;
      test, whether there are $r - 1$ negative and $n - (r - 1)$ positive eigenvalues;
      if not, alter ρ appropriately;
   until $\lambda_{r-1} < \rho < \lambda_r$;
   if $s \neq n$, then
   $\sigma := \tilde{\lambda}_{s+1} - 0.005|\tilde{\lambda}_{s+1}|$;
   repeat
      calculate an LDL^T decomposition of the matrix $A - \sigma B$;
      test, whether there are s negative and $n - s$ positive eigenvalues;
      if not, alter σ appropriately;
   until $\lambda_s < \sigma < \lambda_{s+1}$;
   $m := s - r + 1$;

3. {Spectral shift: bounds are calculated for the residue.}
   $[A] := A - \tilde{\lambda}_j B$;
   $[B] := B$;
   $\rho := \rho - \tilde{\lambda}_j$;
   $\sigma := \sigma - \tilde{\lambda}_j$;
   $u_1 := \tilde{x}_r, \ldots, u_m := \tilde{x}_s$;
   for $i = 1, \ldots, m$
      calculate $[v_i]$ from the equation $[B]\, v_i = [A]\, u_i$;
   $[A_0] := \left(u_i^T [B]\, u_k\right)_{i,k=1,\ldots,m}$;
   $[A_1] := \left(u_i^T [A]\, u_k\right)_{i,k=1,\ldots,m}$;
   $[A_2] := \left([v_i]^T [A]\, u_k\right)_{i,k=1,\ldots,m}$;

   upper bounds:
   if $r = 1$, then
      calculate by bisection bounds

$[\mu_1], \ldots, [\mu_m]$
for the eigenvalues of the problem
$[A_1]\, x = \mu\, [A_0]\, x$;

else
test, whether the matrix $[A_1] - \rho[A_0]$ is positive definite;
if not: stop;
calculate by bisection bounds
$[\mu_1], \ldots, [\mu_m]$
for the eigenvalues of the problem
$([A_2] - \rho[A_1])\, x = \mu\, ([A_1] - \rho[A_0])\, x$;
test, whether $\rho < \underline{\mu}_1$;
if not: stop;

$\overline{\lambda}_r := \tilde{\lambda}_j + \overline{\mu}_1, \ldots, \overline{\lambda}_s := \tilde{\lambda}_j + \overline{\mu}_m$;

lower bounds:
if $s = n$, then
calculate by bisection bounds
$[\mu_1], \ldots, [\mu_m]$
for the eigenvalues of the problem
$[A_1]\, x = \mu\, [A_0]\, x$;

else
test, whether the matrix $-([A_1] - \sigma[A_0])$ is positive definite;
if not: stop;
calculate by bisection bounds
$[\mu_1], \ldots, [\mu_m]$
for the eigenvalues of the problem
$(-[A_2] + \sigma[A_1])\, x = \mu\, (-[A_1] + \sigma[A_0])\, x$;
test, whether $\overline{\mu}_m < \sigma$;
if not: stop;

$\underline{\lambda}_r := \tilde{\lambda}_j + \underline{\mu}_1, \ldots, \underline{\lambda}_s := \tilde{\lambda}_j + \underline{\mu}_m$;
$[\lambda_r] := [\underline{\lambda}_r, \overline{\lambda}_r], \ldots, [\lambda_s] := [\underline{\lambda}_s, \overline{\lambda}_s]$;

Numerical example

Let the matrices A and B be defined by

$$A = (a_{i,k})_{i,k=1,\ldots,18}\,, \quad a_{i,k} := \begin{cases} 0 & \text{if } 19|(i+k) \\ 1 & \text{if } (i+k) \equiv j^2 \mod 19 \\ & \qquad\qquad \text{for a } j \in \mathbb{N} \\ -1 & \text{otherwise} \end{cases}$$

and

$$B = (b_{i,k})_{i,k=1,\ldots,18}\,, \quad b_{i,k} := \begin{cases} 1 & \text{if } i = k \\ 10^{-6} & \text{if } |i-k| = 3 \\ 0 & \text{otherwise} \end{cases} .$$

A is a Hadamard matrix and has the eightfold eigenvalues $\pm\sqrt{19}$ and the simple eigenvalues ± 1 (see [8]). B is a slightly modified identity matrix; therefore, the matrix eigenvalue problem,

$$A\,x = \lambda\,B\,x\,,$$

has two clusters, each containing eight eigenvalues. One obtains the rough bounds

$$\begin{aligned} \lambda_i &\in [-4.3^{54}_{64}\,] && \text{for} \quad i = 1, \ldots, 8\,, \\ \lambda_i &\in [-^{0.9989}_{1.0010}\,] && \text{for} \quad i = 9\,, \\ \lambda_i &\in [\ ^{1.0010}_{0.9989}\,] && \text{for} \quad i = 10\,, \\ \lambda_i &\in [\ 4.3^{64}_{54}\,] && \text{for} \quad i = 11, \ldots, 18\,. \end{aligned}$$

The bounds calculated by theorems 2 and 3 are[3]

$$\begin{aligned} \lambda_1 &\in [-4.358\,906\,293\,914\,31^{77}_{80}\,] \\ \lambda_2 &\in [-4.358\,904\,905\,898\,56^{78}_{81}\,] \\ \lambda_3 &\in [-4.358\,903\,409\,616\,45^{69}_{72}\,] \\ \lambda_4 &\in [-4.358\,900\,453\,991\,24^{19}_{22}\,] \\ \lambda_5 &\in [-4.358\,898\,673\,162\,818^{0}_{4}\,] \\ \lambda_6 &\in [-4.358\,895\,813\,100\,530^{2}_{5}\,] \\ \lambda_7 &\in [-4.358\,893\,709\,563\,09^{37}_{40}\,] \\ \lambda_8 &\in [-4.358\,891\,195\,065\,112^{5}_{9}\,] \end{aligned}$$

$$\begin{aligned} \lambda_9 &\in [-0.999\,999\,333\,334\,141\,9^{6}_{9}\,] \\ \lambda_{10} &\in [\ \ 0.999\,999\,333\,334\,141\,9^{9}_{6}\,] \end{aligned}$$

[3]The bounds have been calculated with FORTRAN-SC on an IBM-3090.

$$\lambda_{11} \in [\ 4.358\,891\,195\,065\,112_5^9\ ]$$
$$\lambda_{12} \in [\ 4.358\,893\,709\,563\,09_{37}^{40}\ ]$$
$$\lambda_{13} \in [\ 4.358\,895\,813\,100\,530_2^5\ ]$$
$$\lambda_{14} \in [\ 4.358\,898\,673\,162\,818_0^4\ ]$$
$$\lambda_{15} \in [\ 4.358\,900\,453\,991\,24_{19}^{22}\ ]$$
$$\lambda_{16} \in [\ 4.358\,903\,409\,616\,45_{69}^{72}\ ]$$
$$\lambda_{17} \in [\ 4.358\,904\,905\,898\,56_{78}^{81}\ ]$$
$$\lambda_{18} \in [\ 4.358\,906\,293\,914\,31_{77}^{80}\ ]$$

3 An eigenvalue problem with an ordinary differential equation

The main area of application for the theorems of Ritz, Lehmann and Goerisch is the treatment of eigenvalue problems with differential equations. These theorems result in matrix eigenvalue problems (the eigenvalues of which yield bounds for the differential equation problem), which have to be solved numerically on a computer. If the procedure proposed in section 2 is used for this purpose, one obtains *guaranteed* eigenvalue bounds. As the matrix elements already have to be calculated with an interval arithmetic, one has to deal with an eigenvalue problem with interval matrices. One then has to determine intervals containing the following sets:

$$\begin{aligned}\{\lambda_j | A\,x = \lambda_j\,B\,x\,,\ \ A \in [A]\,,\ B \in [B]\,,\\ A = A^T\,,\ B = B^T\,,\ B \text{ positive definite}\}\,.\end{aligned} \tag{4}$$

This problem can be solved with an obvious modification of the procedure proposed in section 2.

Inclusion theorem

The inclusion theorem of Goerisch for a positive definite bilinear form on the left hand side of the equation (see [6]) is first stated:

Theorem 4
Let the following assumptions be valid:

A1: *D is a real vector space, $M(.|.)$ und $N(.|.)$ are symmetric bilinear forms on D, $M(.|.)$ is positive definite.*

A2: *There exist a sequence $(\lambda_i)_{i\in J}$ (J index set) of eigenvalues of the eigenvalue problem*

$$M(f|\phi) = \lambda\, N(f|\phi) \text{ for all } f \in D \tag{5}$$

and a sequence $(\phi_i)_{i\in J}$ of corresponding eigenelements such that

$$\begin{array}{ll} M(f|\phi_i) = \lambda_i\; N(f|\phi_i) & \text{for all } f \in D\,,\ i \in J, \\ M(\phi_i|\phi_k) = \delta_{i,k} & \text{for all } i,k \in J \text{ and} \\ N(f|f) = \sum_{i\in J} \lambda_i\; (N(f|\phi_i))^2 & \text{for all } f \in D \end{array}$$

holds true.

A3: *Let $m \in I\!N$, let $u_i \in D$ for $i = 1, \ldots, m$ and $\sigma \in I\!R, \sigma > 0$.*
X is a real vector space,
$b(.|.)$ is a symmetric, positive semidefinite bilinear form on X,
$T : D \to X$ is a linear operator, with $b(Tf|Tg) = M(f|g)$ for all $f, g \in D$.
Elements $w_i \in X$ are chosen such that

$$b(T\,f|w_i) = N(f|u_i) \text{ for all } f \in D,\ i = 1, \ldots, m$$

holds true.

A4: *Let $q \in I\!N$, $q \leq m$; $C \in I\!R^{m\times q}$ is a real matrix. The matrices A_0, A_1, A_2, $\hat{A}$ and $\hat{B}$ are defined by*

$$\begin{aligned} A_0 &:= (M(u_i|u_k))_{i,k=1,\ldots,m}\,, \\ A_1 &:= (N(u_i|u_k))_{i,k=1,\ldots,m}\,, \\ A_2 &:= (b(w_i|w_k))_{i,k=1,\ldots,m}\,, \end{aligned}$$

$$\hat{A} := C^T(A_0 - \sigma\, A_1)\, C \text{ and } \hat{B} := C^T(A_0 - 2\sigma\, A_1 + \sigma^2\, A_2)\, C\,.$$

$\hat{B}$ is positive definite. The matrix eigenvalue problem,

$$\hat{A}x = \mu\, \hat{B}x\,,$$

has the eigenvalues $\mu_1 \leq \mu_2 \leq \ldots \leq \mu_p < 0 \leq \ldots \leq \mu_q$.

Assertion: *For $l = 1, \ldots, p$, the interval*

$$[\sigma - \frac{\sigma}{1-\mu_l}, \sigma)$$

contains at least l eigenvalues of eigenvalue problem (5).

Example

The application of this theorem for the calculation of lower bounds to eigenvalues is demonstrated with a simple example. (Upper bounds are determined with the procedure of Rayleigh–Ritz as usual. This is not further discussed here.) Let

$$D := \{f \in C^1[0,\pi] \mid f(x-\pi) = f(x) \text{ for } x \in [0,\pi],\ f(0) = f(\pi) = 0\},$$

$$\left.\begin{array}{l} M(f|g) := \int_0^\pi f'(x)\, g'(x)\, dx \\[2ex] N(f|g) := \int_0^\pi (1+\alpha \sin(x))\, f(x)\, g(x)\, dx \end{array}\right\} \quad \text{for } f,g \in D, \quad \alpha \in \mathbb{R}$$

The eigenvalue problem is:

" Determine $\phi \in D$, $\phi \neq 0$ and $\lambda \in \mathbb{R}$ such that

$$M(f|\phi) = \lambda\, N(f|\phi) \text{ for all } f \in D." \qquad (6)$$

With these definitions, assumptions A1 and A2 are fulfilled. Problem (6) is equivalent to the following eigenvalue problem:

$$\begin{aligned} -\phi''(x) &= \lambda\,(1+\alpha \sin(x))\, \phi(x) \quad \text{for } x \in [0,\pi], \\ \phi(0) &= \phi(\pi) = 0, \\ \phi(\pi - x) &= \phi(x) \quad \text{for } x \in [0,\pi], \end{aligned}$$

bounds for the eigenvalues can be found in Collatz [3], Fichera [4], Zimmermann [14] and Plum [12].

The trial functions u_i are defined as

$$u_i(x) := \sin(2i-1)x \quad \text{for } x \in [0,\pi], i \in \mathbb{N}.$$

If, furthermore, $X := D$, $b(f|g) := M(f|g)$ for $f,g \in X$, $T : D \to X, Tf := f$, as well as

$$w_1(x) := \frac{\alpha}{8} + \frac{\alpha\pi}{4}x - \frac{\alpha}{4}x^2 + \sin(x) - \frac{\alpha}{8}\cos(2\,x) \quad \text{for } x \in [0,\pi] \text{ and}$$

$$\begin{aligned} w_i(x) := & -\frac{\alpha(2i-1)}{8i^2(i-1)^2} - \frac{\alpha}{8i^2}\cos(2ix) + \frac{\alpha}{8(i-1)^2}\cos(2(i-1)x) \\ & + \frac{1}{(2i-1)^2}\sin((2i-1)x) \quad \text{for } x \in [0,\pi] \text{ and } i \in \mathbb{N}, i \geq 2 \end{aligned}$$

are chosen, one has

$$\begin{aligned} -w_i''(x) &= (1+\alpha\,\sin(x))\,u_i(x) \quad \text{for } x\in[0,\pi] \text{ and} \\ w_i(0) &= w_i(\pi)=0 \end{aligned}$$

and therefore,

$$b(T\,f|w_i) = M(f|w_i) = N(f|u_i)$$

for all $f \in D$ and $i \in I\!N$.

In order to make an appropriate choice for σ, the comparison problem,

$$\begin{aligned} -\phi''(x) &= \Lambda\,(1+\alpha)\,\phi(x) \quad \text{for } x\in[0,\pi]\,, \\ \phi(0) &= \phi(\pi)=0\,, \\ \phi(\pi-x) &= \phi(x) \quad \text{for } x\in[0,\pi]\,, \end{aligned}$$

is considered. For $\alpha > -1$, its eigenvalues are

$$\Lambda_i = \frac{(2i-1)^2}{1+\alpha}\,,\ i \in I\!N\,;$$

if the eigenvalues of problem (6) are also ordered by magnitude ($\lambda_1 \leq \lambda_2 \leq \lambda_3 \leq \ldots$),

$$\Lambda_i \leq \lambda_i \quad \text{for } i \in I\!N$$

is valid. In A4, $q = m$ and $C = I$ (identity matrix) are chosen; the matrices A_0, A_1, and A_2 can now be evaluated with the use of an interval arithmetic. (The primitives of the required integrals can be calculated with a computer algebra system.)

With $\alpha = 1$ and $\sigma = 60.5$ (a lower bound for λ_6), the following eigenvalue inclusions are calculated:[4]

	$m = 14$	$m = 20$
λ_1	$[\,0.540\,318\,859\,5^{63}_{58}\,]$	$[\,0.540\,318\,859\,558^{98}_{56}\,]$
λ_2	$[\,5.448\,636\,1^{53}_{48}\,]$	$[\,5.448\,636\,14^{92}_{88}\,]$
λ_3	$[\,15.312\,608^{41}_{27}\,]$	$[\,15.312\,608\,3^{15}_{04}\,]$
λ_4	$[\,30.114\,98^{53}_{37}\,]$	$[\,30.114\,984^{54}_{41}\,]$

[4]In each case the same matrix dimension m has been used for theorem 4 and for the Rayleigh–Ritz procedure.

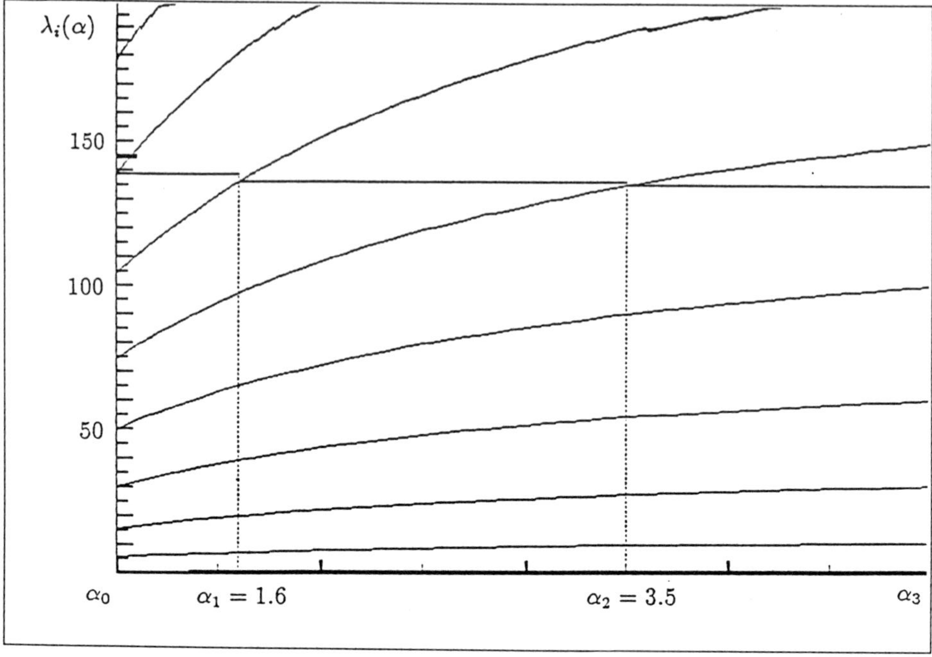

Figure 1: Eigenvalue curves for the step method

Step method

If lower bounds are to be computed for a greater value of the parameter α, for instance for $\alpha = 5$, attention must be paid to the following difficulty: An upper bound for λ_6 is about 29.919, but one has the estimate $28.166 < \frac{169}{6} = \Lambda_7 \leq \lambda_7$; the lower bound derived from the comparison problem is too poor. In order to calculate lower bounds in this situation, a homotopy method (step method) is used (Goerisch [5]). The step method is described with the use of the example.

Two bilinear forms $M_\alpha(.|.)$ and $N_\alpha(.|.)$ are defined by

$$\left.\begin{aligned} M_\alpha(f|g) &:= \int_0^\pi f'(x)\, g'(x)\, dx \\ N_\alpha(f|g) &:= \tfrac{1}{\alpha} \int_0^\pi (1 + \alpha \sin(x))\, f(x)\, g(x)\, dx \end{aligned}\right\} \quad \text{for } f, g \in D, \quad \alpha \in [1,5]\,.$$

The eigenvalue problem,

" Determine $\phi_\alpha \in D$, $\phi_\alpha \neq 0$ and $\lambda(\alpha) \in \mathbb{R}$ such, that

$$M_\alpha(f|\phi_\alpha) = \lambda(\alpha)\, N_\alpha(f|\phi_\alpha) \quad \text{for all } f \in D\,." \tag{7}$$

is now considered. The eigenvalues are ordered by magnitude: $\lambda_1(\alpha) \leq \lambda_2(\alpha) \leq \lambda_3(\alpha) \leq \ldots$. For a fixed α one has the relation $\lambda_i = \frac{1}{\alpha}\lambda_i(\alpha)$ $(i \in \mathbb{N})$ between the eigenvalues of problems (6) and (7). The eigenvalues of problem (7) increase monotonically in α, that is, for all $i \in \mathbb{N}$ and for $\alpha', \alpha'' \in [1,5]$, $\alpha' \leq \alpha''$ the relation $\lambda_i(\alpha') \leq \lambda_i(\alpha'')$ holds true.

If one knows a lower bound $\underline{\lambda_i(\alpha')}$ for $\lambda_i(\alpha')$, $\underline{\lambda_i(\alpha')} \leq \lambda_i(\alpha'')$ holds true. Hence, the following strategy is suggested: Make a sketch of the eigenvalue curves $\lambda_i(\alpha)$ (using, for example, the Rayleigh–Ritz procedure). Then, start with $\alpha_0 = 1$ and $\sigma = 144.5 = \Lambda_9 \leq \lambda_9 = \lambda_9(\alpha_0)$ and calculate a lower bound for $\lambda_8(\alpha_0)$ by means of theorem 4. This yields $138.67 \leq \lambda_8(\alpha_0)$. With the use of the figure, one finds $\alpha_1 = 1.6$ with $\lambda_7(\alpha_1) < 138.67$ and calculates for $\sigma = 138.67$ the lower bound $135.96 \leq \lambda_7(\alpha_1)$. In the same way one gets $\alpha_2 = 3.5$; $\sigma = 135.96$ results in $134.83 \leq \lambda_6(\alpha_2)$. $\sigma = 134.83$ is now a lower bound for $\lambda_6(5)$.

Hence, one can calculate lower bounds for the smallest eigenvalues of problem (7) with $\alpha = 5$ and also for problem (6) with $\alpha = 5$. With $\sigma = \frac{134.83}{5} = 26.966$ this results in the following bounds:

	$m = 15$
λ_1	$[\,0.190\,079\,902\,0^{41}_{37}\,]$
λ_2	$[\,2.111\,279\,13^{34}_{30}\,]$
λ_3	$[\,6.055\,505^{48}_{38}\,]$
λ_4	$[\,12.007\,97^{89}_{79}\,]$

References

[1] H. Behnke. *Die Bestimmung von Eigenwertschranken mit Hilfe von Variationsmethoden und Intervallarithmetik.* Dissertation, Institut für Mathematik, Technische Universität Clausthal, 1989.

[2] J.R. Bunch, L. Kaufmann, and B.N. Parlett. Decomposition of a Symmetric Matrix. *Numer. Math.*, 27:95–109, 1976.

[3] L. Collatz. *Eigenwertaufgaben mit technischen Anwendungen.* Akademische Verlagsgesellschaft Geest & Portig KG: Leipzig, 2. edition, 1963.

[4] G. Fichera. *Numerical and Quantitative Analysis.* Pitman: London-San Francisco-Melbourne, 1978.

[5] F. Goerisch. Ein Stufenverfahren zur Berechnung von Eigenwertschranken. In J. Albrecht, L. Collatz, W. Velte, and W. Wunderlich, editors, *Numerische Behandlung von Eigenwertaufgaben, Band 4. International Series of Numerical Mathematics (ISNM), Volume 83*, pages 104–114, Birkhäuser-Verlag: Basel-Boston, 1987.

[6] F. Goerisch and J. Albrecht. The Convergence of a New Method for Calculating Lower Bounds to Eigenvalues. In J. Vosmanský and M. Zlámal, editors, *Lecture Notes in Mathematics, Volume 1192. Equadiff 6, Proceedings of the International Conference on Differential Equations and their Applications held in Brno, 1985*, pages 303–308, Springer-Verlag: Berlin, Heidelberg, New York, Tokyo, 1986.

[7] F. Goerisch and Z. He. The Determination of Guaranteed Bounds to Eigenvalues with the Use of Variational Methods I. In this volume.

[8] R.T. Gregory and D.L. Karne. *A Collection of Matrices for Testing Computational Algorithms.* Wiley–Interscience: New York-London-Sydney-Toronto, 1969.

[9] N.J. Lehmann. Beiträge zur Lösung linearer Eigenwertprobleme I. *Z. Angew. Math. u. Mech.*, 29:341–356, 1949.

[10] N.J. Lehmann. Beiträge zur Lösung linearer Eigenwertprobleme II. *Z. Angew. Math. u. Mech.*, 30:1–16, 1950.

[11] B.N. Parlett. *The Symmetric Eigenvalue Problem.* Prentice Hall: Englewood Cliffs, 1980.

[12] M. Plum. Eigenvalue inclusions for second–order ordinary differential operators by a numerical homotopy method. To appear in Z. Angew. Math. u. Phys.

[13] H.R. Schwarz. *Methode der finiten Elemente.* Teubner: Stuttgart, 1980.

[14] S. Zimmermann. *Über die Genauigkeit von Eigenwertschranken für selbstadjungierte Operatoren.* Dissertation, Institut für Mathematik, Technische Universität Clausthal, 1989.

Validated Solution of Initial Value Problems for ODE

Hans J. Stetter
Institute for Applied and Numerical Mathematics
Technical University of Vienna
Austria

Abstract

Presently, programs for the validated solution of initial value problems for ODEs require human control and guidance, and they are often very expensive computationally. In this paper, we consider various mechanisms which may lead to a better algorithmic control and to more efficient implemantations of the Taylor series based approach for the above task: We consider the trade-off between effort and accuracy, the generation of an a priori inclusion of the remainder term, the representation of the inclusion sets, and modifications for stiff systems.

Introduction

The solution of initial-value problems for systems of ODEs is one of the most frequently occurring tasks in scientific computation. Therefore it is particularly important to have an efficient and easily useable software product available for the validated execution of this task. The fact that validated results for ODEs may be crucial in many situations has been amply demontrated in the paper by E. Adams ([1]) in this volume.

The history of the problem - as a computational problem - dates back to 1966: A good deal of R.E. Moore's book on 'Interval Analysis' ([2]) has been devoted to that subject. Further landmarks have been the three Ph.D. theses by U. Marcowitz ([3]), P. Eijgenraam ([4]), and R. Lohner ([5]). But to this date, no programs which could truly be called 'scientific software' have been produced, i.e. programs which are easily handled by a standard scientific user.

ISBN 0-12-708245-X

Lohner's program AWA ([6]) represents the state of the art. It can successfully deal with a wide variety of problems - if it is handled by a sufficiently expert user: It requires human control (i.e. adaptive selection) of stepsize and order and a suitable selection of the inclusion set representation. Also it is often very expensive computationally, more expensive than it would seem necessary.

Naturally, the task of computing tight inclusions for solutions of initial value problems for general ODE systems is an intrinsically difficult problem. Even in the class of inversely monotone problems (see e.g. [7]) many of the difficulties remain. Therefore any further progress towards a software product for validated ODE solution will not be easy and require a concentrated effort.

In this paper, we will consider only[1] the Moore-Lohner approach (Taylor series, remainder term inclusion) and discuss some ideas for improving its efficiency and controlling its parameters. Many of these ideas have arisen from discussions with George F. Corliss, Milwaukee, and Rudolf Lohner, Karlsruhe, and with staff members and students in the Technical University of Vienna: Gabriela Schranz-Kirlinger, Monika Kerbl, Veronika Brand. In a cooperation of the groups in Milwaukee, Karlsruhe, and Vienna, we are presently engaged in a joint project to produce software for validated ODE solution.

1 The Method, Areas for Improvement

For reference purpose in the remainder of the paper, we shortly describe the method of Moore-Lohner. We consider the autonomous ODE-system in $\mathbb{R}^s$

$$y'(t) = f(y(t)) \;, \qquad t \in [0,T] \;, \qquad f : \mathbb{R}^s \to \mathbb{R}^s, \tag{1.1}$$

with initial values

$$y(0) \in Y_0 \subset \mathbb{R}^s. \tag{1.2}$$

Where it is essentially that a set in $\mathbb{R}^s$ is an interval, we will use the notation $[\ldots]$. For a set $Y \subset \mathbb{R}^s$, $[Y]$ will denote an interval containing Y. We will also use the notation $y(t; t_\nu, y_\nu)$ for the (exact) value at t of the solution of (1.1) issuing from (t_ν, y_ν) and

$$y(t; t_\nu, Y_\nu) := \cup_{y_\nu \in Y_\nu} y(t; t_\nu, y_\nu) \subset \mathbb{R}^s \tag{1.3}$$

[1]We are well aware of the fact that there are at least two other principal approaches, viz. those of [3] and of [8]; but none of these has been efficiently implemented so far.

for the set of values at t of the solutions of (1.1) issuing from the set Y_ν at t_ν. Furthermore we will use

$$\begin{aligned} f_0(y) &:= f(y) \\ f_j(y) &:= f'_{j-1}(y) \cdot f(y), \qquad j = 1, 2, \ldots \end{aligned} \tag{1.4}$$

so that $\frac{d^j}{dt^j} y(t) = f_{j-1}(y(t))$ for a solution of (1.1).

Assume that, at some $t_\nu > 0$, we have generated the inclusion

$$y(t_\nu; 0, Y_0) \subset \tilde{y}_\nu + S_\nu \cdot [e_\nu] =: Y_\nu \tag{1.5}$$

where $\tilde{y}_\nu \in \mathbb{R}^s$ is a (point) approximation at t_ν,
$[e_\nu] \in \mathbb{IR}^s$ a symmetric interval vector and
S_ν a $s \times s$-matrix,
so that Y_ν is a parallelepipedon centered at $\tilde{y}_\nu$.

An inclusion for $y(t_{\nu+1}; t_\nu, Y_\nu)$ which includes $y(t_{\nu+1}; 0, Y_0)$ because of (1.5), with $t_{\nu+1} = t_\nu + h$, is then computed as follows:

(i) At first, some crude *a priori inclusion* on $[t_\nu, t_{\nu+1}]$ is determined, i.e. some set $\bar{Y}_{\nu+1}$ such that

$$y(t; t_\nu, Y_\nu) \subset \bar{Y}_{\nu+1} \quad \text{for} \quad t \in [t_\nu, t_{\nu+1}]. \tag{1.6}$$

For details, see, e.g., [5]

(ii) Obviously (cf. [2])

$$\begin{aligned} y(t_{\nu+1}; t_\nu, Y_\nu) \in\ & \tilde{y}_\nu + \sum_{j=1}^{p-1} \frac{h^j}{j!} f_{j-1}(\tilde{y}_\nu) \\ +\ & (I + \sum_{j=1}^{p-1} \frac{h^j}{j!} f'_{j-1}(\tilde{y}_\nu + S_\nu[e_\nu])) \cdot S_\nu[e_\nu] \\ +\ & \frac{h^p}{p!} f_{p-1}(\bar{Y}_{\nu+1}) \end{aligned} \tag{1.7}$$

where the a priori inclusion $\bar{Y}_{\nu+1}$ has been used to obtain an inclusion of the remainder term of the Taylor series. The first line on the r.h.s. of (1.7) is a point value whereas the sets on the last two lines must be further enclosed into some parallelepipedon. After adding the center of this parallelepipedon to the first line and subtracting it from the set, we have the inclusion at $t_{\nu+1}$

$$y(t_{\nu+1}; t_\nu, Y_\nu) \subset \tilde{y}_{\nu+1} + S_{\nu+1}[e_{\nu+1}] =: Y_{\nu+1} \tag{1.8}$$

where the choice of $S_{\nu+1}$ will be discussed later. Thus, one forward step of the method consists in properly updating $\tilde{y}_\nu, S_\nu$ and $[e_\nu]$ after an a priori inclusion for the step has been found.

The computation of the $f_j : \mathbb{R}^s \to \mathbb{R}^s$ and $f_j' : \mathbb{R}^s \to \mathbb{R}^{s\times s}$ for point and set arguments is performed via the "automatic differentiation" approach; see, e.g., [9].

In this presentation, we will discuss the following topics which may lead to an improvement of the efficiency and/or the ease-of-use of this method.

- Choice of the stepsize h and order p
- Choice of the a priori inclusion $\bar{Y}_{\nu+1}$
- Choice of the local coordinate system S_ν
- Modifications in the case of stiff systems

2 Accuracy Control

From (1.7) it is clear that – at least in principle – the values of the stepsize h and the order p may be freely chosen in each step. At present, both quantities have to be prescribed by the user and are normally left unchanged over some interval or until something goes wrong. In a software product, we should have an automatic choice which (quasi-)minimizes the computational effort for a prescribed accuracy. With two parameters at hand in each step, one would expect to achieve such a goal quite well. But so far it has not even been clear which strategy should be followed (except for some considerations in [4]).

The strategy to be explained below has been strongly patterned by the established strategy for ordinary ODE-solvers. There (cf. e.g. [10]) the user specifies a scalar parameter, say TOL, and the code attempts to generate an approximate solution $\eta_\nu \approx y(t_\nu)$ such that the global error $\eta_\nu - y(t_\nu)$, measured in some norm which may be manipulated by the user, is *proportional* to TOL:

$$\| \eta_\nu - y(t_\nu) \| \quad \approx \quad c(t_\nu) \cdot TOL \tag{2.1}$$

The function $c : [0, T] \to \mathbb{R}_+$ is unknown and, for a given code, strongly problem dependent; thus the only insight to be gained from (2.1) is through runs of the same problem with different values of TOL.

(2.1) is achieved through a *local* control of stepsize and/or order. In most codes, the control techniques can be rigorously justified only under very narrow assumptions and they contain many heuristic details. Nevertheless, they satisfy the most important requirement, viz. to discover and overcome difficulties in the solution.

Also, in nearly all well-known ODE-codes, only one parameter (step size or order) is changed at a time – except under emergency; so there is no freedom to look for minimal effort at a given tolerance level. There is one notable exception of strong relevance for our case, the Taylor series based code ATSMCC ([11]) which checks the effort needed for more terms in the Taylor series and also checks for bounds on the largest admissible stepsize, in an effort to maintain a given local error at low cost.

In validated ODE-algorithms which produce enclosing sets $Y_\nu \ni y(t_\nu; 0, y_0)$ instead of approximations $y_\nu \approx y(t_\nu; 0, y_0)$, the natural counterpart to the *tolerance proportionality* of (2.1) is

$$\| \; w([Y_\nu]) \; \| \quad \approx \quad c(t_\nu) \cdot TOL \tag{2.2}$$

where $w : \mathbb{IR}^s \rightarrow \mathbb{R}^s$ denotes the *width* of (the smallest enclosing interval of) its argument set. But (2.2) makes sense only for point initial conditions (1.2), with $w(Y_0) = 0$ which implies $w(y(t; 0, Y_0)) = 0$ at all t.

If $w(Y_0) > 0$ and hence $w(y(t; 0, Y_0)) > 0$, the quantity which corresponds to the global error in (2.1) is rather the *overestimation* of Y_ν with respect to $y(t_\nu; 0, Y_0)$, a quantity which is often called the *excess* $[E_\nu]$ of Y_ν:

$$[E_\nu] := [Y_\nu] \dot{-} [y(t_\nu; t_0, Y_0)] \tag{2.3}$$

Here $\dot{-}$ denotes the *interior difference* of the two sets, which is defined only if the second set is a subset of the first:

$$[C] := [A] \dot{-} [B] \Leftrightarrow [A] = [B] + [C] \quad \textit{for} \quad [B] \subset [A] \tag{2.4}$$

Hence, our global goal will be defined as *tolerance proportionality of the excess*:

$$\|w([E_\nu])\| \approx c(t_\nu) \cdot TOL \tag{2.5}$$

Note that $w([E_\nu]) = w([Y_\nu])$ iff $[y(t_\nu; 0, Y_0)]$ is a degenerate interval, i.e. a point solution $y(t_\nu; 0, y_0)$, so that (2.5) and (2.2) are equivalent in this important case.

Now we must find how we may achieve (2.5) by a *local control mechanism*. Note that, locally, we always have set initial conditions because the *local problem* consists in computing an inclusion for the set $y(t_{\nu+1};t_\nu,Y_\nu)$, cf. (1.3) and (1.7), where Y_ν is always a set except possibly at $\nu=0$. The *local excess* l.e. is thus defined as

$$(l.e.)_\nu := Y_{\nu+1} \dot{-} y(t_{\nu+1};t_\nu,Y_\nu) \tag{2.6}$$

In analogy to the well-known theory for ODE approximation, we expect that a control of the *local excess per unit step* $L_\nu = (l.e.)_\nu/(t_{\nu+1}-t_\nu)$ aiming at

$$\|w([L_\nu])\| \approx TOL \tag{2.7}$$

will lead to (2.5). This may be proved under mild technical assumptions since the local excess per unit step may be interpreted as an additive perturbation in the given differential system.

There are two contributions to the local excess where the terms in (1.7) overestimate the corresponding terms in the exact expression for $y(t_{\nu+1};t_\nu,Y_\nu)$: In the second line of (1.7), the rows of the f'_{j-1} should be evaluated at *intermediate* initial values $\tilde{y_\nu}+\lambda S_\nu\varepsilon_\nu$, $\varepsilon_\nu\in[e_\nu], \lambda\in(0,1)$, and in the third line, the use of the rough a priori inclusion $\bar{Y}_{\nu+1}$ leads to an overestimation. Thus a direct estimation of L_ν does not appear feasible.

However, in standard ODE codes, a direct estimation of the local error per unit step l_ν is generally also dismissed for reasons of cost. Instead one controls a quantity which is "proportional" to l_ν, i.e. which is $0(h^p)$ for sufficiently small steps just as l_ν. Since there is an unknown proportionality function c in (2.1) anyway this is justifiable, at least with a good deal of "handwaving".

In the same spirit, when we look for a quantity which should behave like L_ν for sufficiently small steps we are led to the Taylor series remainder term $r_{\nu+1} := \frac{h^p}{p!}f_{p-1}(\bar{Y}_{\nu+1})$. With the standard approach to the determination of the a priori inclusion $\bar{Y}_{\nu+1}$ (cf. [5]) , the excess of this quantity is $0(h^{p+1})$ because the "excess" of $\bar{Y}_{\nu+1}$ over the true intermediate values is $0(h)$. Thus, the asymptotic size $O(h^p)$ of $r_{\nu+1}$ is the same as that of its excess per unit step.

It remains to check whether the contribution of $r_{\nu+1}$ to the local excess is the dominating term. For problems with a point initial condition (1.2), $[e_\nu]$ is $O(h^p)$ and the excess from the second line of (1.7) is $0(h[e_\nu]^2)$ so that we may neglect this contribution. For set initial conditions with $\varepsilon_0 = w([Y_0]) > 0$, however, $[e_\nu] = O(\varepsilon_0)+O(h^p)$ and the excess contribution from the second line of (1.7) may not be small.

Thus it seems that the width of $r_{\nu+1}$ may be a suitable control quantity in the case of point initial conditions only. A more rigorous analysis of the situation is presently made by M. Kerbl in Vienna. In any case, in order to control the accuracy of our computation by the principle of tolerance proportionality of the excess (cf. (2.5)) we must find a quantity which is easily computed in each step and which is $0([L_\nu])$ for sufficiently small steps.

3 Minimizing the Effort

After we have decided that we will apply the local accuracy contral strategy

$$\|w([r_{\nu+1}])\| \approx TOL \quad , \tag{3.1}$$

at least in the case of point initial conditions, we now have to decide about the best mechanism to maintain (3.1) during the validated integration of (1.1) by means of the scheme (1.7):

If we find that $r_{\nu+1}$ is getting too large, should we reduce its size by an increase of p, i.e. use more terms of the Taylor series, or by a decrease of h, i.e. take smaller steps. Both changes increase the total computational effort to cover a given interval of integration, therefore we would like to know which of the two achieves a certain reduction of $r_{\nu+1}$ at less additional cost.

Even more fundamental is the question whether – for a given value of TOL – there exists an *optimal pair* p, h such that the effort for the evaluation of (1.7) is minimal while (3.1) holds. Actually, at such an optimal pair, a given (small) change in TOL costs the same no matter how it is achieved:

Consider two suitably differentiable functions F, G in two variables x, y. A necessary condition for an extremum of G along a curve $F = const$ is

$$F_x G_y - F_y G_x = 0 \tag{3.2}$$

At a pair (x, y) satisfying (3.2), an arbitrary small change in x, y leads to *proportional* changes of F and G

$$\frac{\Delta F}{\Delta G} = \frac{F_x}{G_x} = \frac{F_y}{G_y} \tag{3.3}$$

In order to gain some superficial insight into the relation between the computational effort and the size of the remainder term $r_{\nu+1}$ as p and h vary, we will *model* these two functions by some crude but not completely unrealistic assumptions:

The effort: For nontrivial values of p, the largest part in the evaluation of (1.7) stems from the evaluation of the derivatives f_j and f'_j. According to the mechanisms of automatic differentiation, the effort for the evaluation of higher and higher derivatives

- remains essentially constant for terms in f which are sums,
- grows linearly for terms in f which are products or standard functions.

Thus the computational cost for the evaluation of the coefficients in a Taylor series to order p is a quadratic polynomial in p, cf. e.g. [11]. Locally, we may thus model the *dependence* of the computational effort on p by p^α, with $\alpha \in [1,2]$.

Furthermore, while a change of h does not make the evaluation of (1.7) more costly, the number of steps needed will be proportional to $1/h$. Thus our model for the computational effort $C.E.$ is (with some constant E)

$$C.E.(p,h) = E \cdot \frac{p^\alpha}{h} \quad , \qquad \alpha \in [1,2] \tag{3.4}$$

The remainder term: While the h^p-dependence of the size of the remainder term on h is trivial, its dependence on p will strongly vary with the problem and its solution. However, there are two principal patterns of behavior (for sufficiently large p):

$$\text{geometric series,} \quad i.e. \|r(p,h)\| \;\hat{=}\; (ch)^p\,; \tag{3.5}$$

$$\text{exponential series,} \quad i.e. \|r(p,h)\| \;\hat{=}\; (ch)^p/p!\,. \tag{3.6}$$

The exponential series pattern should dominate if we have nearly the situation of a linear system (1.1) while the geometric series pattern corresponds to a finite radius of convergence $\approx \frac{1}{c}$, due to some singularity of the solution.

With the two functions of p and h represented by (3.5) and (3.6), we may now use (3.2) to look for an optimal point:

1) The combination of (3.4) with the geometric series pattern (3.5) leads to a unique optimal pair

$$p^* \;\hat{=}\; \frac{1}{\alpha}\log(TOL^{-1}) \tag{3.7}$$

$$ch^* \;\hat{=}\; e^{-\alpha} \tag{3.8}$$

The surprising aspect of (3.8) is the independence of h^* from TOL. This would indicate that an optimal control of the stepsize should produce nearly identical stepsize patterns (as a function of t) for different values of TOL. The necessary adjustment of $\|r\|$ to the size of TOL would come about by a linear growth of p with the (negative) order of magnitude of TOL.

2) The combination of (3.4) with the exponential series pattern (3.6), with Stirling's asymptotic expression for $p!$, leads to the relation

$$(\alpha - 1)p + \frac{1}{2}\log\left(\frac{2\pi}{e}p\right) = \log(TOL^{-1}) \tag{3.9}$$

For $\alpha = 1$, which occurs only for rather simple linear ODEs, p^* has to be extremely large for small values of TOL, i.e. the longer the Taylor expansion the better. This is possible because, with (3.6), h may grow even faster than p so that C.E. keeps decreasing, cf. (3.6). This is, of course, unrealistic. For $\alpha > 1$, the linear term in (3.9) will soon dominate and we have

$$p^* \hat{=} \text{ small term} + \frac{1}{\alpha - 1}\log(TOL^{-1}) \tag{3.10}$$

and an h^* which slowly increases with TOL^{-1}.

In all cases it seems that it is important to have the length p of the Taylor series of appropriate size (often quite long) if one wishes to work with a near-optimal combination of p and h.

Naturally, one will always attempt to use h for the "fine-tuning" since the effect of a change of h with constant p is so transparent.

Some preliminary experiments have shown that – for a fixed size of the remainder term – the expected minimum of the computational effort in the p, h-plane actually exists in many cases. Also the expected large values of p^* and h^* for the location of the minimum have been found. More systematic experiments are currently made by V. Brand in Vienna.

But – this is only true if the remainder term is formed with an a priori inclusion which is prescribed and not algorithmically determined. In the ordinary course of Lohner's algorithm, stepsizes near h^* could never be achieved for most problems because the computational generation of the a priori inclusion would not permit them. Therefore we will consider potential improvements of this phase of the inclusion algorithm.

4 A priori Inclusion

As is pointed out by Lohner ([5]) and apparent in the use of AWA, the choice of the stepsize h is presently strongly restricted by the necessity to generate an a priori inclusion $\bar{Y}_{\nu+1}$ for $y(t;t_\nu,Y_\nu)$ on $[t_\nu, t_\nu + h]$, cf. (1.6). The natural and commonly used verification procedure for a suitably chosen trial set $\bar{Y}_{\nu+1}$ and a trial stepsize h is

$$Y_\nu + [0,h] \cdot f(\bar{Y}_{\nu+1}) \subseteq \bar{Y}_{\nu+1}? \tag{4.1}$$

(The choice of trial sets $\bar{Y}_{\nu+1}$ is described in [5], we are not concerned with this question here.)

Clearly, (4.1) requires

$$w(Y_\nu) + hL \cdot w\left(\bar{Y}_{\nu+1}\right) \leq w\left(\bar{Y}_{\nu+1}\right) \tag{4.2}$$

where L is a Lipschitz matrix for f, and (4.2) implies $h\|L\| < 1$. Thus the procedure (4.1) can only work for stepsizes which are approximately as they would be suitable for Euler's method! Accordingly, the use of large orders p and correspondingly large steps h – as suggested by our considerations in section 3 – is not feasible, which makes the use of the inclusion code AWA far more expensive than necessary.

On the other hand, what we really need is an "a priori inclusion" *for the remainder term* $r_{\nu+1}$ of the Taylor series in (1.7) and not for $y([t_\nu, t_{\nu+1}]; t_\nu, Y_\nu)$; the relation

$$r_{\nu+1}(h) \subseteq \frac{h^p}{p!} f_{p-1}(\bar{Y}_{\nu+1}) \, , \tag{4.3}$$

with $\bar{Y}_{\nu+1}$ satisfying (1.6), is merely the most immediate way to obtain such an inclusion.

The exact expression for the remainder term is

$$r_{\nu+1} = \bigcup_{y_\nu \in Y_\nu} \frac{h^p}{(p-1)!} \int_0^1 (1-\lambda)^{p-1} y^{(p)}(t_\nu + \lambda h; t_\nu, y_\nu) d\lambda \tag{4.4}$$

$$= \bigcup_{y_\nu \in Y_\nu} \frac{h^p}{(p-1)!} \int_0^1 (1-\lambda)^{p-1} f_{p-1}(y(t_\nu + \lambda h; t_\nu, y_\nu)) d\lambda \tag{4.5}$$

from which it is obvious that, generally, values of $y(t;t_\nu,y_\nu)$ at some t away from t_ν have very little influence on $r_{\nu+1}$ if p takes larger values. Hence it may happen that (4.3) holds although the set $\bar{Y}_{\nu+1}$ encloses $y(t;t_\nu,Y_\nu)$ only within a beginning section of the interval $[t_\nu, t_{\nu+1}]$. We will call a set $\bar{Y}_{\nu+1}$ *inclusion valid for h* if (4.3) holds.

For the model problem $y'(t) = Ay(t)$ the situation is surprisingly simple:

Theorem: If $\bar{Y}_{\nu+1}$ satisfies

$$Y_\nu + [0, h_0]A\bar{Y}_{\nu+1} \subseteq \bar{Y}_{\nu+1}, \qquad h_0 > 0 \ , \tag{4.6}$$

then $\bar{Y}_{\nu+1}$ is inclusion valid for all $h \in [0, (p+1)h_0]$.

Proof: Inclusion validity of $\bar{Y}_{\nu+1}$ for h is equivalent to

$$y(t_\nu + h; t_\nu, Y_\nu) \quad \subseteq \quad Y^{(0)}_{\nu+1}(h) := \sum_{j=0}^{p-1} \frac{h^j}{j!} A^j Y_\nu + \frac{h^p}{p!} A^p \bar{Y}_{\nu+1} \tag{4.7}$$

By Schauder's fixed point theorem,

$$Y^{(1)}_{\nu+1}(h) := Y_\nu + \int_0^h AY^{(0)}_{\nu+1}(s)ds \subseteq Y^{(0)}_{\nu+1}(h) \tag{4.8}$$

is sufficient for (4.7).

$$\begin{aligned} Y^{(1)}_{\nu+1}(h) &= \sum_{j=0}^{p-1} \frac{h^j}{j!} A^j Y_\nu + \frac{h^p}{p!} A^p Y_\nu + \frac{h^{p+1}}{(p+1)!} A^{p+1} \bar{Y}_{\nu+1} \\ &= \cdots + \frac{h^p}{p!} A^p (Y_\nu + \frac{h}{p+1} A\bar{Y}_{\nu+1}) \\ &\subseteq \cdots + \frac{h^p}{p!} A^p \bar{Y}_{\nu+1} \quad = \quad Y^{(0)}_{\nu+1}(h) \end{aligned}$$

for $h \in [0, (p+1)h_0]$ according to (4.6). □

This theorem shows that a $\bar{Y}_{\nu+1}$ which satisfies the standard requirement (4.1) only for a short step h_0 is likely to be inclusion valid for *much* longer steps h. If the Taylor series behaves like an exponential series, the above result $h \leq (p+1)h_0$ should be qualitatively true.

For an insight into a strongly nonlinear situation with a singularity restricting the stepsize, we consider the scalar model problem $y'(t) = (y(t))^2$, $y(t_\nu) \in Y_\nu = [\underline{y}_\nu, \bar{y}_\nu]$, which has a singularity at $\bar{t} = t_\nu + 1/\bar{y}_\nu$. Take $\bar{Y}_{\nu+1} = [\underline{y}_\nu, \bar{y}_{\nu+1}]$, with some $\bar{y}_{\nu+1} > \bar{y}_\nu$, then

$$\begin{aligned} \bar{Y}_{\nu+1} \text{ satisfies (4.1) for } h &\leq h_0 = (1 - \bar{y}_\nu/\bar{y}_{\nu+1})/\bar{y}_{\nu+1}, \\ \bar{Y}_{\nu+1} \text{ satisfies (1.6) for } h &\leq \bar{h}_0 = (1 - \bar{y}_\nu/\bar{y}_{\nu+1})/\bar{y}_\nu, \qquad (4.9) \\ \bar{Y}_{\nu+1} \text{ is inclusion valid for } h &\leq \bar{h} = (1 - (\bar{y}_\nu/\bar{y}_{\nu+1})^{p+1})/\bar{y}_\nu. \qquad (4.10) \end{aligned}$$

Thus, for this model problem, we have:

(i) For $\bar{y}_{\nu+1} = \bar{y}_\nu(1+\varepsilon)$, ε small,

$$h_0 \approx \bar{h}_0 = \varepsilon/\bar{y}_\nu + O(\varepsilon^2) \tag{4.11}$$
$$\bar{h} = (p+1)\,\varepsilon/\bar{y}_\nu + O(\varepsilon^2) \approx (p+1)h_0, \tag{4.12}$$

i.e. away from the singularity the situation is like in the linear case.

(ii) For $\bar{y}_\nu/\bar{y}_{\nu+1}$ small, $t_\nu + \bar{h}$ gets arbitrarily close to $\bar{t}$ as p increases, i.e. $\bar{Y}_{\nu+1}$ is inclusion valid for steps that nearly reach the singularity.

These results show that the use of long steps in the procedure (1.7) is quite realistic, particularly for large values of p. However – except in the linear case covered by the above theorem – the actual inclusion validity of some set $\bar{Y}_{\nu+1}$ for an h which is not covered by (4.1) must now be *verified* a posteriori! Supposedly this would have to be done by a test analogous to (4.8), with $Y_{\nu+1}^{(0)}(h)$ defined in analogy to (4.7).

Only if such a test may be implemented at reasonable cost, an algorithmic realization of the considerations in this section will be possible. Note that "reasonable cost" has to be measured against stepsize increases which may well be of an order of magnitude. Also, only the attainability of large stepsizes for large p will make the order and stepsize strategies discussed in section 3 feasible.

5 Representation of Inclusion Sets

It has been known for a long time that the direct use of axiparallel intervals in the inclusion of ODE solutions is not feasible (except for quasimonotone systems) because of the so-called *wrapping effect*, cf. e.g. [12]. Moore ([2]) had suggested to use linearly transformed intervals and to update the transformation matrix S_ν in each step, cf. (1.5) and (1.8).

While this eliminates the firstorder wrapping terms, other difficulties appear: The *original coordinate* system still has to be used for the

- computation of an a priori inclusion
- evaluation of the r.h.s. of the differential system (1.1)
- inclusion of the remainder term.

For this purpose, the set $S_\nu[e_\nu]$ has to be included by an axiparallel interval so that some operation C may be performed yielding an axiparallel result interval $[C[S_\nu[e_\nu]]]$. But now this set has to be represented in the local coordinates $S_{\nu+1}$, cf. (1.8), and the contribution to $[e_{\nu+1}]$ of this particular operation becomes

$$[S_{\nu+1}^{-1}[C[S_\nu[e_\nu]]]]. \tag{5.1}$$

Since S_ν and $S_{\nu+1}$ differ only by a factor $I + O(h)$, the operation (5.1) will be very sensitive to an ill-conditioning of S_ν. Unfortunately, the condition of the matrix S_ν usually deteriorates rapidly as the integration proceeds, see e.g. [5].

Various remedies have been suggested for this ill-effect. The orthogonalization of S_ν suggested by Lohner ([5]) combined with other approaches appears to be most efficient; but no approach is generally applicable.

Thus it appears most natural to use a *moving coordinate system* not only for the representation of the inclusion sets Y_ν but for the local representation of the complete problem. This would eliminate the back and forth transformations (5.1) of the inclusion completely.

In the case of the linear model problem $y'(t) = Ay(t)$, the situation is particularly simple because the differential system remains unaltered by the stepwise change of the coordinate system which is to be effected by the Jacobian J_ν of the Taylor sum at t_ν (as in the updating of S_ν, cf. e.g. [5]). But this Taylor sum and its Jacobian are polynomials in A for the above case so that

$$f_{\text{new}}(y) = J_\nu^{-1} A J_\nu y = Ay = f_{\text{old}}(y) \tag{5.2}$$

Some preliminary experiments have established the effectiveness of the approach against the consequences of an ill-conditioned S_ν.

It will have to be seen whether the computational effort for the update of the differential system can be suitably contained so that it pays relative to the tighter inclusions which it permits.

6 Stiff Systems

Even if all the improvements considered in the previous sections could be effectively implemented the inclusion procedure (1.7) would remain inefficient for stiff systems (1.1) because it uses a polynomial approximation. This is well-known from theory and practical experience for approximation algorithms for ODE systems (initial value problems).

Two of the approaches which have been successfully used for approximation methods are

- implicit schemes,
- local linearization.

We will consider how these ideas could possibly be applied for inclusion algorithms.

6.1 Implicit Taylor Series

After several attempts to devise an implicit Taylor series scheme, the following approach appeared as the most suitable one:

As in the approximation case, we consider the solution of the differential equation for decreasing t, i.e. from $t_{\nu+1}$ to t_ν. This means that we propose to find $y_{\nu+1}$ such that

$$y(t_\nu; t_{\nu+1}, y_{\nu+1}) \in Y_\nu. \tag{6.1}$$

Naturally, we must find the set of all values $y_{\nu+1}$ for which (6.1) holds; any set which may be guaranteed to contain all such values may be used as an inclusion set $Y_{\nu+1}$ at $t_{\nu+1}$.

$$Y_{\nu+1} \supset \{y_{\nu+1} : y(t_\nu; t_{\nu+1} y_{\nu+1}) \in Y_\nu\} \tag{6.2}$$

Let $\bar{Y}_{\nu+1}$ be an a-priori inclusion satisfying (1.6). Then the set $Y_{\nu+1}$ of all $y_{\nu+1}$ which satisfy

$$y_{\nu+1} + \sum_{j=1}^{p-1} \frac{(-h)^j}{j!} f_{j-1}(y_{\nu+1}) \in Y_\nu - \frac{(-h)^p}{p!} f_{p-1}(\bar{Y}_{\nu+1}) \tag{6.3}$$

satisfies (6.2) because (6.1) implies

$$y(t_\nu; t_{\nu+1}, y_{\nu+1}) = y_{\nu+1} + \sum_{j=1}^{p-1} \frac{(-h)^j}{j!} f_{j-1}(y_{\nu+1}) + r(y_{\nu+1}) \in Y_\nu \tag{6.4}$$

which, with $r(y_{\nu+1}) \in \frac{(-h)^p}{p!} f_{p-1}(\bar{Y}_{\nu+1})$, implies (6.3).

Note that (6.3) is a (generally nonlinear) system of equations with a set-valued right hand side. For the linear model problem $y'(t) = Ay(t)$, (6.3) becomes

$$\left[I + \sum_{j=1}^{p-1} \frac{(-h)^j}{j!} A^j\right] y_{\nu+1} \in Y_\nu - \frac{(-h)^p}{p!} A^p \bar{Y}_{\nu+1} \tag{6.5}$$

which is a linear system with the point matrix [...] but with a set-valued right hand side. The solution set of such a system, with an interval right hand side, may be enclosed by the procedure LIN0 from the Fortran-SC runtime library. Thus our approach is immediately feasible in the linear case.

From (6.5), the main advantages of implicit Taylor series in the stiff case can be seen. Assume that the spectrum of A is in the complex left half plane, with some eigenvalues of large negative real part. Then

a) For given values of h and p, $\left[I + \sum_{j=1}^{p-1} \frac{(-h)^j}{j!} A^j\right]^{-1}$ is a much better approximation of e^{hA} than $[I + \sum_{j=1}^{p-1} \frac{h^j}{j!} A^j]$.

b) The excess of the a priori inclusion of the remainder term is damped by $[\cdots]^{-1} \approx e^{hA}$.

c) For given p, much larger stepsizes are feasible with (6.5) than with the standard forward inclusion procedure.

All these observations have been confirmed in some preliminary numerical experiments with moderately stiff matrices A.

The solution of (6.3) for nonlinear f by means of a Newton-like procedure has not yet been implemented.

6.2 Local Linearization [2]

If the right hand side f of (1.1) is essentially linear, i.e. if a local representation

$$f(y) = A_\nu y + g_\nu(y) \tag{6.6}$$

exists, with g_ν *small*, then we may set, in $[t_\nu, t_{\nu+1}]$,

$$y(t) = e^{(t-t_\nu)A_\nu} y_\nu + z_\nu(t), \quad y(t_\nu) \in Y_\nu, \quad z_\nu(t_\nu) = 0, \tag{6.7}$$

or

$$y(t) = e^{(t-t_\nu)A_\nu} z_\nu(t), \qquad z_\nu(t_\nu) \in Y_\nu. \tag{6.8}$$

With (6.7), z_ν should remain small in $[t_\nu, t_{\nu+1}]$, with (6.8), it should remain nearly constant.

The exponential e^{hA_ν} may be included via the Padé expansion approach of Bochev and Markov [13].

[2]This section was omitted in the presentation at the Conference due to a shortage of time

Conclusion

In this presentation, we have sketched some ideas that may lead to a significant improvement of the efficiency of programs for the validated solution of inital value problems for systems of ODEs. At the same time, they may permit an automatic control of the essential parameters h and p in each step of an algorithm based upon the Taylor series procedure (1.7). Thus, the implementation of these ideas may lead to the design of a true numerical software package for the indicated task.

The realization of this design will be attempted in close cooperation between researchers in Karlsruhe, Milwaukee, and Vienna.

References

[1] E. Adams: Periodic Solutions: Enclosure, Verification, Applications, this volume.

[2] R.E. Moore: Interval Analysis, Prentice-Hall, 1966.

[3] U. Marcowitz: Fehlerschätzung bei Anfangswertaufgaben für Systeme von gewöhnlichen Differentialgleichungen mit Anwendungen auf das Reentry-Problem, Numer. Math. **24** (1975) 249-275.

[4] P. Eijgenraam: The Solution of Initial Value Problems Using Interval Arithmetic, Math. Centre Tracts 144, 1981.

[5] R. Lohner: Einschließung der Lösung gewöhnlicher Anfangs- und Randwertaufgaben und Anwendungen, Dissertation, Karlsruhe, 1988, resp.
R. Lohner: Enclosing the Solution of Ordinary Initial and Boundary Value Problems, in: E. Kaucher, U. Kulisch, Ch. Ullrich (eds.): Computer Arithmetic, B.G. Teubner, 1987.

[6] R. Lohner: AWA, Program package in Fortran-SC for the inclusion of ordinary initial value problems, Karlsruhe.

[7] W. Walter: Differential and Integral Inequalities, Springer, 1970.

[8] E.W. Kaucher, W.L. Miranker: Self-Validating Numerics in Function Space Problems, Academic Press, 1984.

[9] L.B. Rall: Optimal Implementation of Differentiation Arithmetic, in: E. Kaucher, U. Kulisch, Ch. Ullrich (eds.): Computer Arithmetic, B.G. Teubner, 1987.

[10] H.J. Stetter: Tolerance Proportionality in ODE-codes, in: Symposium on Ordinary Differential Equations, Proceedings, Urbana, 1979.

[11] G. Corliss, Y.F. Chang: Solving Ordinary Differential Equations Using Taylor Series, TOMS **8** (1982) 114-144.

[12] L.W. Jackson: Interval Arithmetic Error-bounding Algorithms, SIAM J. Numer. Anal. **12** (1975) 223-238.

[13] P. Bochev, S. Markov: A Self Validating Numerical Method for the Matrix Exponential, Computing **43** (1989) 59-72.

Guaranteed Inclusions of Solutions of some Types of Boundary Value Problems

L. Collatz
Institut für Angewandte Mathematik
Universität Hamburg
Federal Republic of Germany

Abstract: This survey paper deals with some possibilities to give guarantee for the accuracy of approximate solutions for boundary value problems one has calculated on computers. This is possible for not too complicated problems by combining theoretical inclusions theorems with algorithmic procedures and then with interval arithmetics. Usually the theory is based on monotonicity principles and fixed point theorems; the numerical procedure uses iteration processes, approximation and optimization techniques, and the interval arithmetic is applied if possible to the last step of the algorithmic procedure.

This is illustrated by different types of boundary value problems, and recently calculated numerical test-examples are given.

1. Introduction and operators of monotonic type

A simple inclusion-method is applicable, if monotonicity properties hold. Let B be an open connected domain in the n-dimensional space $\mathbf{R}^n$ of points $x=(x_1, \ldots x_n)$; a wanted function u(x) should satisfy a real differential equation (shortly D.E.)

(1.1) $\quad Lu = r(x) \quad \text{in } B$

and boundary conditions on the boundary ∂B

(1.2) $\quad Su = \gamma(x) \quad \text{on } \partial B.$

Here r, γ are given (f.i. continuous) functions, where S, γ may be vectors. We combine L, S to a vector-operator T and analoguously with r, γ.

(1.3) $\quad Tw = \{Lw, Sw\}\ , \ \ s(x) = (r(x), \gamma(x))\ .$

We introduce an idea of ordering for real-valued functions g(x), h(x):

ISBN 0-12-708245-X

(1.4) $g(x) \leq h(x)$

means, that (1.4) holds for all x of the considered domain; the sign "≤" is used in the sense of classical ordering of real numbers.

Suppose, two functions g, h satisfy (1.4); then we can define the interval J = [g, h] as the set of all functions z(x) "between" g and h, Fig.1

(1.5) $J = [g,h] = \{z(x) \mid g \leq z \leq h\}.$

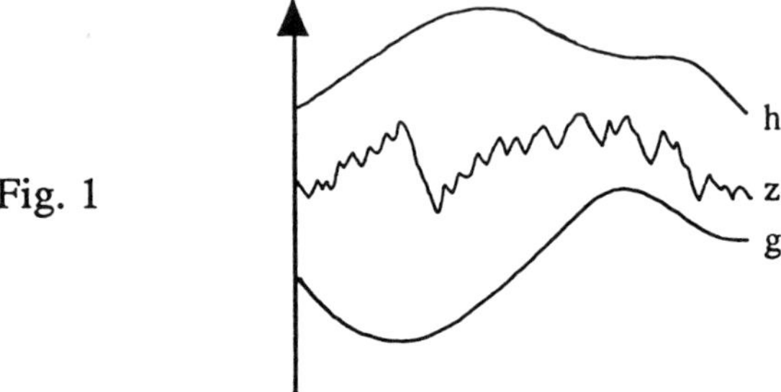

Fig. 1

The fundamental problem of the Numerical Analysis for calculating any wanted real quantity y requires the construction of an interval [g,h] which contains y with guarantee:

$$g \leq y \leq h$$

The operator T is called of "monotonic type", if for every pair v(x), w(x) of functions (for which Tv, Tw are defined) holds (Collatz [52]) :

(1.6) From $Tv \leq Tw$ follows $v \leq w$ in $B \cup \partial B$.

Suppose we can find a function w(x) with $Tu \leq Tw$; then holds

(1.7) From $Tu \leq (Lu,Su) = (r,\gamma) = s(x) \leq Tw$ follows $u \leq w$ in $B \cup \partial B$.

w is a guaranteed upper bound for u; analogeously one tries to find a function v with $Tv \leq Tu$; so u is included in the interval J = [v, w]:

(1.8) From $Tv \leq Tu \leq Tw$ follows $v \leq u \leq w$ in $B \cup \partial B$.

2. Choice of a suitable class of approximating functions

We try to approximate the wanted function u by functions (e.g. compare Meinardus [67], Bredendiek [76], Watson [80], Schwarz [88], Hämmerlin-Hoffmann [89] and many others)

(2.1) $\varphi(x) = w(x, a_1, \ldots, a_p)$

of a class $\Phi = \varphi(x, a)$, in which φ depends of a parameter vector $a = (a_1, \ldots, a_p)$. The suitable choice of the class Φ can be of deciding influence for the numerical success of the method. Often the problem admits an interpretation from the applications, physics, chemistry a.o., and the considered phenomenon suggests to use a class Φ of polynomials or of exponential functions or rational functions for x. The behaviour is comparable with the Ritzprocedure; for which one has also to choose a suitable class as "basis"; if one uses a not appropiate basis, the Ritz-procedure can produce a very bad approximation.

3. The algorithm

One wishes to get an inclusion interval (1.8) as "small as possible". We can often use for the lower and upper bound the same class Φ, but with different values a_j, b_j of the parameters:

(3.1) $v = \varphi(x, a_1, \ldots, a_p), \quad w = \varphi(x, b_1, \ldots, b_p)$

Then we get an optimization problem (semi-infinite) (Hettich-Zencke [82]):

(3.2) $Tv \leq s, \;\; Tw \geq s, \;\; -\delta \leq w - v \leq \delta, \;\; \delta = \text{Min !}$

The restrictions (inequalities) should hold for every x of the domain; but for the calculation on a computer one has the domain to discretize to choose a finite set of points x_σ ($\sigma = 1, \ldots, q$) in the domain and to write the inequalities in (3.2) only for $x = x_\sigma$ ($\sigma = 1, \ldots, q$).

This is a finite (linear or nonlinear) optimization problem (shortly O.P.) with finitely many unknowns a_j, b_j, δ and finitely many restrictions. If the O.P. is linear, then classical subroutines are available. If the O.P. is nonlinear, one has subroutines available only in special cases; otherwise often one can start with only few unknowns and one can use the results from the O.P. with few unknowns as starting values for the O.P. with more unknowns.

4. Interval-Analysis

Usually one can solve the O.P. (3.2) only approximately, and one gets values $\hat{a}_j$, $\hat{b}_j$, $\hat{\delta}$ instead of the solution a_j, b_j, δ of (3.2).

Let be $\hat{v} = \varphi(x, \hat{a}_j)$, $\hat{w} = \varphi(x, \hat{b}_j)$ with these values $\hat{a}_j$, $\hat{b}_j$; we consider now the parameters $\hat{a}_j$, $\hat{b}_j$ as exact values; $\hat{v}$, $\hat{w}$ are then well defined functions $\hat{v}$, $\hat{w}$ and $\hat{\delta}$ satisfy the inequalities (3.2) for all $x = x_\sigma$. The operator T is given and this test of (3.2) has to be done with interval analysis, perhaps with double precision.

Furthermore one has to confirm, that (3.2) is satisfied not only for $x = x_\sigma$, but for all x of the considered domain. If T is given by formulas in not too complicated way, then methods of interval analysis are also available for this test.

5. Some remarks for the numerical computation

The optimization problem (3.2) contains (2p+1) unknowns a_j, b_j, δ. If the problem is nonlinear, one can use instead of (3.2) two problems with only (p+1) unknowns, one problem for an upper bound w

(5.1) $$0 \le Tw - s \le P \cdot \delta, \quad \delta = \text{Min} \quad (\text{with unknowns } b_j, \delta)$$

and one problem for a lower bound v:

(5.2) $$0 \le s - Tv \le \tilde{P} \cdot \tilde{\delta}, \quad \tilde{\delta} = \text{Min} \quad (\text{with unknowns } a_j, \tilde{\delta})$$

The weight-functions P, $\tilde{P}$ are chosen, but fixed functions. Often one chooses $P = \tilde{P} = 1$; but other choices may be better, for instance if in (1.3) S is a vector, containing function values and derivates.

6. Some classes of operators of montonic type

The property of an operator T to be of monotonic type was proved for rather general linear and nonlinear ordinary and partial differential operators of second order of elliptic and parabolic type and for special classes of hyperbolic type (compare Redheffer [67], Nickel [58], Walter [70], Reemtsen [87] a.o.).

We consider as example the first boundary value problem for real linear elliptic operators (for all details see f.i. Collatz [66])

(6.1) $$Lu(x_j) = -\sum_{j,k=1}^{n} a_{jk}(x_l)\,\frac{\partial^2 u}{\partial x_j\,\partial x_k} - \sum_{j=1}^{n} b_j(x_l)\,\frac{\partial u}{\partial x_j} + cu\,.$$

The coefficients a_{jk}, b_j, c are given continuous bounded functions in the domain $B \cup \mathbf{R}^n$. We suppose $c \ge 0$ in B and the matrix $A = (a_{jk})$ as uniformly positive definite in B. The operator S may be $Su = u$, then the vector $Tu = (Lu, Su)$ is an operator of monotonic type; for two functions $v, w \in C^2(B) \cap C(B + \partial B)$ holds:

(6.2) $$\left\{ \begin{array}{l} Lv \le Lw \;\text{ in } B \\ \;\;v \le w \;\text{ on } \partial B \end{array} \right\} \text{ has the consequence } v \le w \text{ in B.}$$

This can be generalized to nonlinear P.D.E. for more general boundary conditions (for integral equations) a.o. (Collatz [66], Bohl [74], Schröder [80], Glashoff-Werner [79] a.o.).

7. Mixed boundary value problems (Kreiss-Lorenz [89])

Often occurs in flow problems the mixed boundary value problem that the values u(x) of the unknown function are prescribed on one part Γ_1 of the boundary ∂B and on another part Γ_2 of the boundary ∂B the values of the derivate $\frac{\partial u}{\partial n}$ with respect to the outer normale n. Then for the Laplacean Operator Δ, there is the operator

$$T(z) = \begin{cases} -\Delta z \text{ in } B \\ z \text{ on } \Gamma_1 \\ \frac{\partial z}{\partial n} \text{ on } \Gamma_2 \end{cases} \tag{7.1}$$

of monotonic type, that means

$$\begin{cases} \text{from } Tv \le Tw & \text{in } B \cup \partial B \\ \text{follows } v \le w & \text{in } B \cup \partial B \end{cases} \tag{7.2}$$

Example: Ideal flow through a channel K.

a) K has a contraction, fig. 2 (compare Collatz).

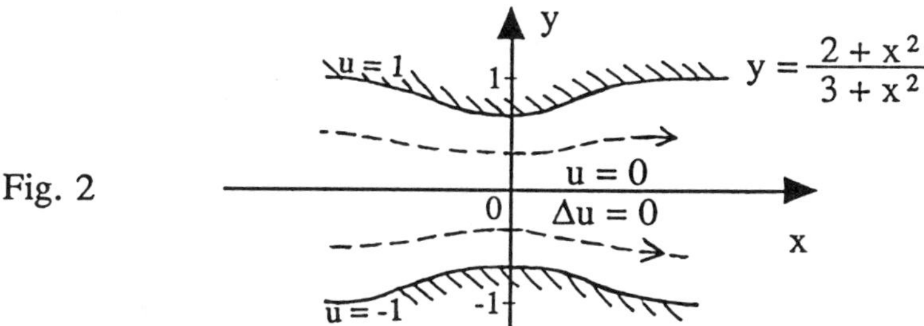

Fig. 2

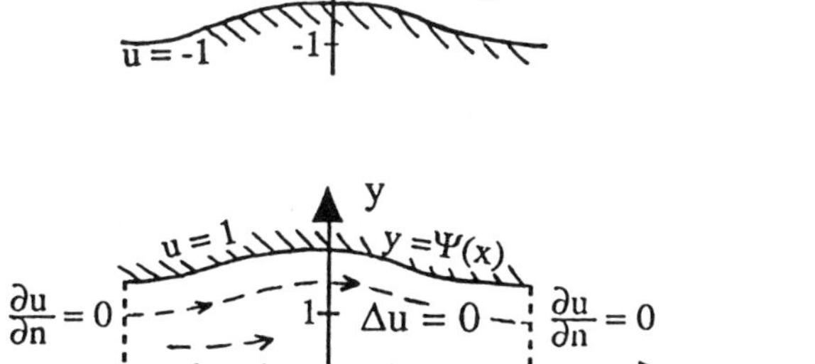

b) K has a dilation, fig. 3.

Fig. 3

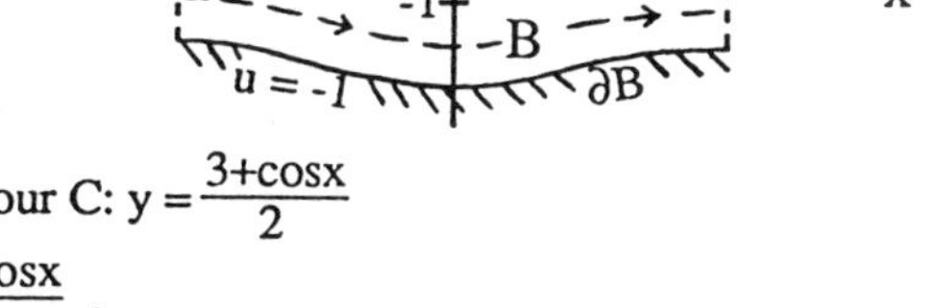

The walls may have the contour C: $y = \frac{3+\cos x}{2}$

domain B: $|x| < \pi,\ |y| < \frac{3+\cos x}{2}$.

$\Delta u = 0$ in B, $u = 1$ on C, $\frac{\partial u}{\partial n} = 0$ on $|x| = \pi$.

Approximate solution $u \approx v = a_0 + \sum_{\nu=1}^{p} a_\nu \cos(\nu x) \sinh(\nu y)$

Guaranteeable error bounds $|\varepsilon| = |v - u| \leq \delta$.

p	0	3	5
δ	0.3333	0.0145	0.0031

I thank Mr. U. Grothkopf for numerical calculation.

(compare Collatz, Proceedings, Numeric-Conference, Praha (CSSR), 21.-25. Aug. 1989, to appear).

8. Discontinuous boundary values, Cavity flow

Often singularities occur in the applications f.i. corners in two- and three dimensions, edges, discontinuities a.o.

Example Cavity-flow. We look on the function u(x, y), which satisfies in the domain $B := \{ (x,y) : |x| < 1, |y| < 1\}$ on a x-y-plane the Laplacean equation

(8.1) $\Delta u = 0$ in B , Fig. 4,

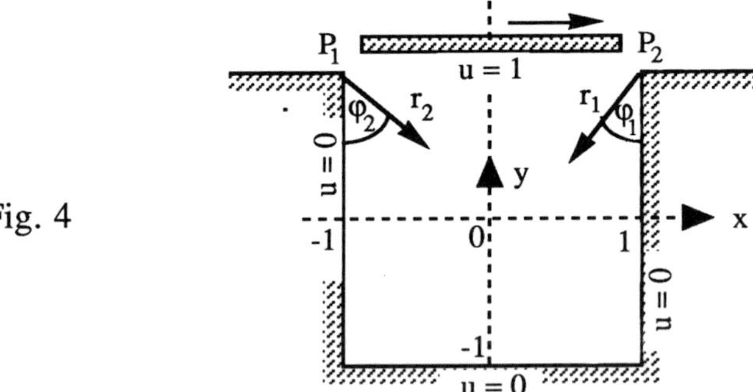

Fig. 4

and the boundary conditions

(8.2) $$u = \begin{cases} 1 & \text{for } y = 1 \\ 0 & \text{otherwise on } \partial B; \end{cases}$$

this corresponds the simplified linearized flow of an ideal liquid in a quadratic hole, fig.4 (Yserentant [89]). We take care on the jumps of the boundary values in the corners P_1, P_2, fig.4, by introducing polar-coordinates r_j, φ_j at P_j (j=1,2) and subtracting the singularity

(8.3) $$\psi = \frac{2}{\pi}(\varphi_1 + \varphi_2)$$

from the solution u. We put an approximate solution v in the form

(8.4) $$u \approx v = \psi + \sum_{\nu=1}^{p} a_\nu v_\nu$$

The functions v_ν are harmonic polynomials satisfying the symmetries of fig.4, f.i. $v_1 = 1$; $v_2 = y$; $v_3 = x^2 - y^2$; $v_4 = 3x^2y - y^3$.

I thank Mr. Th. Schiemann for the calculation on a computer. He got the error bounds in dependence of the number p of polynomials:

p	$\lvert v-u \rvert \leq$
1	0.25
2	0.051
3	0.026
4	0.0027
5	0.0026
6	0.00096
7	0.000075
8	0.000025

The coefficients for p=7:

ν	a_ν
1	0.749999
2	0.219297
3	0.44755 * 10^{-1}
4	0.13390 * 10^{-1}
5	0.13979 * 10^{-6}
6	0.99923 * 10^{-3}
7	0.41298 * 10^{-3}

The framed line for p=6 gives the guarantee for the accuracy $\lvert v-u \rvert \leq 0.00096$, and p=6 was used for drawing the isohypses $u = \text{const.} = \frac{k}{20}$ with $k = 1, 3, 5, \ldots, 19$ in fig.5.

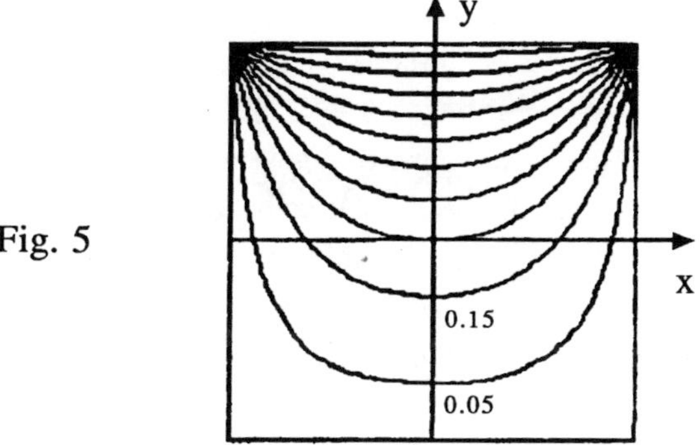

Fig. 5

9. A nonlinear delay equation

We look on the periodic solution u(t) of period 1 of the nonlinear delay equation (9.1) (forced vibrations, elastic force $P(u) = -u - \frac{1}{2}u^3$).

$$\text{(9.1)} \qquad Lu = u''(t) + u(t) + \frac{1}{2}[u(t)]^3 - \frac{1}{2}u(t-1) + \cos(2\pi t) = 0$$

and take (9.2) as approximate solution v(t):

$$\text{(9.2)} \qquad u(t) \approx v(t) = \frac{b_0}{2} + \sum_{\nu=1}^{p} b_\nu \cos(2\pi\nu t) \;.$$

We use different coefficients b_ρ, $\overline{b}_\rho$:

$$\text{(9.3)} \qquad \begin{cases} \underline{b}_\sigma \quad \underline{b}_\nu \quad \text{for the lower bound} \quad \underline{v}(t) \\ \overline{b}_\sigma \quad \overline{b}_\nu \quad \text{for the upper bound} \quad \overline{v}(t) \end{cases}$$

and calculate the arithmetic mean $\hat{v}(t)$:

$$\hat{v}(t) = \frac{1}{2}\left(v(t) + \overline{v}(t) \right);$$

just using only one cosine-term in (9.2) one gets (9.4)

$$\text{(9.4)} \qquad p=1, \; |\hat{v}(t) - u(t)| \le 0.0003$$

10. Generalizations and Outlook

Many other types of problems have been treated with the described methods: eigenvalue problems (Albrecht-Goerisch [87] a.o.), bifurcation problems, free boundary value problems (Hoffmann [78] a.o.), problems with different kind of singularities (Whiteman [85]) a.o. Many numerical examples: Collatz [81], [87], [90], a.o.

If one knows the type and the location of the singularity, the methods of monotonicity are in not too complicated cases often preferable compared with discretization methods, f.i. for unbounded domains, corners in two- and three dimensions, interfaces a.o. There exist also other methods with which one can get guaranteed inclusion intervals for the solution, based on the topological fixed point theorems of Brouwer, Banach, Schauder, Schauder-Leray, Krasnoselskii and other, combined with iteration procedures.

But still much research is necessary to enlarge the fields of applicability of the methods giving guaranteed results. The mentioned methods cover only simple models. One has had great success in the last decades, but not enough success with respect to the applications.

References

Albrecht, J. - F. Goerisch [87] Anwendungen des Verfahrens von Lehmann auf Schwingungsprobleme. Intern. Ser. Num. Math. 83 (1987) 1-9.

Bohl, E. [74] Monotonie, Lösbarkeit und Numerik bei Operatorgleichungen. Springer, 1974, 255p.

Bredendiek, E. - L. Collatz [76] Simultane Approximation bei Randwertaufgaben. Internat. Ser. Num. Math. 30 (1976), 147-174.

Collatz, L. [52] Aufgaben monotoner Art, Arch. Math. Anal. Mech. 3 (1952) 366-376.

Collatz, L. [66] Funktionalanalysis and Numerical Mathematics. Academic Press, 1966, 473 p.

Collatz, L. [78] Approximation methods for boundary value problems with unbounded domains or free boundaries. Proc. Plzen, CSSR. Sept. 1978, ed. J. Marek, p. 51-61.

Collatz, L. [81] Anwendungen von Monotoniesätzen zur Einschließung der Lösungen von Gleichungen. Jahrbuch Überblicke der Mathematik (1981) 189-225.

Collatz, L. [87] Approximation von Eckensingularitäten bei Randwertaufgaben. Intern. Ser. Num. Math. 81 (1987) 44-53.

Collatz, L. [90] Rational and algebraic approx. for initial and boundary value problems. Intern. Ser. Num. Math. 90 (1990) 103-106.

Glashoff, K. - B. Werner [79] Monotonicity in a Stefan Problem for the Heat Equation. Numer. Funct. Anal. and Optimiz. 1 (1979) 431-440.

Hämmerlin, G. - K.H. Hoffmann [89] Numerische Mathematik. Springer, 1989, 448p.

Hettich, R. - P. Zencke [82] Numerische Methoden der Approximation und semi-infiniten Optimierung. Teubner, Stuttgart, 1982.

Hoffmann, K.H. [78] Monotonie bei nichtlinearen Stefan Problemen. Internat. Ser. Num. Math. Vol. 39 (1987) 162-190.

Kreiss, H.O. - J. Lorenz [89] Initial-Boundary value Problems and the Navier-Stokes-Equations. Acad. Press 1989, 402 p.

Meinardus, G. [67] Approximation of Functions, Theory and Numerical Methods. Springer Verlag, 1967, 198 p.

Nickel, K. [58] Einige Eigenschaften von Lösungen der Prandtl'schen Grenzschicht-differentialgleichung. Arch. Rat. Mech. Anal. 4 (1958) 1-31.

Redheffer, R.M. [67] Differentialungleichungen unter schwachen Voraussetzungen. Abhandl. Math. Sem. Univ. Hamburg, 31 (1967) 33-50.

Reemtsen, R. [87] Defect Minimization in operator equations, Theory and Applications. Pitman Research Notes Math. Ser. 1987, 106 p.

Schröder, J. [80] Operator-Inequalities. Academic Press 1980, 367 p.

Schwarz, H.R. [88] Numerische Mathematik. Teubner, Stuttgart 1988, 496 p.

Walter, W. [70] Differential and Integral Inequalities. Springer 1970, 325 p.

Watson, G.A. [80] Approximation Theory and Numerical Methods. John Wiley 1980.

Whiteman, J.R. [85] Singularities on two- and threedimensional elliptic problems and finite element methods for their treatment. Proc. Equadiff 6, Purkyne Univ., Dp. of Math. Brno 1985.

Yserentant [89] Lecture at the GAMM-Meeting on Numerical Methods of fluid dynamics, Kiel, January 1989.

Periodic Solutions: Enclosure, Verification, and Applications

E. Adams
Institut für Angewandte Mathematik
Universität Karlsruhe
Federal Republic of Germany

Abstract: For the finite step sizes of practical numerical work, classical discretizations of nonlinear ordinary differential equations (ODEs) may possess (i) extraneous solutions or (ii) solutions diverting from the neighborhood of one true solution of the ODEs to a neighborhood of a grossly different true solution. This may even occur when Runge-Kutta methods with step size control are employed. In the case of periodic solutions of nonlinear ODEs, the computational requirements for reliable numerical approximations are correspondingly high. The Karlsruhe Enclosure Methods yield a verification of the existence, close bounds for a periodic solution, and (if unknown) its period.

1. Introduction

Systems of autonomous explicit nonlinear ordinary differential equations (ODEs) will be considered:

(1.1) $\quad y' = f(y)$ for $t \geq 0$; $\quad f : D \to \mathbb{R}^n, D \subset \mathbb{R}^n,$ $\quad f$ is sufficiently smooth.

By use of a well-known extension, nonautonomous systems $y' = f(t,y)$ can be reduced to (1.1). Classical solutions of (1.1) will be denoted by $y^* = y^*(t,y(0))$.

The technical preparation of this paper has been very efficiently handled by E.Müller and W.Rufeger.

ISBN 0-12-708245-X

When $y(0) \in D$ is given in the context of an initial value problem (IVP), the existence of y^* is ensured for $t \in [0,t_\infty)$ with a suitable t_∞. The dependency of a solution y^* on t is known for all $t \in \mathbb{R}^+$ in the following special cases: (i) stationary points y^* solving $f(y) = 0$ and (ii) T-periodic solutions y^* with the property

(1.2) $\quad y^*(t+T) = y^*(t)$ for all $t \geq 0$,

where $T \in \mathbb{R}^+$ is fixed but generally unknown. A T-periodic solution y^* of (1.1) with T unknown solves a free boundary value problem (BVP) with a boundary condition of periodicity

(1.3) $y' = f(y)$ for $t \in [0,T]$; $\quad y(T) = y(0)$,

where $[\bullet,\bullet]$ denotes an interval. See Sections 5 and 8 for the practical treatment of (1.3), including the case where $f = f(t,y)$ with $f(t+T,y) = f(t,y)$ for all $(t,y) \in \mathbb{R}^+ \times D$.

With the exception of special cases (particularly for n = 2, e.g. [10] or [14]), the existence of a classical T-periodic solution y^* of (1.1) cannot be guaranteed by means of theorems with conditions depending only on f. The computability of "approximations" $\tilde{T}$ and $\tilde{y}$ of T^*-periodic solutions y^* does not imply the existence of these solutions. That classical numerical methods (e.g. discretizations) for the determination of $\tilde{T}$ and $\tilde{y}$ are unreliable will be shown by use of examples in the present paper. This is obvious when T^* is not small. For sufficiently large T^*, an additional difficulty has to be faced, namely that all bounded difference solutions of evolution problems are periodic provided a digital computer with its bounded set of machine numbers is used (e.g. [33, p.134]). In the case the existence of a periodic solution y^* is asserted qualitatively by a theorem, a computed approximation $\tilde{y}$ may be "far away" from y^*.

The Karlsruhe Enclosure Methods rest on the Kulisch Computer Arithmetic (e.g. [29]). For the present paper, Lohner's enclosure algorithms ([31] and [32]) are particularly important. Pertaining to IVPs or BVPs with ODEs, they yield guaranteed bounds $\underline{T}$, $\overline{T}$ and $\underline{y}$, $\bar{y}$ such that

(1.4) $\quad T^* \in [\underline{T},\overline{T}]$ and $y^*_i(t) \in [\underline{y}_i(t),\bar{y}_i(t)]$ for $i = 1(1)n$ and $t \in [0,T^*]$,

where 1(1)n denotes the set of all integers from 1 to n. If computable, these bounds (i) account automatically for all numerical errors incurred in their determination and (ii) they guarantee the existence of the T^*-periodic solution y^* contained in the enclosures $[\underline{T},\overline{T}]$ and $[\underline{y},\bar{y}]$. Consequently, these enclosure methods are self-validating. Lohner's enclosure algorithm for BVPs is an interval shooting method consisting of (a) enclosures of solutions of auxiliary IVPs and (b) the iterative search by use of an interval Newton method (see [5]) for a shooting vector that leads to satisfying the boundary condition of periodicity. Particularly the close-to-local character of the Newton method confines applications of Lohner's enclosure algorithm for nonlinear BVPs to problems with a sufficiently small number n of dimensions. So far, this algorithm has been executed (i) by Lohner for n = 2 and (ii) by W.Kühn [28] for n = 3, see Section 5 of the present paper. Lohner's enclosure algorithm for nonlinear IVPs has been executed for considerably larger n. For n = 2, the Poincaré-Bendixson theorem (e.g. [10, p.163]) admits a verification of a periodic solution by use of Lohner's enclosure algorithm for IVPs. In [1], this has been executed for the problem of the stiffly coupled Oregonator.

Remark: The co-author D.Straub of [1] has contributed essentially to the present paper's background in the physical theory of dynamics and its applications.

Only as the step size $h \to 0$, the limit of an h-dependent sequence of difference solution is related (by pointwise convergence) to a true solution y^* of an IVP or BVP. This presupposes the consistency and the stability of the discretization. For the finite h employed in practical numerical work, there may be "extraneous" (or "spurious" or "ghost") difference solutions $\tilde{y}$, which are even qualitatively different from any true solution y^* of the underlying ODEs. Unless h is sufficiently small, for example, the Logistic Difference Equation discussed in Section 3 possesses a countable infinity of periodic difference solutions while there are only constant or monotonic true solutions y^* of the underlying Logistic ODE.

Another source of large deviations $y^* - \tilde{y}$ are diversions of $\tilde{y}$ from a small neighborhood of one true solution y^* to a small neighborhood of a grossly different true solution y^*. In Section 5, difference solutions $\tilde{y}$ divert "at" a "stable manifold of the origin". This highly twisted hyperface in phase space depends on the set of (all) solutions. In Section 6, diversions of approximations take place in a neighborhood of a pole of the ODEs.

The computational determination of periodic true solutions y^* of ODEs is closely affected by difficulties denoted by "computational chaos", which has been addressed in numerous recent papers, e.g. [21] and [34]. Extraneous or diverting difference solutions contribute to computational chaos.

Remarks: 1.) In a forthcoming paper [4], it will be discussed whether or not a chaotic set of difference solutions $\tilde{y}$ allows the implication that the set of true solutions y^* of the underlying differential equations is also chaotic.

2.) Extraneous difference solutions were first discussed in literature for certain nonlinear BVPs with ODEs of second order, see Gaines [16] and Spreuer and Adams, [46], [47]. As $h \to 0$, these extraneous difference solutions $\tilde{y}$ approach limiting functions which do not satisfy the ODEs.

3.) See [53] for extraneous fixed points of iterations for nonlinear finite-dimensional equations, e.g. polynomials. This failure is not possible in the case of interval iterations satisfying the conditions of the Brouwer fixed point theorem.

2. Periodic Solutions via Discretizations of ODEs and Discretization Errors

To simplify discussions, (1.1) is now considered for the case of $n = 1$. An equidistant grid $\{0, t_1, ..., t_j, ...\}$ is introduced with $h := t_{j+1} - t_j$. A Taylor-expansion of y^* about t_j is considered, with a remainder term of order two and evaluation at t_{j+1}. This yields

(2.1) $$\frac{y^*_{j+1} - y^*_j}{h} - f(y^*_j) = e^*_j := \frac{h}{2} y^{*''}(\tau_j) \text{ for } j+1 \in \mathbb{N} \text{ with}$$
$$y^*_j := y^*(t_j), \quad \tau_j \in (t_j, t_{j+1}), \quad \text{and } y^*_0 = y_0 \in D.$$

In a graph of τ_j versus t_j for $j \in \mathbb{N}$, the values of τ_j are assigned to the values of t_j. As an interpolation of this graph, a (continuous) linear Spline function $\tau : \mathbb{R}^+ \to \mathbb{R}^+$ may be defined. The explicit Euler discretization of (1.1) [19, p.12],

(2.2) $$\frac{y_{j+1} - y_j}{h} - f(y_j) = 0 \text{ for } j+1 \in \mathbb{N}, \quad y_0 \in D,$$

yields the following extended Comparison-IVP

(2.3) $$\eta' - f(\eta) = \varphi(t) \text{ for } t \geq 0; \quad \eta(0) = y_0; \quad \varphi(t) := -\frac{h}{2} y^{*''}(\tau(t)).$$

This IVP would represent the Euler discretization (2.2) if $y^{*''}$ were replaced by the derivative $\eta^{*''}$ of the classical solution η^* of (2.3). Obviously, $\eta^* \neq y^*$ for $h > 0$ unless $y^{*''} = 0$ for $t \geq 0$. If $y^{*''}$ is oscillating, φ in (2.3) may cause a resonance-like self-excitatory growth of $|\eta^* - y^*|$ as $t \to \infty$. This would be a true self-excitation if $y^{*''}$ in (2.3) were replaced by $\eta^{*''}$.

Remarks: 1.) Whenever a given ODE is discretized, the local discretization error yields an artificial forcing term φ in a Comparison-ODE.

2.) When $h > 0$ is sufficiently small, the phase shift $|\tau_j - t_j|$ is correspondingly small. Then there is a correspondingly small perturbation of the self-excitation.

3.) Self-excitation for ODEs is discussed in [10, p.152-153].

Example 2.4: The following IVP is considered:

(2.5) $$y' - f(y) = 0 \Leftrightarrow \begin{bmatrix} y'_1 \\ y'_2 \end{bmatrix} - \begin{bmatrix} y_2 \\ -y_1 \end{bmatrix} = 0 \text{ for } t \geq 0 \text{ with any } \begin{bmatrix} y_1(0) \\ y_2(0) \end{bmatrix} \in \mathbb{R}^2.$$

Then (i) all solutions with components y^*_1 and y^*_2 can be represented by the functions cos and sin; (ii) for all rational $h > 0$, the explicit Euler discretization (2.2) yields divergent oscillating difference solutions as $j \to \infty$. For this problem, at $t = t_j$, H.Spreuer [48] has derived an explicit representation of the difference solutions in terms of (i) the starting vector and (ii) polynomials in powers of h with degrees 2j or 2j+1. This exhibits the influence of discretization errors concerning self-excitation and the complicated nonlinear dependency of these errors on h. □

Remark: Diekhans [15] applied approximately fifty different discretizations with respect to each one of three linear or nonlinear test-ODEs of the second order with initial conditions. He observed a linear growth of $|y_j - y^*_j|$ as j increases when damping terms were absent in the ODEs. Generally, dissipation damps the influence of preceeding errors incurred in a marching method as applied to an ODE; compare [8] for a counterexample from mechanics.

The execution of a shooting method by means of (uncontrolled) classical numerical methods is subject (a) predominantly to discretization errors affecting the approximations of solutions of the auxiliary IVP and (b) secondarily the truncation error of the iteration. The deceptive nature of these errors is exhibited by the following example concerning the pendulum equation whose periodic solutions have been investigated in literature. Following Holzmüller [23], this ODE here is considered for the case of nonlinear buckling of a beam.

Example 2.6: The nonlinear eigenvalue problem

$$(2.7) \qquad y'' + \lambda \sin y = 0 \text{ for } t \in [0,1]; \qquad y(0) = y'(1) = 0; \qquad \lambda \in \mathbb{R},$$

is considered. Since the solutions of $y'' + \lambda \sin y = 0$ can be represented by means of elliptic integrals, it is known that (2.7) possesses the bifurcation points $\lambda_i = (2i-1)^2\pi^2/4$ for $i \in \mathbb{N}$. As $\lambda_i - 0$ is replaced by $\lambda_i + 0$, a new nontrivial eigensolution of (2.7) is generated. For $\lambda = 417$, which exceeds $\lambda_7 = 416.99...$, Holzmüller executed the shooting method by use of a classical Runge-Kutta method with step sizes $h = 0.02$, $h = 0.01$, and $h = 0.005$. In all three cases, he obtained

$$(2.8) \; y'(1,y'(0)) \begin{cases} < 0 \text{ if } y'(0) = 10^{-12}, \\ > 0 \text{ if } y'(0) = 5 \cdot 10^{-11}. \end{cases}$$

If these inequalities were true for solutions y^* of the ODE, an $y^{*\prime}(0)$ must exist such that $y^{*\prime}(1,y^{*\prime}(0)) = 0$ by use of the true $y^{*\prime}(0)$. By means of Lohner's enclosure method for BVPs, Holzmüller has shown that there is no solution y^* of (2.7) with derivative $y^{*\prime}(0)$ in the interval $[10^{-12}, 5 \cdot 10^{-11}]$. □

Conclusions: (A) Irrespective of the order of a discretization or the employment of a step size control, Example 2.6 implies that classical discretizations are generally unreliable for the purposes of a qualitative and a quantitative assessment of a periodic solution of an ODE.

(B) For this purpose, enclosure methods are the only ones to yield guaranteed bounding one-sided approximations $\underline{y}_i$ and $\bar{y}_i$ with very small distance $\bar{y}_i - \underline{y}_i$, typically, 10^{-6} or less.

(C) The importance of reliability in mathematical modelling is obvious in view of the question: "What are the consequences of a computational failure for a real-world system?"

Because of Conclusion (C), the cost of executing an enclosure method may be insignificant. In the case of nonlinear IVPs, the computability of an enclosure $[\underline{y},\bar{y}]$ of a solution y^* requires a suitably high computational effort. In fact, the enclosure $[\underline{y},\bar{y}]$ will often blow up at a time t shortly after a time t_0 when $\bar{y}(t_0) - \underline{y}(t_0)$ has exceeded a certain value depending on the problem. This situation may be explained by the presence of "extraneous functions" $\hat{y} \in [\underline{y},\bar{y}]$ exerting a "self-excitatory" influence on the continuation of the construction of $[\underline{y},\bar{y}]$; see Appendix for details.

Additionally to Lohner's enclosure algorithms, in literature there are other enclosure methods for periodic solutions, particularly the following ones, which are based on discretizations:

(i) in [55] for the Duffing equation, without any numerical results and

(ii) by use of Poincaré maps (e.g. [20] or [7]), in [17], [18], and [44] with only a few hints concerning numerical results.

Remark: The literature on enclosure methods for inverse-monotone BVPs (e.g. [42]) is not useful here since, generally, inverse-monotone ODEs do not possess periodic or aperiodic oscillatory solutions.

3. The Logistic Equation

The IVP of the Logistic ODE is considered:

(3.1) $y' = \lambda y(k-y)$ for $t \geq 0$; $y(0) = y_0$; $k, \lambda, y_0 \in \mathbb{R}_0^+$.

<u>All</u> solutions of (3.1) can be represented as follows, [22, p.25]:

(3.2) $$y^*(t) = \frac{k}{1 + \left[\frac{k}{y_0} - 1\right] e^{-\lambda k t}} \quad \text{for } t \geq 0.$$

The stationary point $y^* = 0$ of the ODE is unstable, and the stationary point $y^* = k$ is stable. Here and subsequently, (in)stability is defined in the sense of Lyapunov.

According to (2.2), the explicit Euler method yields the following discretization of (3.1):

(3.3) $$\frac{y_{j+1} - y_j}{h} = \lambda y_j(k-y_j) \text{ for } j+1 \in \mathbb{N}.$$

This generates the celebrated Logistic (Difference) Equation

(3.4) $$\eta_{j+1} = a\eta_j(1-\eta_j) \text{ for } j+1 \in \mathbb{N} \text{ with } \eta_j := \frac{h\lambda}{a} y_j \text{ and } a := 1 + h\lambda k.$$

According to the classical paper [35] by R.May, the set of (difference) solutions of (3.4) can be characterized as follows:

<u>(i)</u> there is an infinite sequence $\{a_\nu\}$ of period doubling bifurcations yielding periodic difference solutions, with $a_1 = 3$ and the limit $a_\infty = 3.824\ldots$.

<u>(ii)</u> For $a = a_n$, there exist 2^n periodic (difference) solutions of (3.4) with period n [35, p.153].

<u>(iii)</u> Additionally, for $a \in [3,4]$, there is an uncountable infinity of aperiodic non-monotonic (difference) solutions of (3.4).

For this situation, the expression "(deterministic) chaos" was first introduced in the classical paper [30] by Li and Yorke.

<u>Conclusions:</u> <u>(a)</u> Since all solutions (3.2) of the ODE (3.1) are monotonic or constant, the non-monotonic aperiodic or periodic (difference) solutions must be extraneous.

(b) For all fixed k and λ, the extraneous difference solutions vanish as $h \to 0$ and, thus, $a \to 1$.

4. The Lotka-Volterra Problem

The ODEs of this mathematical model in population dynamics are a generalization of the Logistic ODE (3.1) [38, p.125]:

$$(4.1) \qquad \begin{aligned} x' &= \alpha x - \beta xy \\ y' &= -\gamma y + \delta xy \end{aligned} \qquad \text{for } t \geq 0, \quad \text{with } \alpha, \beta, \gamma, \delta \in \mathbb{R}_0^+.$$

Since x and y represent concentrations, only nonnegative solutions of (4.1) are of interest. According to [25, p.621], all solutions of (4.1) satisfy the integral relation

$$(4.2) \qquad y^{-\beta} e^{\alpha y} = c\, x^{\delta} e^{-\gamma x} \text{ with any } c \in \mathbb{R}^+.$$

The system (4.1) possesses the following solutions (I)-(IV):

(I) the stationary point $(0,0)^T$ which is unstable,

(II) the stationary point $(\gamma/\delta, \alpha/\beta)^T$ which is stable,

(III) the x-axis and the y-axis,

(IV) because of (4.2), a continuous family of periodic solutions in the first quadrant all of which enclose the stationary point (II), [22, p.528-533].

Consequently, (i) no solutions can enter or leave the first quadrant and (ii) (I)-(IV) represent all solutions of (4.1) in this quadrant and on its bounding segments of the axes.

Remarks: 1.) By use of Lohner's enclosure method for free BVPs, Holzmüller [24] has enclosed and verified one of the periodic solutions.

2.) By use of differential inequalities, an enclosure of a periodic solution had already been determined before by W.F.Ames and E.Adams [6].

In [38, p.125-127], Peitgen & Richter define and employ a family of consistent and stable discretizations of (4.1) depending on step size h and a parameter ρ. For $\rho = 0$, this family coincides with the explicit Euler method and for $\rho = 1$ with the Heun method [19, p.15] (in the family of Runge-Kutta methods). For the evaluations, they have assumed that $\alpha = \beta = \gamma = \delta = 1$, $h \in [0,1]$, and $\rho \in [0,1]$.

Results presented in [38] or according to a workshop conducted by the author can be summarized as follows: For suitable choices of h, ρ, x(0), and y(0), there are difference solutions which either

(i) approach solutions of the ODEs of types (II)-(IV), or

(ii) are extraneous and periodic, consisting of finitely many different points, e.g. 9 or 17, or

(iii) are extraneous and diverge to infinity.

Conclusion: Since the set of all solutions of the ODEs (4.1) (with $x \geq 0$ and $y \geq 0$) is known, the extraneous character of the difference solutions (ii) and (iii) of the employed Euler-Heun family is clear.

Subsequently, (a) only selected solutions of the ODEs to be treated are known and (b) discretizations of higher order are used, either without or with step size control. Instead of failures via "extraneous difference solutions", subsequently, failures via "diversions" will be exhibited.

5. The Lorenz Problem

In his classical paper [33], E.N.Lorenz considers a mathematical model for free convection in the earth's atmosphere. Lorenz adopted (Galerkin-type) componentwise truncated series expansions of the solutions. This yielded a system of three ODEs for these coefficients which he named X, Y, and Z:

$$(5.1)\qquad \begin{aligned} X' &= \sigma(-X + Y), \\ Y' &= -XZ + rX - Y, \\ Z' &= XY - bZ, \end{aligned} \qquad \text{for } t \geq 0; \qquad \text{with } b, r, \sigma \in \mathbb{R}^+.$$

In literature, this problem is the paradigm for deterministic chaotic dynamics. C.Sparrow points out in the preface of his monograph [45, p.v] on the Lorenz-ODEs (5.1): "... the number of man, woman, and computer hours spent on them in recent years ... must be truly immense." It will now be summarized what is mathematically known about the true solutions of (5.1).

The origin 0, i.e. the point $(0,0,0)^T$, is a stationary point for all positive values of the parameters b, r, σ. As r goes from $r < 1$ to $r > 1$, a bifurcation yields two new stationary points C_1 and C_2 which are given by $(\pm\alpha,\pm\alpha,r-1)^T$ where $\alpha := \sqrt{b(r-1)}$. Linear stability analysis yields [45, p.11]:

(i) for $r \in (0,1)$, the origin is a globally stable stationary point;

(ii) for $r > 1$, the origin is an unstable saddle point with a two-dimensional "stable manifold of the origin" such that all solutions starting on this sheet are contained in it and tend towards the origin; it is known that this sheet contains the whole z-axis and that it is "twisted in a very strange way" [45, p.13-14];

(iii) for $r \in (1,r_H)$, with $r_H \approx 24.74$...(if $b = 8/3$ and $\sigma = 10$), C_1 and C_2 are stable;

(iv) for $r > r_H$, linear stability analysis of the stationary points C_1 and C_2 yields one negative eigenvalue and a conjugate complex pair of eigenvalues with positive real parts;

(v) according to [45, p.11], ar $r = r_H$, there is a subcritical Hopf bifurcation; i.e., C_1 and C_2 absorb unstable periodic orbits which already existed for $r < r_H$.

According to [45, p.9], there is a bounded ellipsoid in $\mathbb{R}^3$ which all trajectories enter as $t \to \infty$. The divergence of the flow of solutions is

$$\frac{\partial X'}{\partial X} + \frac{\partial Y'}{\partial Y} + \frac{\partial Z'}{\partial Z} = -(\sigma + b + 1) < 0. \tag{5.2}$$

Any compact set S in phase space with volume V(0) is considered. Because of (5.2), as this set evolves in time, its volume changes according to $V(0)\cdot\exp(-(\sigma+b+1)t)$, [33, p.135] or [45, p.9].

For arbitrary positive values of $b = 2\sigma$ and r, W.F.Ames [4] has derived the following equivalent representation of (5.1):

$$X'' + (\sigma+1)X' - \sigma((r-1) - ce^{-2\sigma t})X + (1/2)X^3 = 0, \tag{5.3}$$

$$Z(t) - (1/2\sigma)(X(t))^2 = ce^{-bt} \text{ with } c := Z(0) - (1/2\sigma)(X(0))^2, \tag{5.4}$$

and

(5.5) $\quad Y' + Y = X(t)(r - Z(t)).$

As $t \to \infty$, all solutions of (5.3)-(5.5) are contained in a hyperface; i.e., the solutions of (5.1) cannot be chaotic in this limit for arbitrary positive $b = 2\sigma$ and r. The ODE (5.3) is now considered in the limit as $t \to \infty$. In this limit, the three stationary points 0', C_1', and C_2' of (5.3) are projections onto the X-axis of the three stationary points 0, C_1, and C_2 of the ODEs (5.1). According to the linear variational systems of (5.3) concerning the stationary points, C_1' and C_2' are asymptotically stable for all $r > 0$ and all $\sigma > 0$. The stationary point 0' is asymptotically stable for $r < 1$ and unstable for $r > 1$, in both cases for all $\sigma > 0$. According to the limit as $t \to \infty$ of the energy integral of (5.3), this ODE does not possess periodic solutions.

Remark: For the more special case of $b = 1$, $\sigma = 1/2$, and $r = 0$, the finite relationship (5.4) is referred to in [52, p.347], and X is expressed via an implicit equation which can be solved by use of Jacobi elliptic functions. In [45, p.185], special fixed choices of b, σ, (and r) are listed for which (5.1) is completely integrable.

This completes the discussion of what is rigorously known about the solutions of the Lorenz-ODEs (5.1). Covering approximately 176 out of a total of 269 pages in [45], Sparrow's investigations of (5.1) are based on (i) quantitative approximations by use of discretizations and (ii) qualitative discussions of subsets of the set of solutions in the phase space $\mathbb{R}^3$. According to [45, p.21], there is a "strange invariant set" consisting of a countable infinity of periodic orbits, an uncountable infinity of aperiodic trajectories, and an uncountable infinity of trajectories terminating at the origin. Sparrow also deduces that there is an infinite sequence $\{r_\nu\}$ of values of r such that a period doubling bifurcation takes place at r_ν [45, p.113]. Periodic orbits of this kind have been found in literature by use of numerical approximations, see Appendix E in [45]. Concerning the invariant set,

Sparrow points out [45, p.3], that "the general form ... does not depend ... on our choice of initial conditions ... or on our choice of integration routine ... . The details ... depend crucially on both factors ... . As a consequence of this, it is not possible to predict the details of how the trajectories will develop over anything other than a very short time interval." This is truly a devastating jugdement concerning the performance of classical numerical methods!

For $b = 8/3$, $r = 28$, and $\sigma = 6$, W.Kühn [28] has

(I) enclosed and verified (separately) three T-periodic solutions of (5.1), making use of the representation (1.3) of this problem,

(II) enclosed and verified several aperiodic solutions for certain intervals of time t, and

(III) approximated the solutions addressed in (ii) by use of several Runge-Kutta methods with different step sizes and with or without step size control.

Concerning (I), Kühn chose and employed the following transformation of (1.3) with T unknown and q denoting the transposition of a vector:

(5.6) $\quad \tau := t/T$ and $\eta : \mathbb{R} \to \mathbb{R}^3$ with $\eta(\tau) := y(\tau T) := (X(\tau T), Y(\tau T), Z(\tau T))^q$.

This yields the following non-free BVP as an equivalent representation of (1.3):

(5.7) $\quad \eta' = Tf(\eta)$ for $\tau \in [0,1]$ with $\eta(1) = \eta(0)$.

Kühn's search for a T-periodic solution η^* of $\eta' = Tf(\eta)$ consisted of the following steps:

(a) the choice of a hyperplane $Z = c \in \mathbb{R}$ in $\mathbb{R}^3$ to be selected such that it is intersected by the desired periodic solution η^*;

(b) the following determination of T and of the remaining components X(0) and Y(0) of the initial vector $(X(0), Y(0), c)^q$;

(c) $s := (X(0), Y(0), T)^q$ is the vector to be determined by means of a shooting method;

(d) this shooting method consists of the auxiliary IVP

(5.8) $\eta' = Tf(\eta)$ for $\tau \in [0,1]$ with $\eta(0) = s$ to be chosen

(a solution to be enclosed is denoted by $\eta^* = \eta^*(\tau,s)$) and

(e) an interval Newton-method employed for an enclosure of a solution s^* of

(5.9) $F(s) = 0$ with $F(s) := \eta^*(1,s) - s$;

(f) the derivatives $\partial F/\partial X(0)$ and $\partial F/\partial Y(0)$ as required for the execution of (e) were determined as solutions of corresponding linear variational ODEs of $\eta' = Tf(\eta)$;

(g) the derivative $\partial F/\partial T$ was determined by use of

$$\frac{\partial F(s)}{\partial T} = \frac{\partial \eta^*(\tau,s)}{\partial T}\Big|_{\tau=1} - \frac{\partial s}{\partial T} = \tau \frac{\partial y^*(\tau T,s)}{\partial(\tau T)}\Big|_{\tau=1} - e_3 =$$

$$(5.10) \quad \frac{\partial y^*(t,s)}{\partial t}\Big|_{t=T} - e_3 = f(y^*(T,s)) - e_3 = f(\eta^*(1,s)) - e_3 \quad \text{with}$$

$$e_3 := (0,0,1)^q.$$

Remarks: 1.) Kühn's algorithm pertaining to (b)-(g) corresponds essentially to Lohner's algorithm for non-free BVPs.

2.) The choice of the transformation (5.6) avoids Lohner's extension ([31] or [32]) of ODEs such as (5.1) by means of additional equations $dT(t)/dt = 0$ and $T(0) = T$.

3.) Subsequent to his applications, Kühn noticed that the transformation (5.6) had already been employed by E.Hopf.

Concerning (II), Kühn employed Lohner's enclosure algorithm for IVPs which was also used by him for the enclosure of the solutions $\eta^* = \eta^*(\tau,s)$ of the auxiliary IVP (5.8). Because of the hardware and software available to him for enclosure purposes, Kühn decided to implement Lohner's enclosure algorithm and numerous required auxiliary algorithms in the programming language C.

Concerning (III), the details of the algorithm will be given subsequently.

Now, Kühn's major results can be summarized. Kühn verified three periodic solutions of (5.1) which are depicted in Figures 5.11-5.13 by use of their projections onto the X-Y-coordinate plane. The widths of the computed enclosures do

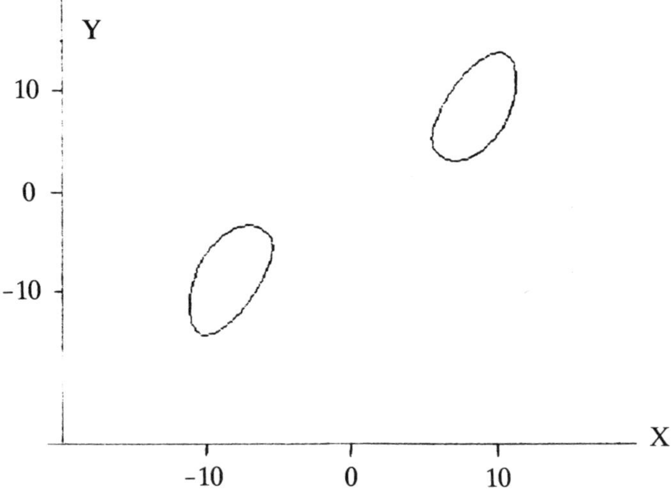

Figure 5.11: Projection of Periodic Solution $\eta^*_{(1)}$ of Lorenz ODEs (5.1). There is no periodic solution other than $\eta^*_{(1)}$ whose "initial values" are $(6.8311189693^{6}_{3}, 3.221312211^{5}_{4}, 27)^{q}$.

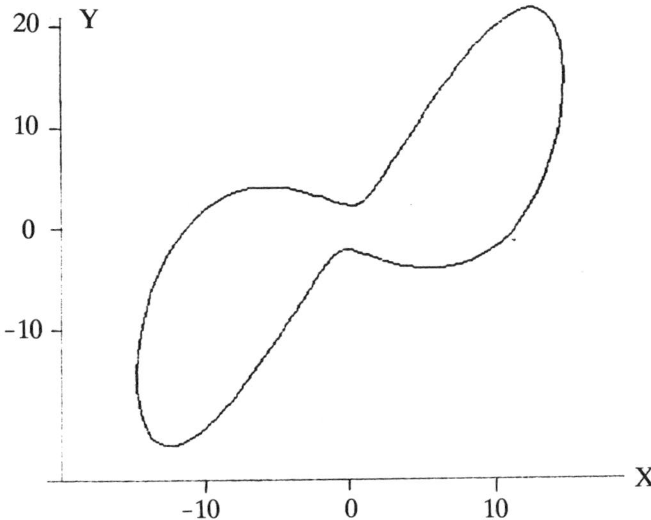

Figure 5.12: Projection of Periodic Solution $\eta^*_{(2)}$ of Lorenz ODEs (5.1). There is no periodic solution other than $\eta^*_{(2)}$ whose "initial values" are $(0.5121927^{4}_{3}, 2.2498373^{8}_{7}, 15)^{q}$.

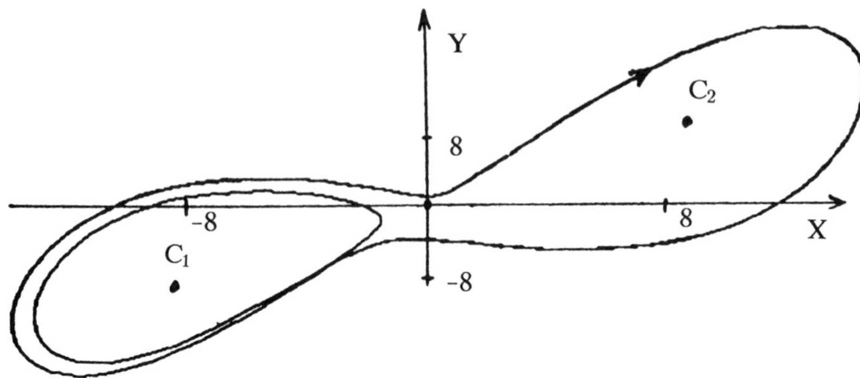

Figure 5.13: Projection of Periodic Solution $\eta^*_{(3)}$ of Lorenz ODEs (5.1). "Initial values" of $\eta^*_{(3)}$ are contained in the interval vector $(10.952283216^{4}_{0}, 21.716013685^{7}_{5}, 20)^{q}$.

not exceed 10^{-8}. By use of simple shooting, Figure 5.11 exhibits two realizations of periodic solutions $\eta^*_{(1)}$ with period T^*_1 where $T^*_1 \in 0.6899186868^{28}_{15}$. By use of simple shooting, the periodic solution $\eta^*_{(2)}$ depicted in Figure 5.12 was obtained with period $T^*_2 \in 1.75168488^{41}_{27}$. By use of multiple shooting with two intermediate times, the periodic solution $\eta^*_{(3)}$ depicted in Figure 5.13 was obtained with period $T^*_3 \in 2.5942776279^{90}_{03}$. For the computational determination of the enclosures of $\eta^*_{(1)}$, $\eta^*_{(2)}$, and $\eta^*_{(3)}$, the step size $h = 2^{-5}$ and the order $p = 15$ were used in Lohner's enclosure algorithm.

With $y' = f(y)$ representing (5.1), functions $\xi_{(i)} := y - \eta^*_{(i)} : \mathbb{R} \to \mathbb{R}^3$ are considered for a fixed $i \in \{1,2,3\}$. With respect to $y' = f(y)$ and $\eta^*_{(i)}$, the linear variational system concerning variations of $\eta(0)$,

(5.14) $\quad \xi'_{(i)} = A_{(i)}(t)\xi_{(i)}$, with $i \in \{1,2,3\}$ arbitrary but fixed,

can be derived where $A_{(i)} : \mathbb{R} \to L(\mathbb{R}^3)$ possesses the period of $\eta^*_{(i)}$. In the context of the Floquet theory, then $\mu = 1$ is one of the characteristic factors of (5.14), e.g., [10, p.98]. Then y' = f(y) is said to possess asymptotic orbital stability if the two remaining characteristic factors of (5.14) have negative real parts [10, p.98]. This is not true for (5.1) and $\eta^*_{(1)}$ or $\eta^*_{(2)}$ or $\eta^*_{(3)}$, as has been shown by Kühn. Consequently, each one of these periodic solutions of (5.1) is locally repellent within a manifold in $\mathbb{R}^3$ containing the periodic solution $\eta^*_{(i)}$ under consideration.

Subsequently in Section 5 and in Section 6, it is to be shown that classical discretization methods are unreliable for the purpose of approximating periodic or aperiodic trajectories.

<u>Figures 5.15-5.17</u> are concerned with a solution starting close to $\eta^*_{(3)}$. In these figures, boxes demarcate (symbolically) the projection of the enclosure $[\underline{y},\bar{y}]$ whose width is less than 10^{-8}. Here, $h = 2^{-5}$ and p = 15 were used in Lohner's enclosure algorithm for IVPs. The solid lines represent projections of approximations $\tilde{y}$ due to a classical Runge-Kutta method (with order four) and the following step sizes: <u>Figure 5.15</u>: h = 1/64, <u>Figure 5.16</u>: h = 1/256, and <u>Figure 5.17</u>: h = 1/1024. According to Figures 5.15 and 5.17, $\tilde{y}$ and $[\underline{y},\bar{y}]$ are coincident within graphical accuracy for the interval of time for which $\tilde{y}$ and $[\underline{y},\bar{y}]$ were computed. This is not so in Figure 5.16: here $\tilde{y}$ begins to <u>deviate grossly</u> from $[\underline{y},\bar{y}]$ as $\tilde{y}$ and $[\underline{y},\bar{y}]$ approach the origin. In a comparison of Figures 5.15, 5.16, and 5.17, it is remarkable that this gross deviation occurs for the case of the intermediate step size h.

The deviation in Figure 5.16 will now be explained. Starting at a position $y(0) \in \mathbb{R}^3$ close to $\eta^*_{(3)}$, the enclosed true solution y^* of (5.1) and its approximation $\tilde{y}$ come close to the stable manifold of the origin at a time $t_1 > 0$. Since this manifold consists of solutions of (5.1), it cannot be pierced by y^*. Because of computational errors, however, $\tilde{y}$ for h = 1/256 does pierce this manifold. For

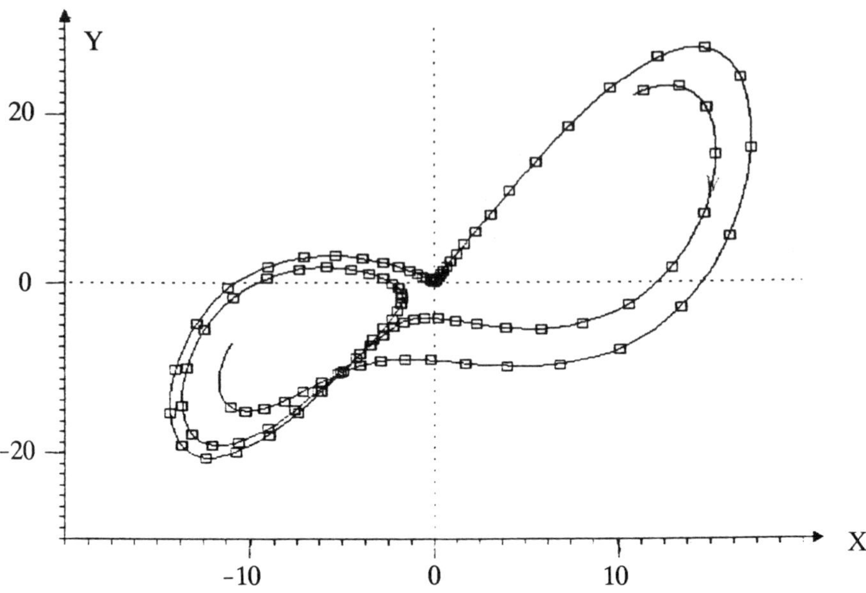

Figure 5.15: Solution of Lorenz ODEs (5.1), Starting Close to $\eta^*_{(3)}$

□ □ □ Enclosure Method ——— Runge-Kutta method with h = 1/64

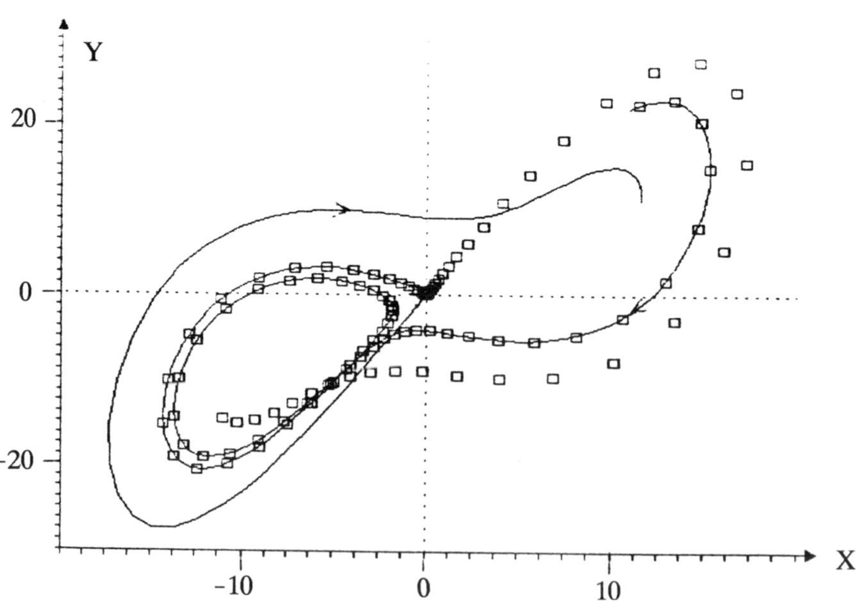

Figure 5.16: Just Like Figure 5.15, However, h = 1/256

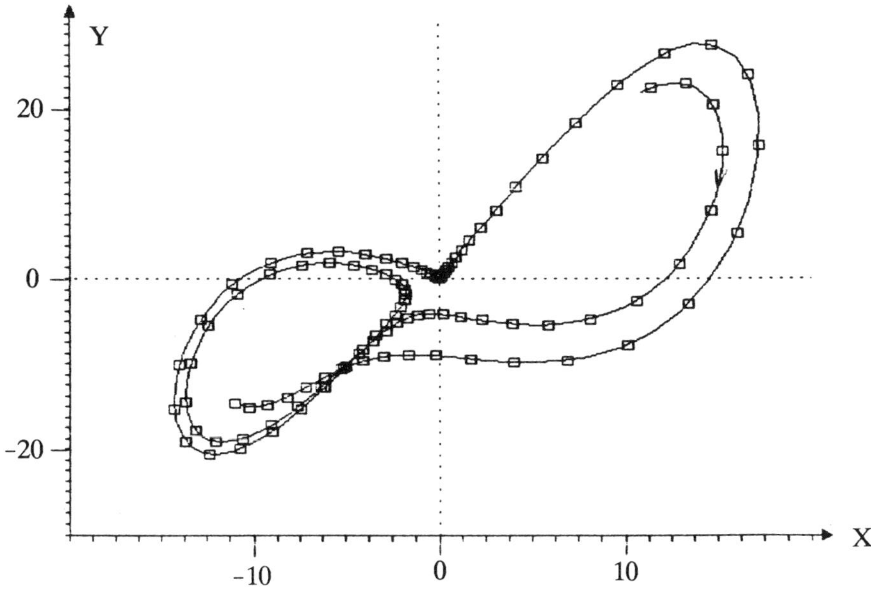

Figure 5.17: Just Like Figure 5.15, However, h = 1/1024

$t \in (t_1,t_2)$, y^* and $\tilde{y}$ are almost parallel to one another with a small distance, but on different sides of the manifold referred to before. As y^* and $\tilde{y}$ approach the origin at $t \approx t_2$, they begin to move in opposite directions, roughly perpendicular to the manifold. This is a consequence of the fact that the origin is a saddle point which is (i) attractive within the manifold referred to and (ii) repellent in a perpendicular direction. For $t > t_2$, $\tilde{y}$ is a close approximation of a solution y^{**} with $y^{**} \neq y^*$. As $t > t_2$ increases, there is the gross deviation of y^* and $\tilde{y}$ which can be seen in Figure 5.16. In the situation just outlined and in comparable cases, the difference approximation $\tilde{y}$ is said to divert from one true solution of the ODEs, y^*, to another true solution of the ODEs, y^{**}.

Remark: In [45, p.216], numerical difficulties are reported: "when a trajectory passes near the origin ... we get 'extreme' sensitive dependence on initial conditons."

Figures 5.18-5.21 are concerned with another solution starting close to $\eta^*_{(3)}$. These figures present the Euclidean distance d in $\mathbb{R}^3$ of various Runge-Kutta approximations and the (one) enclosure of this solution, which were determined for $t \in [0,24.3]$. Within this interval of time, the width of the enclosure is smaller than 10^{-8}. The data of the employed Runge-Kutta approximations are as follows, with RK denoting a classical method (of order four) and RKC a method of order four or five with step size control.

Figure 5.18: RK with $h = 0.01$,

Figure 5.19: RKC with precision $\epsilon = 10^{-7}$, yielding an average step size of 0.01,

Figure 5.20: RK with $h = 0.0025$,

Figure 5.21: RKC with precision $\epsilon = 10^{-10}$, yielding an average step size of 0.0025.

Figures 5.18-5.21 exhibit the following situation:

(i) for $t \in [0,t_1]$ with $t_1 \approx 14$, the Runge-Kutta approximations $\tilde{y}$ follow closely the true (enclosed) solution y^* of (5.1);

(ii) $\tilde{y}$, beginning at t_1, deviates grossly from y^* such that d reaches values of up to 50; this is approximately the maximum diameter of the periodic solution $\eta^*_{(3)}$.

This situation is again explained by diversions of the approximations $\tilde{y}$ as y^* and $\tilde{y}$ come close to the stable manifold ot the origin.

Conclusions: (a) With or without step size control, Runge-Kutta approximations $\tilde{y}$ of a true periodic or aperiodic solution y^* of the Lorenz-ODEs (5.1) are unreliable. They may exhibit gross deviations from y^*.

(b) The quoted "crucial dependency" of all computed details on the employed classical integration routine [45, p.3] may be related to the diversions of $\tilde{y}$ and y^* as shown in Figures 5.16 and 5.18-5.21.

(c) Generally for systems of nonlinear autonous ODEs in $\mathbb{R}^n$ with $n \geq 3$, the existence of stable manifolds of stationary points may be expected, provided these points are saddle points.

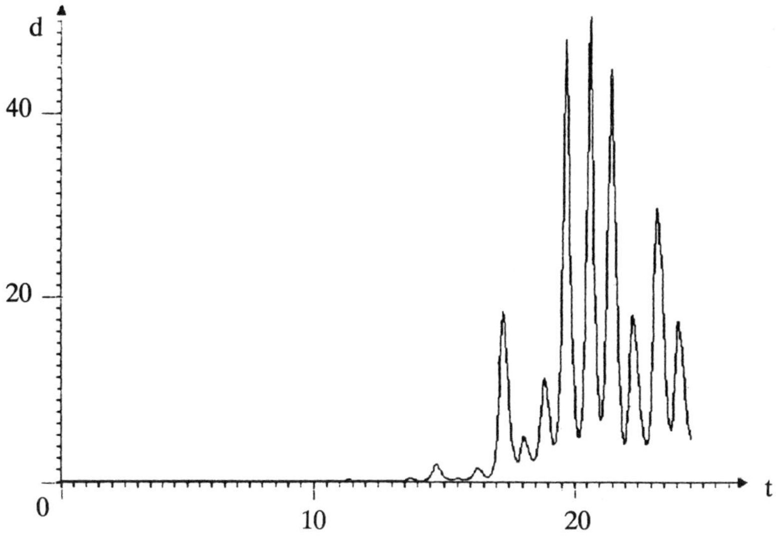

Figure 5.18: Euclidean Distance d of Enclosure and Runge-Kutta Trajectory Computed with h = 0.01

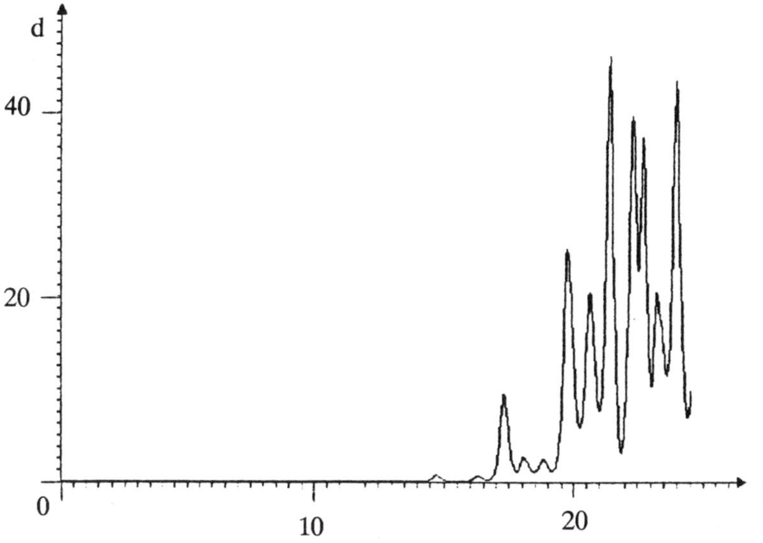

Figure 5.19: Euclidean Distance d of Enclosure and Runge-Kutta Trajectory Computed with Step Size Control ($\epsilon = 10^{-7}$)

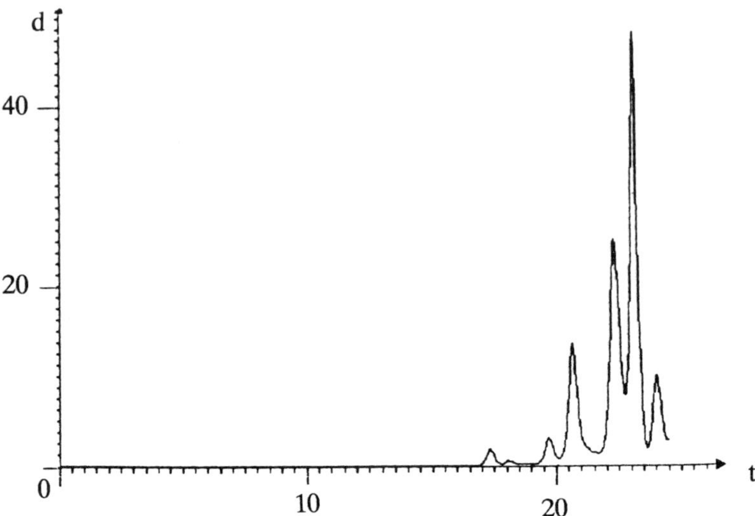

Figure 5.20: Euclidean Distance d of Enclosure and Runge-Kutta Trajectory Computed with $h = 0.0025$

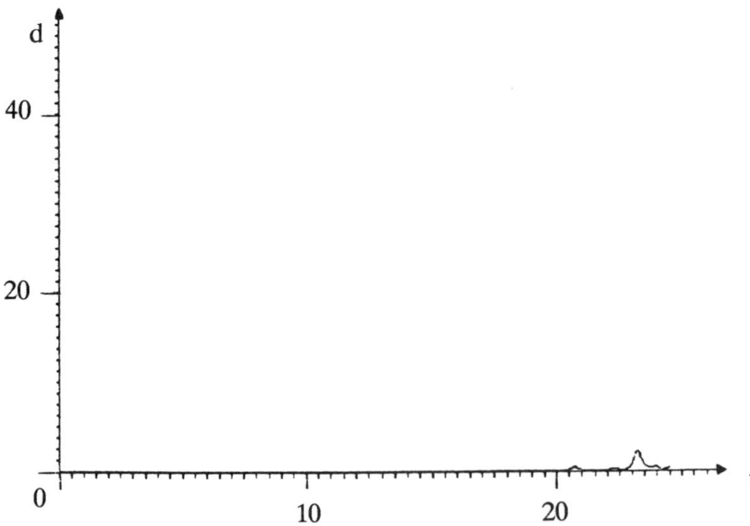

Figure 5.21: Euclidean Distance d of Enclosure and Runge-Kutta Trajectory Computed with Step Size Control ($\epsilon = 10^{-10}$)

6. The Restricted Three Body Problem of Celestial Mechanics

The following idealization of celestial mechanics is considered [54, p.5]:

(i) the orbits of the earth E and the moon M are confined to a plane in $\mathbb{R}^3$;

(ii) in this plane, there is a suitable rotating Cartesian y_1-y_2-basis whose origin is attached to the center of gravity, C, of E and M; the points E, C, and M are on the y_1-axis;

(iii) the position of C relative to E and M is determined by the ratio $\mu = 1/82.45$ of the masses of M and E; consequently, $-\mu$ is the location of E and $\lambda := 1 - \mu$ is the location of M on the y_1-axis;

(iv) in the y_1-y_2-plane, trajectories of a small satellite S are to be determined;

(v) for theses trajectories, the phase space possesses the Cartesian coordinates y_1, y_2, $y_3 := y'_1$ and $y_4 := y'_2$.

For the restricted three body problem defined by (i), (ii), and (iv), (v), the equations of motion are as follows [54, p.5] in the employed rotating basis:

$$
\begin{aligned}
&y'_1 = y_3, \qquad y'_2 = y_4,\\
(6.1)\qquad &y'_3 = y_1 + 2y_4 - \frac{\lambda(y_1 + \mu)}{P} - \frac{\mu(y_1 - \lambda)}{Q},\\
&y'_4 = y_2 - 2y_3 - \frac{\lambda y_2}{P} - \frac{\mu y_2}{Q}\\
&\text{where } P := ((y_1 + \mu)^2 + y_2^2)^{3/2} \text{ and } Q := ((y_1 - \lambda)^2 + y_2^2)^{3/2}.
\end{aligned}
$$

Subsequently, the closed orbit of S in Figure 6.3 will be investigated which has been approximated by Bulirsch and Stoer [9] and Stroud [50]. In [9] and [50], the following data are listed for an initial position at time $t = 0$ on this orbit and for its period T:

$$
\begin{aligned}
(6.2)\qquad &y_1(0) = 1.2, \qquad y_2(0) = y_3(0) = 0, \qquad y_4(0) \approx -1.049\,357\,509\,83,\\
&T \approx 6.192\,169\,331\,396\ldots\,.
\end{aligned}
$$

The discretization methods developed in [9] rest on extrapolation techniques such that an (automatic) control of step size h is executed. In an application to the problem (6.1), (6.2), the closed orbit was covered in 48 time steps involving 5218

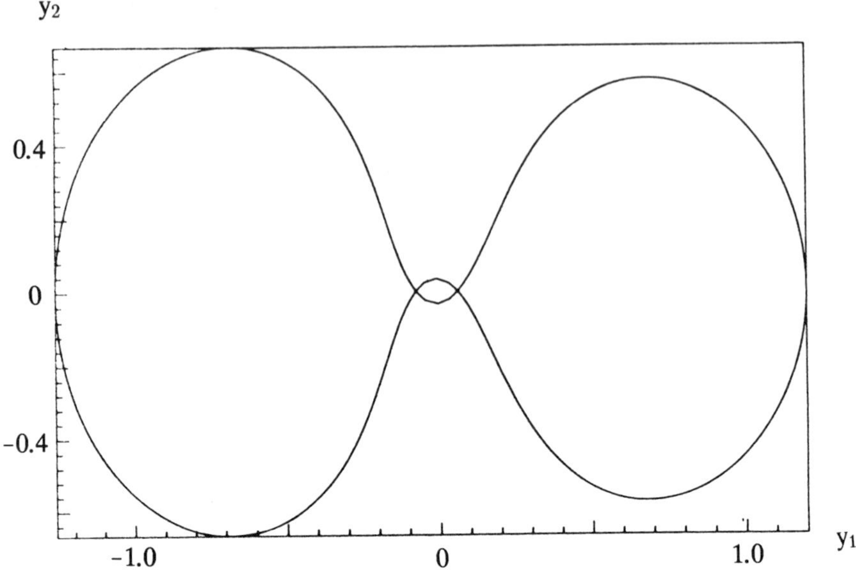

Figure 6.3: Periodic Solution of Restricted Three Body Problem by Use of Classical Runge-Kutta Method with $h = 0.001$

evaluations of the right hand sides of the ODEs (6.1). At the end of one revolution, there was an agreement of approximately ten mantissa digits for the final and the initial values of y_i with $i = 1(1)4$. In [50], a Runge-Kutta method of order five with control of h was used whose execution yielded the minimum value $h = 0.0001...$ of h. Without a verification of the orbit determined by (6.1) and (6.2), this orbit has been enclosed as follows: (A) by Scheu and Adams [41] and (B) by Lohner [31] and [32]. On the basis of differential inequalities, Scheu and Adams made use of the trapezoidal rule for the marching determination of bounds $\underline{y}_i$ and $\bar{y}_i$ for $i = 1(1)4$. They employed (a) $h = 0.125$ and (b) $h = 0.007815$. Enclosing the solution for one revolution of the orbit, they obtained the following maximum distances $\bar{y}_i(T) - \underline{y}_i(T)$ for $i = 1(1)4$ at the time T of one period: $(3.1) \cdot 10^{-4}$ in case (a) and $(1.2) \cdot 10^{-6}$ in case (b). For the execution of his enclosure algorithm for IVPs, Lohner chose a sequence of step sizes on the basis of the ones which had been generated by the execution of a Runge-Kutta method with (automatic) step size control. For i = 1(1)4 and $t \in [0,T]$, Lohner's results for $\bar{y}_i(t)$ and $\underline{y}_i(t)$ show an agreement of at least nine mantissa digits.

Poincaré (see [51, p.61]) has shown the existence of a countable infinity of periodic solutions of (6.1). Additionally, Poincaré and Birkhoff (e.g. [7, p.631]) have investigated the stability of periodic orbits solving (6.1). By means of usual numerical methods, Strömgren et al. (e.g. [51, p.61]) have computed approximations of numerous periodic solutions of (6.1). As compared to the relatively simple structure of the orbit in Figure 6.3, in literature sequences of closed orbits are shown which possess an increasing number of loops. In view of physics, a loop is "activated" when an orbit comes close to one of the poles of the ODEs (6.1), at either E or M. Close to such a pole, a high sensitivity of the solutions is to be expected with respect to positional errors of the trajectory, compare Section 5.

Assuming that the closed orbit in Figure 6.3 represents a periodic solution of (6.1) and (6.2), approximations were determined by W.Rufeger [40] employing a family of Runge-Kutta methods as given by Grigorieff [19]. Additionally, variations of h were carried out, or an (automatic) control of h was used. The execution of this control was essentially identical to the one employed in the case of Figures 5.19 and 5.21.

W.Rufeger obtained results of the following kinds:

(I) Within graphical accuracy, the orbit in Figure 6.3 was reproduced when a fixed value of $h = 10^{-3}$ was used or when controls of h were used, even with a (crude) prescribed precision $\epsilon = 10^{-2}$. In a neighborhood of the pole at E, the controls yielded values $h < 10^{-3}$.

(II) When h was fixed and larger than 10^{-3}, diversions of the computed approximation took place in a neighborhood of E.

Now selected results of type (II) are presented because of their relevance for the problem of reliability of classical numerical methods for the determination of approximations of periodic or aperiodic solutions.

Because of the identical starting position, all trajectories to be presented should be identical with the one in Figure 6.3, provided there are no computational errors. Figure 6.4 shows a trajectory which was computed by use of a classical Runge-Kutta method with $h = 5 \cdot 10^{-3}$. A sequence of "extraneous loops" can be seen whose size shrinks as t increases. For a smaller neighborhood of E, Figure 6.5 exhibits a continuation of the sequence of loops, terminating in an "ejection" from this neighborhood. For another enlarged neighborhood of E, Figure 6.6 shows the loops depicted in Figure 6.4 with, additionally, ten points (1)...(10) demarcated, which are ordered as time t increases. Trajectories were determined as follows in a piecewise fashion: (A) from the initial point as given in (6.2) to one of the points (1)...(10) by use of the Runge-Kutta method (without step size control) employed for Figure 6.4, followed by (B) a continuation by use of a Runge-Kutta method with control of h and a prescribed precision $\epsilon = 10^{-9}$, starting from the chosen point in the set {(1),...,(10)}, with additional consideration of the values of y_3 and y_4 at this point as determined by means of (A). According to the result (I) referred to before, the numerical method (B) may be expected to be reliable in the problem under consideration.

Figure 6.7 exhibits the results for the choice of the starting point (1) as indicated in Figure 6.6. Here, the continuation does not possess extraneous loops, and it is almost closed. Figure 6.8 shows the corresponding results for the choice of the starting point (2). The distance of (1) and (2) in Figure 6.6 is just one time step of length h of the execution of method (A). Figure 6.8 shows three extraneous loops and there would be further loops upon continuation of the analysis. Figure 6.9-6.11 show the corresponding results for the starting points (3)-(5), respectively. In Figure 6.6, these points are adjacent time steps of the execution of

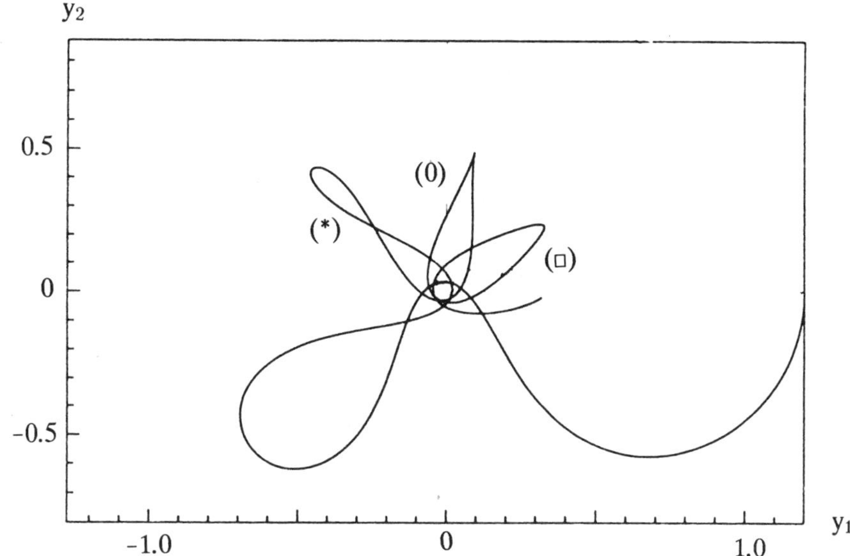

Figure 6.4: Trajectory of Restricted Three Body Problem by Use of Classical Runge-Kutta Method with h = 0.005

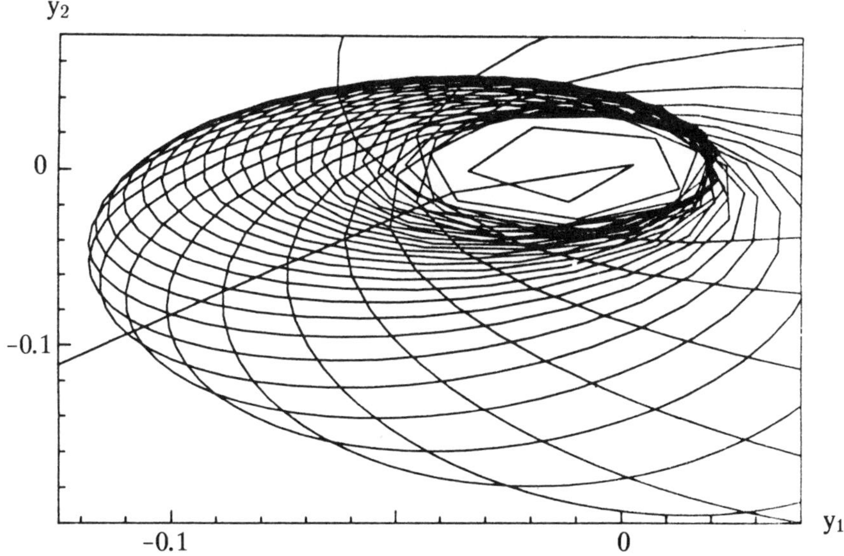

Figure 6.5: Continuation of Trajectory Shown in Figure 6.4

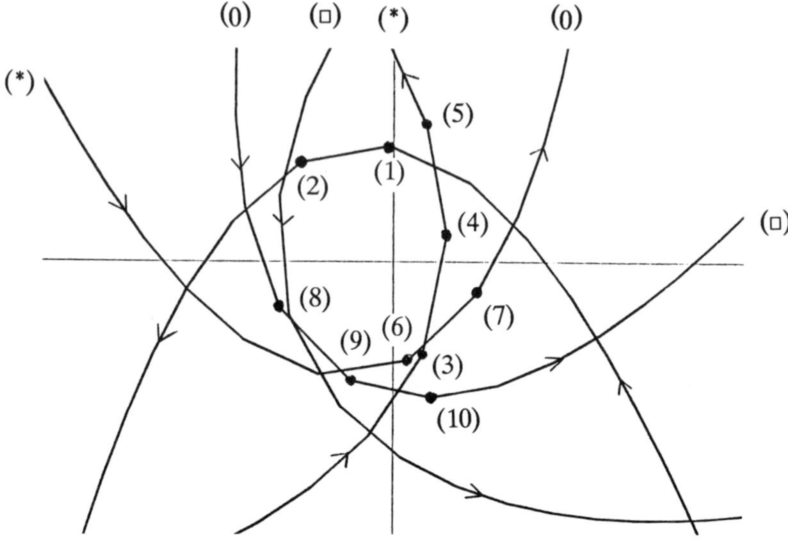

Figure 6.6: Trajectory Shown in Figure 6.4 with Points (1),...,(10) Demarcated on this Trajectory

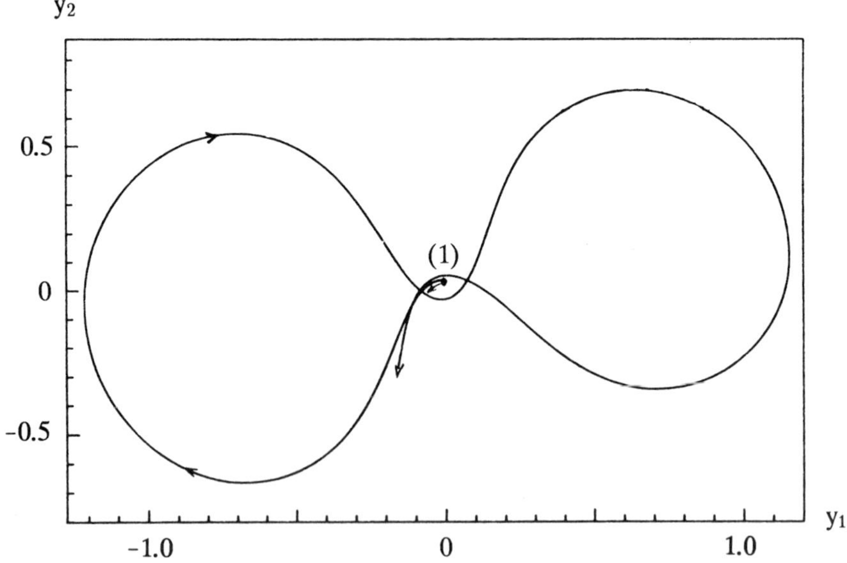

Figure 6.7: Consecutive Determination of Trajectory by Use of Runge-Kutta Methods (A) and (B), with (B) Starting at (1)

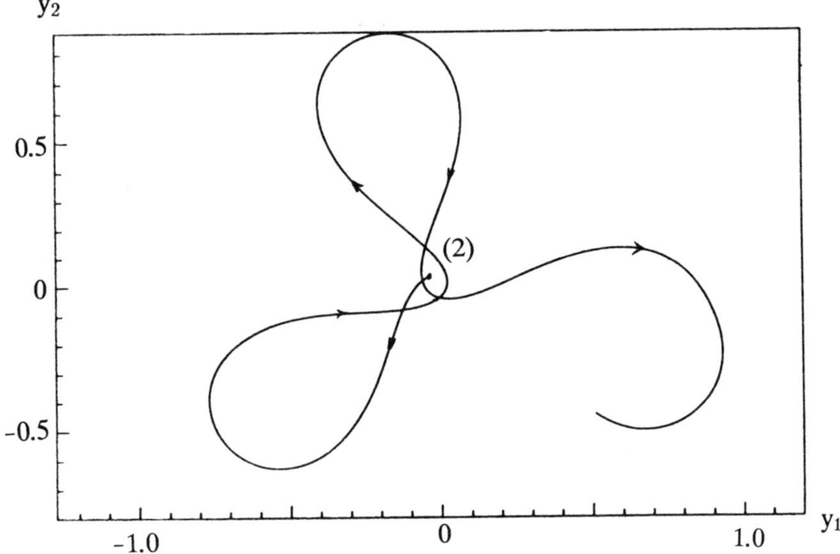

Figure 6.8: Consecutive Determination of Trajectory by Use of Runge-Kutta Methods (A) and (B), with (B) Starting at (2)

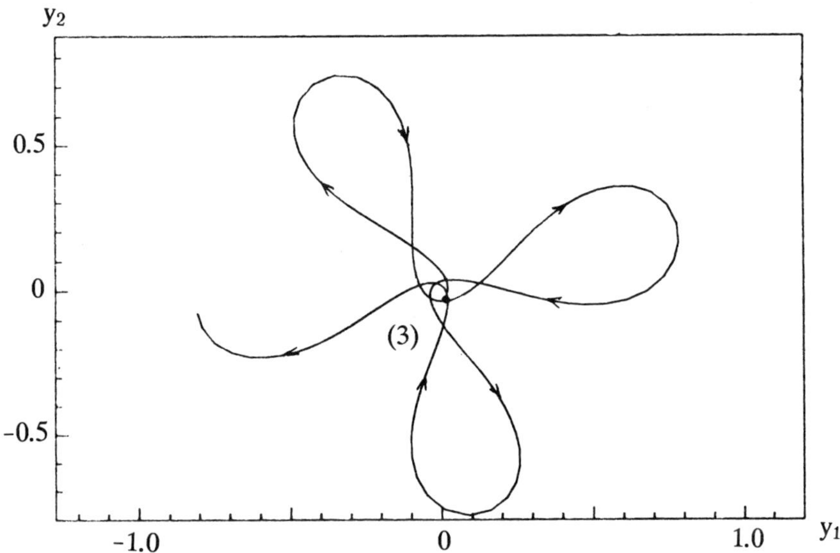

Figure 6.9: Consecutive Determination of Trajectory by Use of Runge-Kutta Methods (A) and (B), with (B) Starting at (3)

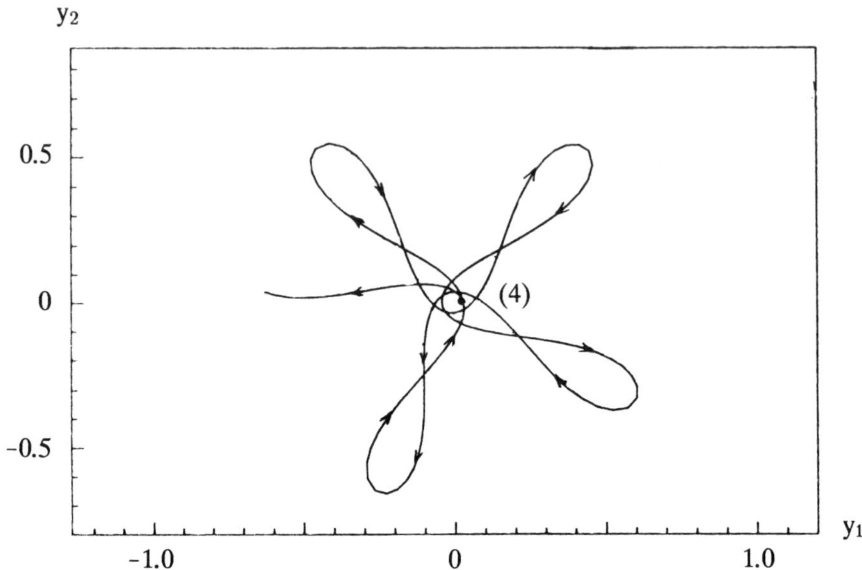

Figure 6.10: Consecutive Determination of Trajectory by Use of Runge-Kutta Methods (A) and (B), with (B) Starting at (4)

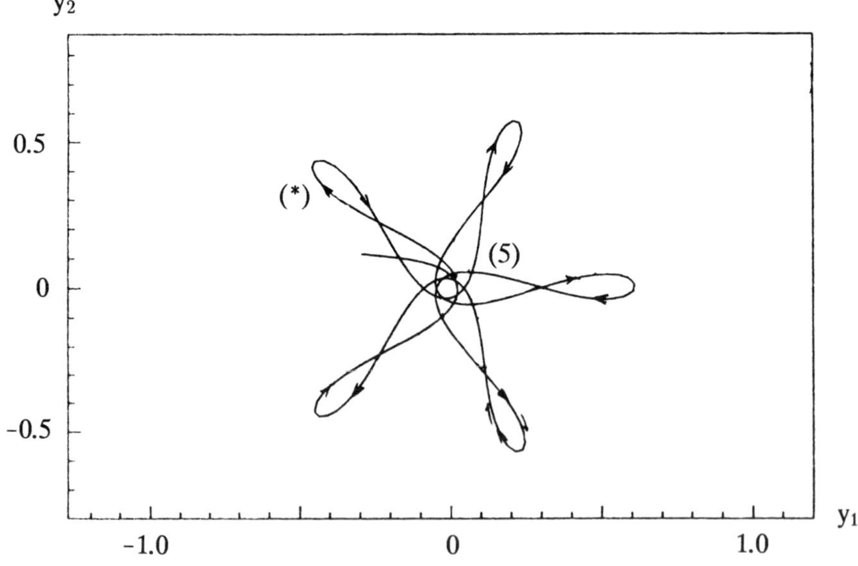

Figure 6.11: Consecutive Determination of Trajectory by Use of Runge-Kutta Methods (A) and (B), with (B) Starting at (5)

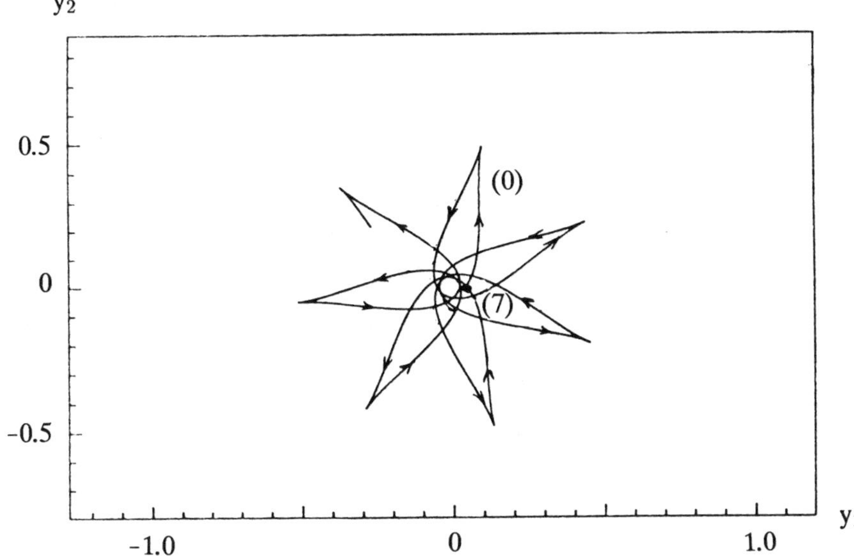

Figure 6.12: Consecutive Determination of Trajectory by Use of Runge-Kutta Methods (A) and (B), with (B) Starting at (7)

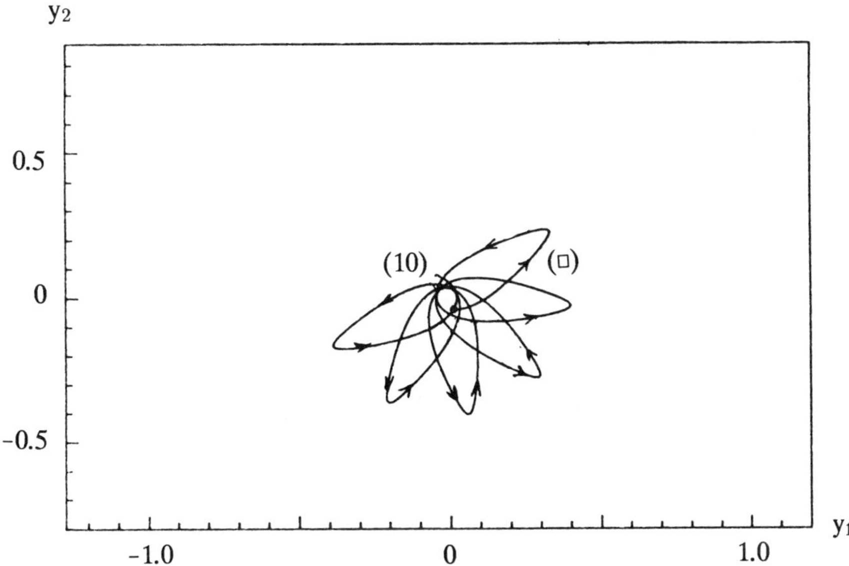

Figure 6.13: Consecutive Determination of Trajectory by Use of Runge-Kutta Methods (A) and (B), with (B) Starting at (10)

method (A). Figures 6.12 and 6.13 show the extraneous loops for the starting points (7) and (10), respectively. Figures 6.7-6.13 allow the following observations:

(a) the patterns of extraneous loops are essentially different when method (B) is started at adjacent time steps of method (A);

(b) within graphical accuracy, the three loops demarcated by the symbols (*), (O), or (□) reappear as follows: (*) in Figures 6.4 and 6.11; (O) in Figures 6.4 and 6.12; (□) in Figures 6.4 and 6.13;

(c) the sizes of the extraneous loops decrease as one goes from Figure 6.8 to Figure 6.13, which is consistent with Figure 6.5.

These results concerning extraneous loops have not been verified by use of enclosure methods; however, they appear to be reliable because of the properties (I) and (II) of controlled or uncontrolled Runge-Kutta methods, respectively. With these reservations in mind, Figures 6.7-6.13 allow the following interpretation (or at least conjecture):

Almost every entry or reentry of the approximation $\tilde{y}$ computed by use of method (A) into a certain vicinity of the pole at E leads to a diversions of $\tilde{y}$ from a neighborhood of one solution y^* of (6.1) to a neighborhood of some other solution y^{**} of (6.1), with $y^* \neq y^{**}$, compare Section 5.

Concerning periodic or aperiodic solutions of (6.1), in literature there are many graphical representations of sequences of loops, e.g., [38, p.2] or [13, p.133-134]. Could it be that some of these patterns of loops are due to (perhaps multiple) diversions?

If sufficiently intricate, sequences of loops as discussed in Figures 6.8-6.13 have been called "chaotic" in literature, e.g., [38, p.2]. Recently, in literature, the problem of

"Chaos or Instability in the Solar System"

has been investigated and discussed frequently, making use of clasical numerical methods, e.g., [27]. In this context in literature, instability is assumed to begin when a computed trajectory starts to "wiggle", generally in an aperiodic oscillatory fashion. Figures 5.18-5.21 concerning the Lorenz-ODEs (5.1) suggest that this may be due to discretization errors, causing an "unusual character" of the computed approximations. This would be computational chaos rather than "real chaos" for the solutions of the governing ODEs. The ruling (however misguided) opinion in literature is revealed by the following excerpts from [27]:

(i) "Advent ... of faster computers is pushing analyses of solar system stability to longer and longer time spans until chaos is appearing in the behavior of planets..."

(ii) Concerning errors in the analysis: "... only the most minor of approximations such as ignoring general relativity" have been made.

Remarks: 1.) A relationship between computational instability and computational chaos has been suggested in a recent paper [34] by E.N.Lorenz.

2.) For solutions of the ODEs (6.1) coming close to the pole at E, Lohner's enclosure algorithms must be supplemented by an automatic step size control; compare Stetter's plea [49] for the development of this kind of control.

Conclusions: For the computational determination of periodic or aperiodic trajectories in celestial mechanics, classical discretization methods are unreliable. They may yield diversions to extraneous loops or to extraneous aperiodic oscillations of trajectories.

7. Periodic Solutions of Mathematical Models for Gear Drive Vibrations

The ususal mathematical simulation of gear drive vibrations employs a system of linear nonhomogeneous ODEs possessing T-periodic coefficients, with 1/T as the tooth engagement frequency:

(7.1) $\quad \eta' - (A+B(t))\eta = a$ for $t \in \mathbb{R}^+$; $\quad a \in \mathbb{R}^n$; $\quad A \in L(\mathbb{R}^n)$;

$B : \mathbb{R}^+ \to L(\mathbb{R}^n), \quad B(t+T) = B(t)$ for all $t \in \mathbb{R}^+$,

where $L(\mathbb{R}^n)$ denotes the set of real n×n-matrices. In engineering literature, the following two properties are of predominant interest: (a) the asymptotic stability of

(7.2) $\quad \eta' - (A+B(t))\eta = 0$ for $t \in \mathbb{R}^+$

and (b) for $a \neq 0$, the existence and the determination of a globally unique T-periodic solution η^*_{per} of (7.1), provided there is one. According to the Floquet theory [10, p.58], property (a) is true if and only if $\rho(\phi(T)) < 1$ where $\phi : \mathbb{R} \to L(\mathbb{R}^n)$ is the fundamental matrix of (7.2) and ρ denotes the spectral radius of a matrix. D.Cordes ([11] or [12] or [3]) has derived a T-periodic semi-explicit representation of η^*_{per} in terms of a, ϕ, and $(\phi(T))^{-1}$. If $\rho(\phi(T)) < 1$, (α) $(\phi(T))^{-1}$ exists and (β) η^*_{per} is globally unique. By use of Cordes' enclosure methods,

(A) $\rho(\phi(T)) < 1$ can be verified by means of a sufficient but nonconservative algorithmic test and

(B) η^*_{per} can be enclosed and verified.

Details of (A), (B), and applications to gear drive vibrations are presented in [2]. U.Schulte ([43] and [2]) has carried out applications of Cordes' analysis with respect to (I) a class "CL" of mathematical models of gear drive vibrations which has been frequently employed in engineering literature and (II) a modified class "MCL" of models, with negligible or absent dynamical implications of the transfer from "CL" to "MCL".

Concerning class "CL", Schulte discovered that $\phi(T)$ always possesses an eigenvalue $\lambda = 1$ with an algebraic multiplicity of two and a geometric multiplicity of one. According to the Floquet theory [10, p.58], (7.2) then possesses unbounded solutions as $t \to \infty$. Because of the nonexistence of $(\phi(T))^{-1}$, η^*_{per} then does not exist.

In engineering literature, this existence had always been tacitly assumed for class "CL". Consequently, in this literature, (7.1) with a suitably estimated initial vector $y(0) \in \mathbb{R}^n$ was always approximately solved by use of Runge-Kutta methods, either without or with step size control. Because of this tacit assumption, these marching methods were terminated at times $t > 0$ when the computed approximation appeared to be almost T-periodic.

For several mathematical models in class "MCL", Schulte has executed Cordes' enclosure methods and thus enclosed and verified η^*_{per}.

Conclusions: Enclosure methods have been

(α) instrumental in the discovery of the non-existence of T-periodic solutions of (7.1) for class "CL" when $a \neq 0$ and

(β) practically applicable for the verification of the desired properties (A) and (B) for class "MCL".

In [3] and [26], a in (7.1) is replaced by a+b where $b : \mathbb{R}^+ \to \mathbb{R}^n$ is a suitable stochastic process for the simulation of pitch errors of the gears as fabricated. Then there is a set of solutions of (7.1) with a computable enclosure $[\underline{y},\bar{y}]$. Under suitable conditions, $\underline{y}$ and $\bar{y}$ are periodic. Very small pitch errors may exert a surprisingly large influence on the dynamic response of the model, both concerning $[\underline{y},\bar{y}]$ and marginal distribution functions over component intervals $[\underline{y}_i,\bar{y}_i]$.

8. Perturbation Method for the Enclosure of Periodic Solutions of Nonlinear Periodic ODEs

Problem (7.1) is reconsidered with the following nonlinear extension:

(8.1) $y' = (A + B(t))y + \epsilon g(y) + a$ for $t \in \mathbb{R}$; $\epsilon \in \mathbb{R}$; $g : D \to \mathbb{R}^n$, $D \subset \mathbb{R}^n$, g is sufficiently smooth, etc as in (7.1);

T-periodic solutions are denoted as follows: $\eta^*_{per} = \eta^*_{per}(t)$ in the case of (7.1) and $y^*_{per} = y^*_{per}(t,\epsilon)$ in the case of (8.1). By use of Cordes' enclosure algorithm as outlined in Section 7, it is assumed that (i) asymptotic stability has been verified for (7.2) and (ii) an enclosure $[\eta^*_{per}(0)]$ has been determined. Because of (i), then $\eta' - (A + B(t))\eta = 0$ does not possess a T-periodic solution. Consequently and according to a theorem given in [10, p.137], (a) for all sufficiently small $|\epsilon|$, (8.1) possesses one and only one T-periodic solution y^*_{per} and (b) as $\epsilon \to 0$, η^*_{per} is the limit of y^*_{per} for all $t \in [0,T]$. Additionally, a theorem in [10, p.97] asserts (c) that y^*_{per} then is asymptotically stable, i.e. locally attractive.

In order to determine y^*_{per}, a function F is introduced:

(8.2) $F(s) := y^*(T,s) - s \overset{!}{=} 0$ where $y^* = y^*(t,s)$ solves (8.1) for $t \in [0,T]$ with $y^*(0) =: s$.

For numerical applications, F is replaced by the following interval function:

(8.3) $F([s]) := [y(t,[s])] - [s]$ for $t \in [0,T]$ with $y^*(t,s) \in [y(t,[s])]$ for $s \in [s]$.

By use of an interval Newton method [5], a sequence of intervals can be determined:

(8.4) $[s]_\nu$ for $\nu \in \mathbb{N}$ with $[s]_0 := [\eta^*_{per}(0)]$.

Provided the enclosure condition

(8.5) $[s]_{\bar\nu+1} \subset [s]_{\bar\nu}$ for a fixed $\bar\nu \in \mathbb{N}$

is satisfied, the Brouwer fixed point theorem [37, p.161] asserts that

(8.6) $s^* := y^*_{per}(0,\epsilon) \in [s]_{\bar\nu+1}$.

In a practical execution, (8.4) is replaced as follows:

(i) starting with $[s]_0$, only $[s]_1$ is determined,

(ii) if (8.5) is not satisfied for $\nu = 0$, then a new starting interval $[s]_{00}$ is chosen, either as

(iii) the convex hull of $[s]_0$ and $[s]_1$ or as

(iv) a suitably selected interval $[s]_{00}$ containing $[s]_0$.

If $|\epsilon|$ is sufficiently small, the theorem in [10, p.137] asserts that

$\underset{t}{\mathrm{Max}}\,|y^*_{per}(t,\epsilon) - \eta^*_{per}(t)|$ is correspondingly small for $t \in [0,T]$. Starting with $[s]_0 = [\eta^*_{per}(0)]$, (8.5) will be satisfied by an iterated interval denoted by $[s]_\mu$. By construction, then

(8.7) $\quad y^*_{per}(t,\epsilon) \in [y(t,[s]_\mu)]$ for $t \in [0,T]$.

Remarks: 1.) Because of the periodicity of y^*_{per}, it is generally not possible to satisfy the enclosure condition (8.5) when the Newton operator is replaced by F as defined in (8.2) or (8.3). This is possible, however, in the case of an asymptotically vanishing solution of (8.1), as has been shown by D.Cordes [12, p.117-123] for a nonlinear Mathieu equation.

2.) A perturbation enclosure method can be executed analogously for autonomous ODEs (8.1) with B absent. The period $T = T(\epsilon)$ is then unknown. Its enclosure can be carried out by use of the transformation (5.6) and on the basis of a theorem in [10, p.139].

3.) Applications of Lohner's enclosure algorithm for BVPs require highly accurate starting approximations for the Newton interval iteration. This algorithm has not yet been executed successfully for nonlinear BVPs when the (total) order of the system of ODEs exceeds three. The perturbation enclosure method presented here may be useful for the determination of an approximation of this kind when the order exceeds three.

Because of a lack of space, R.Heck's numerous completed examples for the presented perturbation enclosure method cannot be treated here.

9. Conclusions and Final Remarks

Outside of mathematics, there is still an almost universal tacit belief in classical discretizations of ODEs to yield practically negligible errors in their computational assessment of true solutions of ODEs. Recently, however, some physicists and engineers are becoming aware of the possibility of "computational chaos". Whereas the problem of the existence of a certain solution is relatively mild in the case of (i) IVPs, this is not so concerning (ii) BVPs and, especially, (iii) periodic solutions of ODEs. For problems (ii) or (iii), the unreliability of classical discretizations is particularly grave. Because of their self-validating character, enclosure methods (if applicable in view of available supporting hardware and software) are totally reliable. In the case of ODEs, this is due to the fact that they establish a controlled constructive bridge between (a) the function space of the ODEs and (b) the Euclidean space of the discretization, whose dimension in practical work is necessarily bounded together with the step size. This paper is closed by

(A) a plea for more advanced hardware and software supporting enclosure methods and

(B) concerning these computational tools, an appreciation of the remarkable progress achieved by U.Kulisch and coworkers who laid the foundations of the Karlsruhe Enclosure Methods.

Appendix: Self-Excitation of Enclosures for IVPs

Lohner's enclosure algorithm for IVPs is considered for the system (1.1) of ODEs with a given initial vector $y(0) \in D$. The algorithm employs the equidistant grid points $t_j := jh$ for $j+1 \in \mathbb{N}$. The algorithm yields an enclosure $[y] = [y(t)] := [\underline{y}(t),\bar{y}(t)] \ni y^*(t)$ for the true solution y^* of the IVP. The execution of the algorithm proceeds stepwise; i.e., $[y(t)] \subset D$ is computed for $t \in (t_j,t_{j+1}]$ provided $[y(t)] \subset D$ has already been determined for $t \in (0,t_j]$. According to Stetter's discussion [49], the choice and the employment of an a priori interval $[y^{(0,j+1)}] \subset D$ (containing $y(t_j)$) is the bottleneck for a fast execution of the algorithm. The candidate

(A1) $[y^{(0,j+1)}] \ni y^*(t)$ for $t \in [t_j,t_{j+1}]$

for a preliminary enclosure is (i) verified by means of the Banach fixed point Theorem [54, p.45] (if this is possible) and then (ii) improved. The bounds defining $[y^{(0,j+1)}]$ do not depend on $t \in [t_j,t_{j+1}]$. Additionally to the true solution y^*, the intervals in the sequence

(A2) $S^{(0)} := \{ [y^{(0k)}] \}$ with $k = 1(1)j+1$

contain the range of an uncountable set $\{\hat{y}\}$ of "extraneous functions" $\hat{y}$. The stepwise execution of Lohner's enclosure algorithm is consistent with the admission of a subset $\{\hat{y}_{st}\}$ of $\{\hat{y}\}$ containing all step curves with range in $\{\hat{y}\}$ and discontinuities only at the grid points $t_k = kh$ for $k \in \mathbb{N}$. It is assumed that the ODEs (1.1) possess oscillatory solutions. Extraneous step curves $\hat{y}_{st}$ can then be chosen so as to exert the maximum self-excitation which is consistent with the range of $\hat{y}$ available in the "function strip" $S^{(0)}$. For the discontinuities of these step functions $\hat{y}_{st}$, the jump times t are chosen as integer multiples of h such that a situation of parameter or combination resonance is approximated as closely as possible; compare [10] or [3] for these resonances. The maximum excitation takes place when there is a number $m \in \mathbb{N}$ such that $1/mh$ coincides with one of the resonance frequencies. This situation is explained by the following demonstration-type Example A3.

By assumption, the enclosure $[y] = [y(t))]$ has been computed for $t \in [0,t_{j+1}]$. This enclosure then contains (i) the true solution y^* of the IVP, consisting of $y' = f(y)$ and a given vector $y(0)$ and (ii) the piecewise smooth solutions $\hat{y}^*$ of an uncountable number of modified IVPs whose ODEs $y' = \hat{f}(t,y)$ possess interval extensions contained in the ones for $y' = f(y)$ with $t \in [t_k,t_{k+1}]$ and $k = 0(1)n$.

Example A3: The following ODEs and initial data are chosen:

$$\text{(A4)}\quad \begin{array}{lll} y_1'' + (a+by_2)y_1 = 0, & & y_1(0) = 1,\ y_1'(0) = 0, \\ y_2' + cy_2 = 0, & \text{for } t \in [0,t_\infty] \text{ with} & y_2(0) \in [\underline{y}_2(0),\overline{y}_2(0)]; \\ a, b, c \in \mathbb{R}^+; \quad 0 < b < a; & c = 1/t_\infty, \quad t_\infty >> 1; & \underline{y}_2(0) > 0. \end{array}$$

These IVPs are equivalently represented as follows by use of $y_3 := y_1'$:

$$\text{(A5)}\quad \left.\begin{aligned} y_1' &= f_1(y) := y_3, \\ y_2' &= f_2(y) := -cy_2, \\ y_3' &= f_3(y) := -(a+by_2)y_1, \end{aligned}\right\} \text{etc. as in (A4) with } y := \begin{bmatrix} y_1 \\ y_2 \\ y_3 \end{bmatrix}.$$

Since properties of Lohner's enclosure algorithm are to be discussed, it is ignored that the solutions y_2^* of $y_2' = -cy_2$ can be represented explicitly. In the construction of $[y] = [y(t)]$ for $t \in [t_j, t_{j+1}]$, the following interval evaluation $f_3([y^{(0,j+1)}])$ of f_3 has to be carried out:

$$\text{(A6)}\quad f_3([y^{(0,j+1)}]) := -(a+b[y_2^{(0,j+1)}])[y_1^{(0,j+1)}].$$

Since all solutions $y_2^* = y_2^*(t,y_2(0))$ are positive, the interval evaluation in (A6) may be executed for the following intervals:

$$\text{(A7)}\quad \begin{aligned} &[y_i^{(0,j+1)}] = [\underline{y}_i^{(0,j+1)}, \bar{y}_i^{(0,j+1)}] \text{ with } i = 1 \text{ or } 2 \text{ and} \\ &\underline{y}_1^{(0,j+1)} \lessgtr 0, \quad \underline{y}_1^{(0,j+1)} \le \bar{y}_1^{(0,j+1)}, \quad \text{and } \underline{y}_2^{(0,j+1)} > 0. \end{aligned}$$

Because of interval arithmetic (e.g. [5] or [36]),

$$-a[y_1^{(0,j+1)}] - b[y_2^{(0,j+1)}][y_1^{(0,j+1)}] =: [\underline{z}_3^{(j+1)}, \bar{z}_3^{(j+1)}] \text{ with}$$

$$\text{(A8)}\quad \bar{z}_3^{(j+1)} = -a\underline{y}_1^{(0,j+1)} \begin{cases} -b\bar{y}_2^{(0,j+1)}\underline{y}_1^{(0,j+1)} & \text{if } \bar{y}_1^{(0,j+1)} < 0, \\ -b\underline{y}_2^{(0,j+1)}\underline{y}_1^{(0,j+1)} & \text{if } \underline{y}_1^{(0,j+1)} > 0, \\ -b\bar{y}_2^{(0,j+1)}\underline{y}_1^{(0,j+1)} & \text{if } \underline{y}_1^{(0,j+1)} \le 0 \le \bar{y}_1^{(0,j+1)}, \end{cases}$$

$$\underline{z}_3^{(j+1)} = \ldots$$

correspondingly. Because of $c = 1/t_\infty$, the solutions $y_1^* = y_1^*(t,y_1(0),y_2(0))$ are aperiodically oscillating with a local frequency $\sqrt{a + by_2^*(t,y_2(0))}$ which <u>slowly</u> decreases as t increases. If $\bar{y}_2(0) - \underline{y}_2(0)$ is sufficiently small, the overestimate of the true set of solutions y_1^* by the enclosure $[\underline{y}_1, \bar{y}_1]$ is correspondingly small; $\underline{y}_1$ and $\bar{y}_1$ then are oscillating almost like the true solutions y_1^*. Therefore, the sequence $\{\bar{z}_3^{(j+1)}\}$ is determined by the step function

$$\text{(A9)}\quad \hat{y}_{2st} := \begin{cases} \bar{y}_2^{(0k)} & \text{if } \bar{y}_1^{(0k)} < 0, \\ \underline{y}_2^{(0k)} & \text{if } \underline{y}_1^{(0k)} > 0, \\ \bar{y}_2^{(0k)} & \text{if } \underline{y}_1^{(0k)} \le 0 \le \bar{y}_1^{(0k)}, \end{cases} \qquad \text{for } t \in (t_k, t_{k+1}] \text{ with } k = 0(1)j.$$

Concerning $y_1'' + (a+by_2)y_1 = 0$ in (A4), the function $\hat{y}_{2st}$ is automatically accounted for in the construction of the enclosure. The discontinuous transfer from $\bar{y}_2{}^{(0k)}$ to $\underline{y}_2{}^{(0,k+1)}$ (or from $\underline{y}_2{}^{(0k)}$ to $\bar{y}_2{}^{(0,k+1)}$) <u>may</u> take place at equidistant times t_ℓ such that $\hat{y}_{2st}$ yields the following Mathieu-like ODE with a close-to-periodic coefficient $b(\hat{y}_{2st}(t) - \hat{y}_{2m}(t))$:

(A10) $$y_1'' + ((a + b\hat{y}_{2m}(t)) + b(\hat{y}_{2st}(t) - \hat{y}_{2m}(t)))y_1 = 0 \text{ with}$$
$$\hat{y}_{2m}(t) := (\underline{y}_2{}^{(0k)} + \bar{y}_2{}^{(0k)})/2 \text{ for } t \in (t_k, t_{k+1}] \text{ with } k = 0(1)j.$$

For the Mathieu equation [10, p.65-66],

(A11) $$\eta'' + (\alpha + \beta\cos 2t)\eta = 0 \text{ for } t \in \mathbb{R}, \text{ with } \alpha, \beta \in \mathbb{R}^+,$$

the Ince-Strutt stability chart exhibits <u>(i)</u> the parameter resonances $\alpha = n^2$ for all $n \in \mathbb{N}$; <u>(ii)</u> the domain of stability where all solutions decrease exponentially as $t \to \infty$, and <u>(iii)</u> the domain of instability where some solutions grow exponentially as $t \to 0$. For $\alpha \neq n^2$, $(\alpha,\beta)^q$ is in the domain (ii) provided $\beta \in \mathbb{R}^+$ is sufficiently small. Correspondingly, (A10) is in a domain of stability (or instability) if

(A12) $$\underset{k \in \mathbb{N}}{\text{Max}} (\bar{y}_2{}^{(0k)} - \underline{y}_2{}^{(0k)})$$

is sufficiently small (or large). □

<u>Conclusions:</u> <u>(i)</u> This type of self-excitation built into the computational determination of an enclosure is confined to nonlinear ODEs with oscillatory solutions and

<u>(ii)</u> its influence increases in a nonlinear fashion as the width of the computed enclosure grows.

<u>Remarks:</u> <u>1.)</u> Conclusion (ii) is consistent with numerical experience concerning Lohner's enclosure algorithm for nonlinear IVPs. According to this experience, the timewise rate of growth of the enclosure is a strongly nonlinear increasing function of the width of the enclosure.

<u>2.)</u> Whereas the self-excitation discussed here is related to parameter or combination resonances, the one treated in Section 2 is concerned with classical resonance (by means of forcing terms).

3.) Generally, the excitation of any type of resonance can be reduced by means of damping terms in the ODEs with suitable damping coefficients, however, compare [8].

4.) Extraneous difference solutions are not necessarily a consequence of any one of these types of resonance, as has been shown by use of the case of the monotonic solutions (3.2) of the logistic ODE (3.1).

List of References

[1] E.Adams, A.Holzmüller, D.Straub, The Periodic Solutions of the Oregonator and Verification of Results, p.111-121 in: Scientific Computation with Automatic Result Verification, Comp. Suppl. 6, ed.: U.Kulisch, H.J.Stetter, Springer, Wien, 1988.

[2] E.Adams, Enclosure Methods and Scientific Computation, p.3-31 in: Volume 1: Num. and Appl. Math., ed.: W.F.Ames, IMACS Annals on Computing and Applied Mathematics, J.C.Baltzer, Basel, 1989.

[3] E.Adams, D.Cordes, H.Keppler, Enclosure Methods as Applied to Linear Periodic ODEs and Matrices, accepted for publication in ZAMM.

[4] E.Adams, W.F.Ames, W.Kühn, W.Rufeger, Chaos: Computational Errors or Reality, in preparation.

[5] G.Alefeld, J.Herzberger, Introduction to Interval Computation, Academic Press, New York, 1983.

[6] W.F.Ames, E.Adams, Monotonically Convergent Numerical Two-Sided Bounds for a Differential Birth and Death Process, p.135-140 in: Interval Mathematics, ed.: K.Nickel, Springer, Berlin, 1975.

[7] J.Argyris, H.P.Mlejnek, Chapter 25: Eine erste Fibel über Chaos, p.555-708, in: Die Methode der finiten Elemente, vol.III: Dynamik, F.Vieweg, Braunschweig, 1988.

[8] L.Becker, Experimentelle und numerische Untersuchung von Kombinationsresonanzen, Dissertation, Karlsruhe, 1972.

[9] R.Bulirsch, J.Stoer, Numerical Treatment of Ordinary Differential Equations by Extrapolation Methods, Num. Math. 8, 1966, p.1-13.

[10] L.Cesari, Asymptotic Behavior and Stability Problems in Ordinary Differential Equations, 2nd edit., Springer, Berlin, 1962.

[11] D.Cordes, E.Adams, Test for Uniform Boundedness for Dynamic Problems Admitting Parameter and Combination Resonance, p.115-132 in: Computer Arithmetic, ed.: E.Kaucher, U.Kulisch, Ch.Ullrich, B.G.Teubner, Stuttgart, 1987.

[12] D.Cordes, Verifizierter Stabilitätsnachweis für Lösungen von Systemen periodischer Differentialgleichungen auf dem Rechner mit Anwendungen, Dissertation, Karlsruhe, 1987.

[13] J.W.Daniel, R.E.Moore, Computation and Theory in Ordinary Differential Equations, Freeman, San Francisco, 1970.

[14] H.T.Davis, Introduction to Nonlinear Differential and Integral Equations, Dover Publ., New York, 1962.

[15] G.Dieckhans, Numerische Simulation von parametererregten Getriebeschwingungen, Dissertation, Aachen, 1981.

[16] R.Gaines, Difference Equations Associated With Boundary Value problems for Second Order Nonlinear Ordinary Differential Equations, SIAM J. Num. Anal. 11, 1974, p.411-433.

[17] S.De Gregorio, E.Scoppola, B.Tirozzi, A Rigorous Study of Periodic Orbits by Means of a Computer, J. of Statistical Physics 32, 1983, p.25-33.

[18] S.De Gregorio, The Study of Periodic Orbits of Dynamical Systems. The Use of a Computer, J. of Statistical Physics 38, 1985, p.947-972.

[19] R.D.Grigorieff, Numerik gewöhnlicher Differentialgleichungen, Vol.1, B.G.Teubner, Stuttgart, 1972.

[20] J.Guckenheimer, P.Holmes, Nonlinear Oscillations, Dynamical Systems and Bifurcations of Vector Fields, Springer, New York, second printing, 1986.

[21] B.M.Herbst, M.J.Ablowitz, Numerically Induced Chaos in the Nonlinear Schrödinger Equation, Phys. Rev. Letters 62, 1989, p.2065-2068.

[22] H.Heuser, Gewöhnliche Differentialgleichungen, B.G.Teubner, Stuttgart, 1989.

[23] A.Holzmüller, Einschließung der Lösung linearer oder nichtlinearer gewöhnlicher Randwertaufgaben, Diplomarbeit, Karlsruhe, 1984.

[24] A.Holzmüller, unpublished communication.

[25] E.Kamke, Differentialgleichungen Lösungsmethoden und Lösungen, Vol.1: Gewöhnliche Differentialgleichungen, 3rd edition, Akademische Verlagsgesellschaft, Leipzig, 1944.

[26] H.Keppler, Numerische Behandlung von Zufallsschwingungen, insbesondere bei Getrieben, Diplomarbeit, Karlsruhe, 1988.

[27] R.A.Kerr, Does Chaos Permeate the Solar System? Science 244, 1989, p.144-145.

[28] W.Kühn, Über Einschließungen, das Lorenz-Problem und eine Klasse stochastischer Differentialgleichungen, Diplomarbeit, Karlsruhe, 1990.

[29] U.W.Kulisch, W.L.Miranker, The Arithmetic of the Digital Computer: A New Approach, SIAM Review 28, 1986, p.1-40.

[30] T.-Y.Li, J.Yorke, Period Three Implies Chaos, Amer. Math. Monthly 82, 1975, p.985-992.

[31] R.J.Lohner, Enclosing the Solutions of Ordinary Initial and Boundary Value Problems, p.255-286 in: Computer Arithmetic, ed.: E.Kaucher, U.Kulisch, Ch.Ullrich, B.G.Teubner, Stuttgart, 1987.

[32] R.Lohner, Einschließung der Lösung gewöhnlicher Anfangs- und Randwertaufgaben und Anwendungen, Dissertation, Karlsruhe, 1988.

[33] E.N.Lorenz, Deterministic Nonperiodic Flow, J. of the Atmospheric Sciences 20, 1963, p.130-141.

[34] E.N.Lorenz, Computational Chaos - A Prelude to Computational Instability, Physica D 35, 1989, p.299-317.

[35] R.M.May, Simple Mathematical Models With Very Complicated Dynamics, Nature 261, 1976, p.459-467.

[36] R.E.Moore, Methods and Applications of Interval Analysis, SIAM, Philadelphia, 1979.

[37] J.M.Ortega, W.C.Rheinboldt, Iterative Solution of Nonlinear Equations in Several Variables, Academic Press, New York, 1970.

[38] H.O.Peitgen, P.W.Richter, The Beauty of Fractals, Springer,Berlin, 1986.

[39] H.O.Peitgen, Symmetrie im Chaos - Selbstähnlichkeit in komplexen Systemen, p.30-49 in: Symmetrie in Geistes- und Naturwissenschaften, ed.: R.Wille, Springer, Berlin, 1988.

[40] W.Rufeger, Numerische Ergebnisse der Himmelsmechanik und Entwicklung einer Schrittweitensteuerung des Lohner'schen Einschließungs-Algorithmus, Diplomarbeit, Karlsruhe, 1990.

[41] G.Scheu, E.Adams, Zur numerischen Konstruktion konvergenter Schrankenfolgen für Systeme nichtlinearer, gewöhnlicher Anfangswertaufgaben, p.279-287 in: Interval Mathematics, ed.: K.Nickel, Springer, Berlin, 1975.

[42] J.Schröder, A Method for Producing Verified Results for Two-Point Boundary Value Problems, p.9-22 in: Comp. Suppl. 6, ed.: U.Kulisch, H.J.Stetter, Springer, Wien, 1988.

[43] U.Schulte, M.Kölle, E.Adams. Dynamische Zahnkräfte, ZAMM 69, 1989, p.T350-T352.

[44] Ja.G.Sinai, E.B.Vul, Discovery of Closed Orbits of Dynamical Systems with the Use of Computers, J. of Statistical Physics 23, 1980, p.27-47.

[45] C.Sparrow, The Lorenz Equations: Bifurcations, Chaos, and Strange Attractors, Springer, New York, 1982.

[46] H.Spreuer, E.Adams, Pathologische Beispiele von Differenzenverfahren bei nichtlinearen gewöhnlichen Randwertaufgaben, ZAMM 57, 1977, p.T304-T305.

[47] H.Spreuer, E.Adams, on Extraneous Solutions With Uniformly Bounded Difference Quotients for a Discrete Analogy of a Nonlinear Ordinary Boundary-Value problem, J. Engin. Math. 19, 1985, p.45-55.

[48] H.Spreuer, unpublished communication.

[49] H.J.Stetter, Validated Solutions of Initial Value Problem for ODE, this volume.

[50] A.H.Stroud, Numerical Quadrature and Solution of Ordinary Differential Equations, Springer, New York, 1974.

[51] K.Stumpff (edit.) Astronomie, 4, Das Fischer Lexikon, Fischer, Frankfurt am Main, 1957.

[52] M.Tabor, Chaos and Integrability, J.Wiley, New York, 1989.

[53] E.R.Vrscay, W.J.Gilbert, Extraneous Fixed Points, Basin Boundaries and Chaotic Dynamics for Schröder and König Rational Iteration Functions, Numerische Mathematik 52, 1988, p.1-16.

[54] W.Walter, Gewöhnliche Differentialgleichungen, Springer, Berlin, 1972.

[55] M.Yamaguti, H.Yoshihara, T.Nishida, Periodic Solutions of Duffing Equation, p.80-95 in: Self-Validating Algorithms and Applications, Lecture Note 673, Research Institute for Mathematical Science, Kyoto University, 1988.

Numerical Algorithms for Existence Proofs and Error Estimates for Two-Point Boundary Value Problems

Johann Schröder
Mathematisches Institut der Universität zu Köln
Weyertal 86-90, D-5000 Cologne 41
Federal Republic of Germany

Abstract: We describe a general method for constructing EB-algorithms, i.e., algorithms for proving the existence of a solution of a given (nonlinear boundary value) problem and calculating bounds for this solution.

1. Introduction

The method for constructing EB-algorithms considered here will be explained by discussing vector-valued two-point boundary value problems of the form

$$-U''(x) + F(x, U(x)) = 0 \quad (0 < x < 1) \quad , \quad U(0) = r_0 \quad , \quad U(1) = r_1 \tag{1.1}$$

with $U(x) = (U_i(x)) \in \mathbf{R}^n$, where $F(x, y)$ is twice continuously differentiable for $0 \leq x \leq 1$, $y \in \mathbf{R}^n$. Our approach can also be used for other problems.

Each of the EB-algorithms considered is composed of two procedures (subalgorithms) **A** and **E**, with **A** denoting a procedure for calculating approximate solutions of certain boundary value problems, and **E** a procedure for calculating bounds for certain continuous functions. The idea is to carry out procedure **A** in real arithmetic, and procedure **E** in interval arithmetic. In this way, the approximate solution of the problem considered is separated from the application of interval arithmetic.

As a theoretical tool for the existence proof we use here the fixed-point theorem of Schauder. For this purpose we transform the given problem into an equivalent fixed-point equation

$$u \in X \quad , \quad u = Tu \tag{1.2}$$

in a Banach space X, and construct a suitable set $D \subset X$ such that

$$\text{(C)} \qquad TD \subset D \quad .$$

When this *Contraction Property* (C) holds, there is a fixed-point $u^* = Tu^* \in D$. (All further assumptions of the Schauder fixed-point theorem are satisfied in the cases considered.)

Computer Arithmetic and Self-Validating Numerical Methods

ISBN 0-12-708245-X

Our final goal is the construction of EB-algorithms which do not require further assumptions concerning smoothness, or growth properties of the nonlinear function F in (1.1) (except those necessary for the procedures **A** and **E** which one wants to apply). The general approach is described in sections 2, 3, and 4. The essential tool used there is a condition on a suitable linearization A of the differential operator in (1.1) (called condition (L)). *This condition enables one to estimate terms involving A^{-1} without calculating this inverse operator explicitly.* In the cases discussed in section 7, A must be inverse-positive, or have an inverse-positive majorant. Such restrictions can be avoided, when transformations are used, as discussed in sections 8, 9, and 10. The introduction of *breakpoints* in section 10, for example, yields a method of two-sided bounds (and also a method of pointwise norm bounds), which does not require any monotonicity property of the given differential operator, or its linearization.

Algorithms of the type considered here have been presented in [22, 23, 24], [6]. The (somewhat less simple) theory in the report [22] is concerned with more general two-point boundary value problems for $n \geq 1$, and also with more general types of estimates. In [6], problems with $n = 1$, a nonlinearity $F(x, U(x), U'(x))$, and nonlinear boundary conditions are treated, using breakpoints. This paper presents also a *programmed* EB-algorithm (including procedures **A** and **E**), which was developed in cooperation with M. Göhlen and M. Plum.

For problems with $n = 1$, a differential equation $-U'' + bU' + F(x, U) = 0$, and more general linear boundary conditions, M. Plum [19] uses an algorithm of the same structure (and with essentially the same procedures **A** and **E**). However, instead of applying the theory of (pointwise) differential inequalities, he applies inequalities in function spaces for avoiding the calculation of A^{-1}. These inequalities can, for example, be verified by calculating bounds of eigenvalues as described in [18].

Concerning other methods for obtaining verified results for two-points boundary value problems see [7,8,9,10,13,14,15,16,17], and the discussion in [23].

Notation: Let $\mathbf{R}^n[0,1]$ be the space of all functions $u : [0,1] \to \mathbf{R}^n$, and $S_0^n[0,1]$ the subspace of all $u \in \mathbf{R}^n$ which are piecewise continuous. More precisely, we require of $u = (u_i) \in S_0^n[0,1]$ that each component u_i is continuous on $[0,1]$, except at finitely many (jump) points $\xi \in (0,1)$, where $u_i(\xi) = u_i(\xi - 0)$ holds, $u_i(\xi + 0)$ exists, and $u_i(\xi - 0) \neq u_i(\xi + 0)$ is allowed. By $S_2^n[0,1]$ we denote the space of all continuously differentiable $u \in \mathbf{R}^n[0,1]$, which have a second derivative $u'' \in S_0^n[0,1]$. (At the jump points u_i'' is a one-sided derivative.)

Then define $R = S_2^n[0,1]$ and denote by S the space of all $u \in \mathbf{R}^n[0,1]$ such that $u(x) = v(x)$ for $0 < x < 1$ with some $v \in S_0^n[0,1]$. In particular, let R' and S' denote these spaces with $n = 1$. Moreover, define X to be the Banach space of all continuous $u \in \mathbf{R}^n[0,1]$, endowed with the norm $\|u\|_\infty$ of uniform convergence.

For $u, v \in \mathbf{R}^n[0,1]$ the terms $|u| \in \mathbf{R}^n[0,1]$, and $u \leq v$ are defined point- and componentwise:

$$|u|(x) = (|u_i(x)|); \quad u \leq v \quad \Leftrightarrow \quad u_i(x) \leq v_i(x) \quad \text{for } 0 \leq x \leq 1; i = 1, 2, \ldots, n.$$

The given problem can briefly be written as

$$U \in R \quad , \quad MU = 0 \tag{1.3}$$

with

$$\begin{array}{ll} MU(x) = -U''(x) + F(x, U(x)) \quad \text{for} \quad 0 < x < 1 & , \\ MU(0) = U(0) - r_0 \quad , \quad MU(1) = U(1) - r_1 & . \end{array}$$

2. The construction of T

Using procedure **A** we calculate an approximate solution $\omega \in R$ with *defect* d, briefly:

$$\omega \in R \quad , \quad M\omega \approx 0 \quad , \quad d := -M\omega \quad .$$

By Taylor expansion we obtain

$$MU = M(\omega + u) = -d + Au + Nu$$

for the *error function* $u = U - \omega \in R$, with a linear operator $A = M'(\omega) : R \to S$, and a "very small" nonlinear operator $N : X \to S$. Then the given problem is equivalent to

$$u \in R \quad , \quad Au + Nu = d \quad . \tag{2.1}$$

Here and in the succeeding section we *assume* that the operator A is invertible. (This property will be a consequence of condition (L) required in section 4.) Under this assumption we define T on X by

$$v = Tu \quad , \quad Av + Nu = d \quad .$$

Then the problems (1.2) and (2.1) are equivalent, and the contraction property is equivalent to:

$$\text{(C)} \qquad (u \in D \ , \ Av + Nu = d) \quad \Rightarrow \quad v \in D \quad .$$

In the present case, the operators A and N have the following form:

$$\begin{array}{llll} Au(x) = -u''(x) + c(x)u(x) & , \quad Nu(x) = f(x, u(x)) & \text{for} & 0 < x < 1 \\ Au(x) = u(x) & , \quad Nu(x) = 0 & \text{for} & x \in \{0, 1\} \end{array}$$

with the $n \times n$ matrix $c(x) = F_y(x, \omega(x))$ and a well defined f such that, uniformly in x,

$$\|f(x, y)\|_2 = O(t^2) \quad \text{for} \quad \|y\|_2 \leq t \quad , \quad t \to 0 \quad . \tag{2.2}$$

Remark: In general, the operator A^{-1} will not be known explicitly. One could use a simpler splitting $A + N$ instead of the one above, defining, for example, $Au = -u''$ on $(0, 1)$. Then A^{-1} could be calculated; but, in general Nu would not be "very small", and one had to expect difficulties in constructing D.

3. The choice of D

We will mainly consider sets D which can be written in the form

$$D = \{u \in X \ : \ \| u(x) \| \leq \psi(x) \ \text{ for } 0 \leq x \leq 1\} \quad .$$

In particular, the cases 1, 2, 3, 4 described as follows (with $y \in \mathbf{R}^n$) will be treated.

Case 1: $n = 1$, $\psi \in R'$, $\| y \| := |y|$.

Case 2: $n \geq 1$, $\psi \in R'$, $\| y \| := \|y\| = \langle y, y \rangle^{\frac{1}{2}}$ with an inner product $\langle \ , \ \rangle$.

Case 3: $n \geq 1$, $\psi \in R'$, $\| y \| := \|y\| = \|y\|_\infty$.

Case 4: $n \geq 1$, $\psi \in R$, $\| y \| := |y| = (|y_i|) \in \mathbf{R}^n$.

For brevity we write

$$\| u \| \leq \psi \ :\Leftrightarrow \ |u(x)| \leq \psi(x) \quad (0 \leq x \leq 1) \quad .$$

Then

$$D = \{u \in X \ : \ \| u \| \leq \psi\} \quad , \tag{3.1}$$

and the contraction property is equivalent to:

(C) $\qquad (u \in X \ , \ \| u \| \leq \psi \ , \ Av + Nu = d) \quad \Rightarrow \quad \| v \| \leq \psi \quad .$

We have to solve the nonlinear problem of finding a bound function ψ *such that* (C) *holds.*

Remark: One could also consider sets D of the form (3.1) with $\| u \|$ denoting any generalized norm of u, i.e, $\| \ \|$ denotes a function with norm properties and values in an ordered linear space. (In particular, $\| \ \|$ may be a norm in a function space, and $\psi \in \mathbf{R}$.)

For treating the above cases simultaneously we define

$$\begin{array}{llll} \widetilde{R} = R' \ , & \widetilde{S} = S' & \text{in the cases} & 1,2,3 \ ; \\ \widetilde{R} = R \ , & \widetilde{S} = S & \text{in the case} & 4 \ , \end{array}$$

so that, for example, $\psi \in \widetilde{R}$ in all these cases.

4. Constructing ψ by linearization

Using procedure **E** we calculate a $\delta \in \widetilde{S}$ such that $\delta \not\equiv 0$ and

$$\| d \| \leq \delta \quad . \tag{4.1}$$

In the case of a linear problem (1.1), with Nu vanishing, property (C) holds, if $\| Av \| \leq \delta \;\Rightarrow\; \| v \| \leq \psi$. In order to take into account non-vanishing ("very small") Nu, we require that the operator A *satisfies* the following *linear condition* (L) with known $\psi_0 \in \widetilde{R}$, $\delta_0 \in \widetilde{S}$, $\delta_0(x) < \delta(x)$ $(0 \leq x \leq 1)$:

(L) $$(w \in R\,,\ \| Aw \| \leq \delta - \delta_0) \quad \Rightarrow \quad \| w \| \leq \psi_0 \quad .$$

If (L) *holds, then A on R is invertible.* For, let $Aw = 0$, then $A(\lambda w) = 0$ and hence $\| \lambda w \| \leq \psi_0$ for each $\lambda > 0$, so that $w = 0$.

We define

$$\psi = (1+\varepsilon)\psi_0 \qquad \text{with a small } \varepsilon > 0 \text{ (say, } \varepsilon = 0.01) \quad ,$$

and *use procedure* **E** for calculating a $k \in \widetilde{S}$ such that

$$\| Nu \| \leq k \qquad \text{for all} \qquad u \in X \text{ with } \| u \| \leq \psi \quad . \tag{4.2}$$

Finally we check whether the following *Simple Final Condition* holds:

(SFC) $$\varepsilon\delta \geq k + (1+\varepsilon)\delta_0 \quad .$$

Result: *Let* (L) *and* (SFC) *hold. Then* (C) *is satisfied, and hence the given problem has a solution U^* such that $\| U^* - \omega \| \leq \psi$.*

Proof. The relations $Av = d - Nu$, $\| u \| \leq \psi$, together with (4.1), (4.2), and (SFC) imply that $\| Av \| \leq \delta + k \leq (1+\varepsilon)(\delta - \delta_0)$, so that $\| v \| \leq (1+\varepsilon)\psi_0 = \psi$, due to (L).

Condition (SFC), in general, will be satisfied, if procedures **A** and **E** of sufficient accuracy have been applied. (See, e.g., the treatment of the cases 1 through 4 in section 7, and also formula (4.3) below.)

Thus, the problem of verifying (C) for a suitable ψ essentially has been reduced to the problem of veryfying (L) for suitable δ_0 and ψ_0.

Remarks. α) Terms of the form $\delta - \delta_0$ with "very small" δ_0 will occur in a natural way when one of the methods in section 7 is applied. If only constant bounds are used, the above formulas can be simplified. For example, (L) may then be replaced by a condition of the form $\| Aw \| \leq 1 \Rightarrow \| w \| \leq \varphi \cdot 1$ (with a suitable "unit function" 1). In this case $\psi_0 = (\delta - \delta_0)\varphi$. In general, however, such a function ψ_0 need not belong to $\widetilde{R}$.

β) Instead of calculating k in (4.2) numerically with procedure **E** one can often use *majorizing functions* for N constructed by analytical means. To explain the use of such a majorizing function $\mathcal{G}$, we consider case 1 and assume, for simplicity, that δ, δ_0, and k are constant, $\delta_0 = \Delta\delta$. Then condition (L) is equivalent to

$$(w \in R \ , \ |Aw| \leq 1 - \Delta) \quad \Rightarrow \quad |w| \leq \varphi$$

with $\psi_0 = \delta\varphi$.

Now let $\mathcal{G} : [0, \infty) \to [0, \infty)$ denote a non-decreasing function which satisfies $\mathcal{G}(t) = 0(t^2)$ for $t \to 0$, and

$$|Nu| \leq \mathcal{G}(t) \quad \text{for} \quad u \in X, |u| \leq t$$

(or, equivalently, $|f(x, y)| \leq \mathcal{G}(|y|)$ for $0 \leq x \leq 1$, $y \in \mathbb{R}$).

Then one can use $k = \mathcal{G}((1 + \varepsilon)\delta\nu_0)$ with $|\varphi| \leq \nu_0 \in \mathbb{R}$, where ν_0 is *calculated with procedure* **E**, and (SFC) assumes the form

$$\varepsilon\delta \geq \mathcal{G}((1 + \varepsilon)\delta\nu_0) + (1 + \varepsilon)\Delta\delta \quad . \tag{4.3}$$

Instead of choosing ε a priori one may here calculate ε such that the latter inequality holds, for example, by carrying out a simple iteration step (see [6]).

5. On the procedures A and E

We will make a few remarks on procedures **A** and **E** of the type used in [6], where $n = 1$. These remarks are helpful in understanding some of the discussions in the sections below. Analogous procedures can also be developed for the case $n \geq 1$. Of course, we cannot explain here all details of the practical application. Also, the general EB-algorithms discussed in this paper do not require the use of procedures **A** and **E** of the particular type considered here. It is helpful, however, to have some concrete procedures **A** and **E** in mind, when discussing the general algorithm.

The procedure **A** used in [6] for solving linear boundary value problems is a collocation method with piecewise polynomial functions and "Chebyshev collocation points" in each subinterval where the functions are polynomial. For calculating the approximate solution ω of the given nonlinear problem, this method is combined with a Newton iterative procedure. That means, in the m-th step (for calculating u_m) the above collocation method is applied to a problem which is obtained by linearizing the given problem at u_{m-1}.

For starting this iterative procedure one needs an initial approximation u_0. In the examples presented in [6] we could use a homotopy method for calculating u_0. (All examples treated there contain a certain parameter λ.) For the general case we suggest to use, for example, a starting procedure of the following type.

Starting procedure: a) Calculate a discrete (or continuous) approximate solution $\tilde{\omega}$ of the given problem *with any method which is suitable for this particular problem* (e.g., a difference method, a shooting method, or a collocation method as described in [1]). Suppose that this method yields approximate values $\tilde{\omega}_k$ (of a solution) at certain points x_k.

b) Apply a collocation method with piecewise polynomial functions and the collocation points x_k to the problem which is obtained by linearizing the given problem at $\tilde{\omega}$. (No values $\tilde{\omega}(x)$ other than $\tilde{\omega}_k = \tilde{\omega}(x_k)$ are needed for this purpose.)

c) Use the approximate solution obtained in b) as an initial approximation u_0. (If $\tilde{\omega}$ is defined on $[0,1]$, one may use $u_0 = \tilde{\omega}$. Then b) need not be carried out.)

If the nonlinearity $F(x,y)$ in the given problem is polynomial in x and y (or piecewise polynomial), all functions to be estimated in the EB-algorithm of [6] are piecewise polynomial. For this case, procedure **E** in [6] is based on the following result of Ehlich and Zeller [4] (see also [20]).

If P_m is a polynomial of degree $\le m$, then

$$\max\{|P_m(x)| : x \in [-1,1]\} \le C \cdot \max\{|P_m(x_j)| : j = 0,1,\dots,N\} \tag{5.1}$$

with $N > m, \quad C = C(\frac{m}{N}) = [\cos \frac{\pi m}{2N}]^{-1}, \quad x_j = \cos \frac{j\pi}{N}$.

This result yields also

$$P_m(x) \ge \frac{1}{2}\{(C+1)p_\ell - (C-1)p_u\} \quad \textit{for} \quad -1 \le x \le 1 \tag{5.2}$$

with p_ℓ, p_u denoting any numbers such that

$$p_\ell \le P_m(x_j) \le p_u \quad \textit{for} \quad j = 0,1,2,\dots,N \quad .$$

These inequalities remain true, if $[-1,1]$ is replaced by any compact interval $[a,b]$, and $x_j = \frac{1}{2}(a+b) + \frac{1}{2}(b-a)\cos\frac{j\pi}{N}$.

If, for example, $g \in C[0,1]$ is piecewise polynomial, one can apply the above results for computing piecewise constant bounds for g.

The above inequalities can also be applied for calculating bounds of functions of several variables. If, for example, $P(x,y)$ is a polynomial of degree m_1 in x, and a polynomial of degree m_2 in y, then one can, for example, on a compact rectangle derive an inequality of the type

$$\max_{x,y}|P(x,y)| \leq C_1 C_2 \max_{j,k}|P(x_j,y_k)| \quad \text{with } C_1 = C(\frac{m_1}{N_1}), \quad C_2 = C(\frac{m_2}{N_2}) \quad . \tag{5.3}$$

We remark, furthermore, that inequalitites analogous to (5.1), (5.2) hold also for trigonometric polynomials (with equidistant x_k, see [4]). As a consequence one can also obtain bounds for functions of several variables which in each of their variables are either a polynomial, or a trigonometric polynomial. For example, if $P(x,y)$ is a polynomial in x, and a trigonometric polynomial in y, one can derive an inequality of the form (5.3).

Of course, one is also interested in obtaining bounds for functions which are not polynomials (or trigonometric polynomials), so that one can treat boundary value problems with non-polynomial nonlinearity $F(x,y)$.

As already remarked above, the use of the general EB-algorithms discussed here does not require to apply procedures **A** and **E** of the type considered above. On the other hand, U. Gärtel [5] has developed methods for obtaining estimates similar to (5.1), (5.2) for more general classes of functions (functions which are composed of certain elementary functions such as sin, cos, log, ... etc.). These methods, however, have not yet been systematically applied in this context.

6. Some results on differential inequalities

As a theoretical tool the notion of an inverse-positive differential operator, and related concepts will play an important role in our methods. (Such *operators of monotonic type* were already investigated and applied by Collatz [2,3].) For later use we will here formulate some corresponding results, which are of a very simple form.

Let $B : R \to S$ denote an operator given by

$$Bu(x) = \begin{cases} -u''(x) - P(x)u(x) & \text{for} \quad 0 < x < 1 \\ u(0) & \text{for} \quad x = 0 \\ u(1) & \text{for} \quad x = 1 \end{cases} \tag{6.1}$$

where $P(x)$ is an n×n-matrix such that

$$p_{ik}(x) \geq 0 \quad \text{for} \quad i \neq k; \quad i,k = 1,2,\ldots,n; \quad 0 < x < 1 \quad .$$

Theorem I. *If a function $z \in R$ exists such that*

$$z \geq 0, \quad \textit{and} \quad (Bz(x))_i > 0 \quad \textit{for} \quad 0 \leq x \leq 1; \quad i = 1,2,\ldots,n, \tag{6.2}$$

then B is inverse-positive on R, i.e., for each $u \in R$, $Bu \geq 0 \Rightarrow u \geq 0$.

This statement is an easy consequence of the abstract Monotonicity Theorem in [21] (Thm. 1.2 on p.31, see also p. 64). For later reference, however, we will briefly explain the idea of the proof. First we prove

Lemma I'. *The inequalities* (6.2) *imply that*

$$z_i(x) > 0 \quad for \quad 0 \leq x \leq 1; \quad i = 1, 2, \ldots, n \quad . \tag{6.3}$$

To see this suppose that $z_j(\xi) = 0$ for some $\xi \in (0,1)$ and some index j. Then $z_j'(\xi) = 0$ and $z_j'' \geq 0$. These relations at ξ, and $z \geq 0$ yield $(Bz(\xi))_j \leq 0$, which contradicts (6.1).

Now let $Bu \geq 0$, but $u \not\geq 0$, for some $u \in R$. Because of (6.3), there is a smallest $\lambda > 0$ such that $v := u + \lambda z \geq 0$. The inequalities (6.2) hold also for v, instead of z. Consequently, also (6.3) holds for v, instead of z. This contradicts the property of λ to be the smallest value with $u + \lambda z \geq 0$.

□

Remark. If it is known a priori that the function u considered satisfies $u_j(\xi) > 0$ for some j and some ξ, then these values j and ξ need not be considered in the proof. Consequently, the inequality $(Bu(\xi))_j \geq 0$ contained in $Bu \geq 0$ is not needed for proving $u \geq 0$, and thus the differentiability properties required of u_j at ξ are not needed either.

Now consider an operator $A : R \to S$ of the form described in section 2, and let $P(x)$ in (6.1) be a matrix such that

$$c_{ii}(x) = -p_{ii}(x) \quad , \quad |c_{ik}(x)| \leq p_{ik}(x) \quad \text{for} \quad i \neq k \tag{6.4}$$

and $0 \leq x \leq 1$; $i, k = 1, 2, \ldots, n$.

Theorem II. *If a function* $z \in R$ *exists such that* (6.2) *holds, then for* $w, \Psi \in R$

$$|Aw| \leq B\Psi \quad \Rightarrow \quad |w| \leq \Psi \quad .$$

This theorem is a simple special case of results on two-sided bounds. (See [21], Section V.2, for example.) The proof uses arguments similar to those in the above proof of Theorem I.

7. Veryfying (L) in some special cases

Suppose that a bound δ has been calculated in one of the cases considered in section 3. We discuss possibilities for constructing ψ_0 and δ_0, such that the corresponding condition (L) holds. For simplicity we assume that $\delta(x) > 0$ for

$0 \le x \le 1$ in case 1, 2 and 3, and that in case 4 each component δ_i of δ satisfies such an inequality.

This assumption on δ can be weakened, but the proofs then become somewhat less simple. For example, in [6] defect bounds were considered which may vanish at the boundary points. As a consequence, we had to apply there a result on inverse-positivity which is somewhat less simple than Theorem I.

The methods described in this section will only work, if the function ψ_0 calculated satisfies $\psi_0 \ge 0$. If this inequality cannot be verified, one may carry out a transformation as described below (e.g., by introducing breakpoints).

Case 1: *Using procedure* **A** we calculate ψ_0 such that

$$\psi_0 \in R \quad , \quad A\psi_0 \approx \delta \quad .$$

Then, *applying procedure* **E**, we compute $\delta_0 \in S$ such that

$$|A\psi_0 - \delta| \le \delta_0 \quad , \tag{7.1}$$

and check whether $\psi_0 \ge 0$.

In general, the values of δ_0 will be "very small". Here, we need only assume that $\delta_0 \le \alpha\delta$ with some constant $\alpha < 1$.

Remark. *The inequality $\psi_0 \ge 0$ holds if and only if the operator A is inverse-positive on R.*

Proof. Inequality (7.1) yields $A\psi_0(x) > 0$ for $0 \le x \le 1$. By applying Theorem I to $B = A$ and $z = \psi_0$ we see that A is inverse-positive, if $\psi_0 \ge 0$. On the other hand, if A is inverse-positive, then $A\psi_0 \ge 0$ implies $\psi_0 \ge 0$.

Result. *If $\psi_0 \ge 0$, then* (L) *holds.*

Proof. Suppose that (7.1) holds, $\psi_0 \ge 0$, and $|Aw| \le \delta - \delta_0$. Then $Aw \le \delta - \delta_0 \le A\psi_0$ and hence $w \le \psi_0$ because of the inverse-positivity of A. In an analogous way the inequality $-\psi_0 \le w$ is proved, so that $|w| \le \psi_0$.

Case 2. Here we first introduce a *majorizing operator* A_0 in the following way.

Assume that Λ is a function in $S_0[0,1]$ such that

$$\langle \eta, c(x)\eta \rangle \ge \Lambda(x) \quad \text{for all } \eta \in \mathbb{R}^n \text{ with } \|\eta\| = 1 \quad , \tag{7.2}$$

and all $x \in [0,1]$. In other words, $\Lambda(x)$ is assumed to be a lower bound of all eigenvalues of the symmetric part $c_S = \frac{1}{2}(c + c^*)$ of c, where $c^*(x)$ denotes the adjoint matrix of $c(x)$ with respect to the given inner product $\langle\ ,\ \rangle$. (Observe that $\langle \eta, c(x)\eta \rangle = \langle \eta, c_S(x)\eta \rangle$.)

Then define $A_0\varphi$ for $\varphi \in R'$ by

$$\begin{aligned} A_0\varphi(x) &= -\varphi''(x) + \Lambda(x)\varphi(x) \quad \text{for } 0 < x < 1 \quad , \\ A_0\varphi(x) &= \varphi(0) \quad , \quad A_0\varphi(1) = \varphi(1) \end{aligned} \tag{7.3}$$

Using procedure **A** we calculate ψ_0 such that

$$\psi_0 \in R' \quad , \quad A_0\psi_0 \approx \delta \quad .$$

Then, *applying procedure* **E**, we compute $\delta_0 \in S'$ such that

$$|A_0\psi_0 - \delta| \leq \delta_0 \tag{7.4}$$

and check whether $\psi_0 \geq 0$. (Again we assume that $\delta_0 \leq \alpha\delta$ with $\alpha < 1$.)

Remark. *The inequality $\psi_0 \geq 0$ holds if and only if the operator A_0 is inverse-positive.*

Result. *If $\psi_0 \geq 0$, then* (L) *holds.*

Proof. Lemma I' applied to $B = A_0$, $z = \psi_0$ yields $\psi_0(x) > 0$ for $0 \leq x \leq 1$. If (7.4) and $\|Aw\| \leq \delta - \delta_0$ hold, then

$$\psi_0(0) \geq \|w(0)\| \quad , \quad \psi_0(1) \geq \|w(1)\| \quad , \tag{7.5}$$

and

$$\begin{aligned} &-\psi_0''(x) + \langle \eta, c(x)\eta \rangle \psi_0(x) \geq \|Aw(x)\| \\ &\text{for all } x \in (0,1) \text{ and all } \eta \in \mathbf{R}^n \text{ with } \|\eta\| = 1 \;. \end{aligned} \tag{7.6}$$

In order to verify, for example, the latter inequality, observe that for $\psi_0 \geq 0$

$$-\psi_0''(x) + \langle \eta, c(x)\eta \rangle \psi_0(x) \geq A_0\psi_0(x) \geq (\delta - \delta_0)(x) \geq \|Aw(x)\| \quad .$$

According to a theory on pointwise norm inequalities, the relations (7.5), (7.6) together with $\psi_0 \geq 0$ imply that $\|w\| \leq \psi_0$. (Apply, for example, Theorem V.4.1 in [21] to $\psi_\lambda = (1+\lambda)\psi_0$, and $f(x,y,p) = c(x)y$, observing $\psi_0(x) > 0$ $(0 \leq x \leq 1)$.) □

The result can also be proved by applying Theorem I to A_0, and observing the remark following this theorem.

One writes $w(x)$ in the form $w(x) = \rho(x)\eta(x)$ with $\rho(x) = \|w(x)\|$, $\|\eta(x)\| = 1$. Then for all $x \in (0,1)$ with $\rho(x) > 0$:

$$\begin{aligned} A_0\rho(x) &\leq (-\rho'' + \langle \eta', \eta' \rangle \rho + \langle \eta, c\eta \rangle \rho)(x) \\ &= \langle \eta, Aw \rangle (x) \leq \|Aw(x)\| \leq A_0\psi_0(x) \quad , \end{aligned}$$

and $A_0\rho(x) = \rho(x) = \|w(x)\| \leq \psi_0(x) = A_0\psi_0(x)$ for $x \in \{0,1\}$. Thus $A_0(\psi_0 - \rho)(x) \geq 0$ for all $x \in [0,1]$, except possibly for those $x \in (0,1)$ for which $\rho(x) = 0$. For such values x, however, the inequality $(\psi_0 - \rho)(x) \geq 0$ is true, a priori. Therefore, $\rho \leq \psi_0$ follows. (For more details of this approach see [24].)

There are several possibilities for calculating Λ in (7.2) numerically. We will describe here two ways in which a piecewise constant bound can be obtained in the case of a piecewise polynomial $c(x)$. For explaining the ideas involved it suffices to consider a polynomial $c(x)$ and constant bounds Λ.

Thus we assume, for simplicity, that all elements $c_{ik}(x)$ of $c(x)$ are polynomials (of degree $\leq m$) and that we want to calculate a constant Λ satisfying (7.2).

1) For each fixed η with $\|\eta\| = 1$ formula (5.2) (with $N > n$, $C = C(\frac{m}{N})$) yields

$$\langle \eta, c(x)\eta\rangle \geq \frac{1}{2}\{(C+1)p(\eta) - (C-1)q(\eta)\} =: \Lambda(\eta) \quad \text{for} \quad 0 \leq x \leq 1 \quad ,$$

where $p(\eta)$ denotes the minimum and $q(\eta)$ the maximum of all values $\langle \eta, c(x_j)\eta\rangle$ ($j = 0, 1, \ldots, N$) with $x_j = \frac{1}{2}(1 + \cos \frac{j\pi}{N})$. Thus, (7.2) is satisfied for $\Lambda = \Lambda^0 := \min\{\Lambda(\eta) : \|\eta\| = 1\}$.

By further estimates we see that (7.2) holds also with

$$\Lambda = \tfrac{1}{2}\{(C+1)\lambda_1 - (C-1)\lambda_n\} \quad ,$$
$$\lambda_1 = \min\{\lambda_1(x_j) : j = 0, 1 \ldots, N\} \quad , \quad \lambda_n = \max\{\lambda_n(x_j) : j = 0, 1, \ldots, N\} \quad ,$$

where $\lambda_1(x_j)$ is the smallest and $\lambda_n(x_j)$ is the largest eigenvalue of the matrix $c_S(x_j)$.

Observe, for example, that

$$\min_\eta p(\eta) = \min_\eta \min_j \langle \eta, c(x_j)\eta\rangle = \min_j \min_\eta \langle \eta, c(x_j)\eta\rangle \quad ,$$

and

$$\min_\eta \langle \eta, c(x_j)\eta\rangle = \min_\eta \langle \eta, c_s(x_j)\eta\rangle = \lambda_1(x_j)$$

(where, of course, only values η with $\|\eta\| = 1$ are to be considered).

2) Each $\eta \in \mathbf{R}^n$ with $\|\eta\| = 1$ can be described as a trigonometric function of parameters $\alpha_1, \ldots, \alpha_{n-1}$. Thus, for polynomial $c(x)$, the function $F(x, \alpha_1, \ldots, \alpha_{n-1}) = \langle \eta, c(x)\eta\rangle$ is a polynomial in x, and a trigonometric polynomial in each of the variables α_i. A lower bound of such a function can be obtained with the tools explained in section 5. (See, for example, formula (5.3), and the remarks following this formula.)

Case 3. Again, we introduce a majorizing operator A_0 on R' of the form (7.3). Here, however, $\Lambda \in S_0[0,1]$ is assumed to have the following property:

$$c_{ii}(x) - \sum_{\substack{k=1 \\ k \neq i}}^{n} |c_{ik}(x)| \geq \Lambda(x) \quad \text{for} \quad 0 \leq x \leq 1; \quad i = 1, 2, \ldots, n \quad . \tag{7.7}$$

Then, ψ_0 and δ_0 are calculated as in case 2. Moreover:

The remark and the result formulated in case 2 hold also in case 3.

Proof of the result. If $\|Aw\| \leq \delta - \delta_0$, then for each index i

$$|w_i(0)| \leq \psi_0(0) \quad , \quad |w_i(1)| \leq \psi_0(1) \quad ,$$
$$|(Aw)_i(x)| \leq (\delta - \delta_0)(x) \leq A_0\psi_0(x)$$
$$\leq (-\psi_0''(x) + (c_{ii}(x) - \sum_{k \neq i} |c_{ik}(x)|)\psi_0(x) \quad \text{for} \quad 0 < x < 1 \quad .$$

Now let $\Psi = (\Psi_i) \in R$ with $\Psi_i = \psi_0$ $(i = 1, 2, \ldots, n)$, and define B as in (6.1) with $p_{ii} = -c_{ii}$, and $p_{ik} = |c_{ik}|$ for $i \neq k$ (see (6.4)). Then the above inequalities yield $|Aw| \leq B\Psi$. Thus, since $z = \Psi$ satisfies (6.2), it follows from Theorem II that $|w| \leq \Psi$, i.e., $\|w\| \leq \psi_0$.

□

As in case 2, we describe a possibility of calculating a piecewise constant $\Lambda(x)$ for piecewise polynomial $c(x)$. For simplicity, we assume again that all elements $c_{ik}(x)$ of $c(x)$ are polynomials (of degree $\leq m$) and that we want to calculate a constant Λ.

For constant Λ, (7.7) is equivalent to $\Lambda_i \geq \Lambda$ $(i = 1, 2, \ldots, n)$ and

$$\sum_{k=1}^{n} c_{ik}(x)\eta_k \geq \Lambda_i \quad \text{for} \quad 0 \leq x \leq 1 \quad , \quad \eta \in P_i \quad ;$$

$$P_i = \{\eta = (\eta_k) \in \mathbb{R}^n \ : \ \eta_i = 1, |\eta_k| = 1 \ (k = 1, 2, \ldots, n)\} \quad .$$

For each fixed $\eta \in P_i$ formula (5.2) (with $N > m$, $C = C(\frac{m}{N})$) yields

$$\sum_{k=1}^{n} c_{ik}(x)\eta_k \geq \Lambda_i(\eta) \quad \text{with} \quad \Lambda_i(\eta) = \tfrac{1}{2}\{(C+1)p_i(\eta) - (C-1)q_i(\eta)\} \quad ,$$

where $p_i(\eta)$ denotes the minimum and $q_i(\eta)$ the maximum of all values

$$\sum_{k=1}^{n} c_{ik}(x_j)\eta_k \quad (j = 0, 1, 2, \ldots, N) \quad \text{with} \quad x_j = \tfrac{1}{2}(1 + \cos \tfrac{j\pi}{N}) \quad . \tag{7.8}$$

Consequently, *inequality* (7.7) *holds with*

$$\Lambda = \min\{\Lambda_i : i = 1, 2, \ldots, n\} \quad , \quad \Lambda_i = \min\{\Lambda_i(\eta) : \eta \in P_i\} \quad .$$

In general the computation of the values $c_{ik}(x_j)$ (for $c(x) = F_y(x, \omega(x))$ is the most time consuming part of the calculation of this values Λ. Calculating the various linear combinations in (7.8) will require less work. By further estimates, however, we can obtain a value Λ which can be computed without calculating such linear combinations.

Inequality (7.7) *holds also with*

$$\begin{aligned} &\Lambda = \tfrac{1}{2}\{(C+1)p - (C-1)q\} \quad , \\ &p = \min\{c_{ii}(x_j) - \sum_{k \neq i} |c_{ik}|(x_j) : j = 0, 1, \ldots, N; i = 1, 2, \ldots, n\} \quad , \\ &q = \max\{c_{ii}(x_j) + \sum_{k \neq i} |c_{ik}|(x_j) : j = 0, 1, \ldots, N; i = 1, 2, \ldots, n\} \quad . \end{aligned}$$

Case 4. Here, too, we need a majorizing operator A_0. Define $A_0 = B$ on R with B in (6.1), where $P(x)$ is a matrix such that (6.4) holds. Then ψ_0 and δ_0 are calculated using the same formulas as in case 2, except that now $\psi_0 \in R$, $\delta_0 \in S$. Moreover:

The remark and the result formulated in case 2 *hold also in case* 4.

The proof of the result is similar to that in case 3, which is a special case of case 4.

The practical calculation of ψ_0 with procedure **A** could cause difficulties, if one chooses $p_{ik} = |c_{ik}|$ for $i \neq k$, since these functions p_{ik} may not have the right smoothness properties. There are various possibilities to construct suitable functions p_{ik}, such as piecewise smooth p_{ik} which have discontinuities only at given knots. This subject, however, shall not be discussed here.

8. Transformation, general remarks

Suppose that, in one of the cases 1 through 4 considered, we cannot verify (L), or find that too difficult. Then one may try to transform problem (2.1) into a problem of an analogous form

$$u \in \mathcal{R} \quad , \quad \mathcal{A}u + \mathcal{N}u = \vartheta \quad , \tag{8.1}$$

such that the methods explained above are applicable to this transformed problem, and, in particular, condition (L) holds for the transformed operator $\mathcal{A}$.

We will mainly consider transformations which are obtained by applying a linear operator G to the equation in (2.1). The idea is to choose G in such a way that $\mathcal{A}u$ has a certain prescribed form (e.g., $\mathcal{A}u = u$), while $\mathcal{N}u$ is "very small" for "small" u.

For $u \in R$, the transformed quantities will have the form

$$\mathcal{A}u = GAu - \mathcal{L}u \quad , \quad \mathcal{N}u = GNu + \mathcal{L}u \quad , \quad \vartheta = Gd \tag{8.2}$$

with a "very small" linear operator $\mathcal{L}$. This operator has to be introduced since, in general, we will not be able to construct G in such a way that GA itself has the form required of $\mathcal{A}$.

For example, to achieve $\mathcal{A}u = u$ one will choose G to be an approximation of A^{-1} (assuming A is invertible). In this case, $\mathcal{N}u = GNu + \mathcal{L}u$ with $\mathcal{L}u = GAu - u$, for $u \in R$.

In order to obtain a problem of the form (8.1) on an appropriate space $\mathcal{R} \supset R$, the transformed operators will be extended to $\mathcal{R}$ in a suitable way (e.g., by applying partial integration to $\mathcal{L}u$). Finally, the equivalence of the problems (2.1) and (8.1) has to be verified.

In all cases considered we may then proceed by applying the ideas described in sections 2, 3, and 4 to the transformed problem (8.1) instead of (2.1). That means we have to replace there $A : R \to S$, $N : X \to S$, d,R', S', $\widetilde{R}$, $\widetilde{S}$ by $\mathcal{A} : \mathcal{R} \to \mathcal{S}$, $\mathcal{N} : X \to \mathcal{S}$, ϑ, $\mathcal{R}'$, $\mathcal{S}'$, $\widetilde{\mathcal{R}}$, $\widetilde{\mathcal{S}}$, respectively. Here, the spaces $\mathcal{R}$ and $\mathcal{S}$ will be defined, separately, for each type of transformation. Then $\mathcal{R}'$ denotes the space of all $\varphi \in [0,1]$ such that $\varphi e \in \mathcal{R}$ (with $e = (e_i) \in \mathbf{R}^n$, $e_i = 1$ for all i). Moreover, $\widetilde{\mathcal{R}} = \mathcal{R}'$ in cases 1, 2, and 3; $\widetilde{\mathcal{R}} = \mathcal{R}$ in case 4. For given $\mathcal{S}$, the spaces $\mathcal{S}'$ and $\widetilde{\mathcal{S}}$ are defined analogously.

In particular, condition (L) assumes the form

$$\text{(L)} \qquad (w \in \mathcal{R} \ , \ \|\mathcal{A}w\| \le \delta - \delta_0) \quad \Rightarrow \quad \|w\| \le \psi_0 \quad .$$

If $\mathcal{A}u = u$, for example, this condition holds with $\delta_0 \equiv 0$, if $\delta \le \psi_0$. More generally, if $\mathcal{A}u = u - \mathcal{B}u$ with some operator $\mathcal{B}$ such that $\|\mathcal{B}w\| \le \Delta \|w\|$ for $w \in \mathcal{R}$ (with "small" $\Delta \in S'$), then (L) holds with $\delta_0 = \Delta\delta$, if $\delta \le \psi_0$.

The transformations discussed will only work, if the non-transformed operator A is invertible. This will be the case if ω is a sufficiently good approximation of the solution U^* (whose existence shall be verified), and if the linearization of M in (1.3) at U^* is invertible. (Compare corresponding discussions in [6], section 7.3.)

Remark. Under special assumptions one can use the transformation in order to verify (L) for the non-transformed operator A in (2.1), by exploiting the first relation in (8.2). Suppose, for example, that $n = 1$, δ is constant, $\mathcal{A}u = u$, and that $|Gu| \le \gamma\|u\|_\infty$, $|\mathcal{L}u| \le \Delta\|u\|_\infty$ for $u \in C_0[0,1]$ with constants γ, Δ. Then condition (L) for A holds with $\delta_0 = \Delta\delta$, and $\psi_0 = \gamma\delta$. (Here, the terms $\mathcal{N}$, $\vartheta, \mathcal{R}$ are not needed.)

9. Transformation into an integral equation

For simplicity, we consider here only the case $d(0) = d(1) = 0$, i.e., we assume that ω satisfies the given boundary conditions exactly. Let $d(x) = \bar{d}(x)$ $(0 < x < 1)$ with $\bar{d} \in S_0^n[0,1]$

We want to achieve a transformed problem of the form (8.1) with $\mathcal{R} = \mathcal{S} = X$, and $\mathcal{A}u = u$, or more generally, $\mathcal{A}u(x) = \sigma(x)u(x)$ $(0 \le x \le 1)$ with suitable non-singular matrices $\sigma(x)$. For this purpose we apply an integral operator of the form

$$Gv(x) = \int_0^1 g(x,s)v(s)\,ds$$

with

$$g(x,s) = \begin{cases} \Phi(x)\beta^T(s) & \text{for} \quad 0 \le x \le s \le 1 \\ \Psi(x)\alpha^T(s) & \text{for} \quad 0 \le s < x \le 1 \end{cases} ,$$

where $\Phi(x)$, $\Psi(x)$, $\alpha(x)$, $\beta(x)$ are n×n-matrices, such that

$$\Phi(0) = \Psi(1) = \alpha(0) = \beta(1) = 0 \quad , \tag{9.1}$$

and all components of these matrices belong to $S_2[0,1]$.

Partial integration applied to $\int_0^1 g(x,s)L[u](s)\,ds$ with $L[u] = -u'' + cu$, and $u \in R$ with $u(0) = u(1) = 0$ yields for $0 \le x \le 1$

$$(\sigma_{01}u - \sigma_{00}u')(x) - \int_0^1 g(x,s)L[u](s)\,ds + \mathcal{L}u(x) = 0 \tag{9.2}$$

with

$$\begin{gathered} \sigma_{00} = \Psi\alpha^T - \Phi\beta^T \quad , \quad \sigma_{01} = \Psi(\alpha')^T - \Phi(\beta')^T \quad , \\ \mathcal{L}u(x) = \Psi(x)\int_0^x (L^*[\alpha](s))^T u(s)\,ds + \Phi(x)\int_x^1 (L^*[\beta](s))^T u(s)\,ds \end{gathered} \tag{9.3}$$

where $L^*[u] = -u'' + c^T u$.

In order to obtain a "very small" operator $\mathcal{L}$, one will calculate the quantities α and β such that they are approximate solutions of $L^*[u] = 0$, i.e., $L^*[\alpha] \approx 0$, $L^*[\beta] \approx 0$.

In the following we assume that

$$\sigma_{00}(x) = 0 \quad , \qquad \text{and } \sigma_{01}(x) \text{ is nonsingular} \quad , \quad \text{for } 0 \le x \le 1 \quad .$$

Then we consider problem (8.1) with

$$\mathcal{R} = \mathcal{S} = X \quad , \quad \mathcal{A}u(x) = \sigma(x)u(x) \quad , \quad \sigma := \sigma_{01} \quad ,$$
$$\mathcal{N}u(x) = \int_0^1 g(x,s)f[u](s)\,ds + \mathcal{L}u(x) \quad , \quad f[u](x) = f(x,u(x)) \quad ,$$
$$\vartheta(x) = \int_0^1 g(x,s)\bar{d}(s)\,ds \quad , \quad \text{and } \mathcal{L} \text{ in (9.3)} \quad .$$

This is an integral equation.

Equivalence statement. *Problem* (8.1), *as defined above, and problem* (2.1) *are equivalent, if, for each* $r \in S_0^n[0,1]$

$$\int_0^1 g(x,s)r(s)\,ds = 0 \quad (0 \le x \le 1) \quad \Rightarrow \quad r(x) = 0 \quad (0 \le x \le 1) \quad . \tag{9.4}$$

Proof. Due to identity (9.2), each solution u of problem (2.1) is also a solution of problem (8.1).

Now let u be a solution of problem (8.1). Then $u(0) = u(1) = 0$, since (9.1) holds and the matrices $\sigma(0)$ and $\sigma(1)$ are non-singular. If $u \in R$, one obtains

$$\int_0^1 g(x,s)r(s)\,ds = 0 \quad (0 \le x \le 1) \quad \text{for } r = L[u] + f[u] - \bar{d}$$

by subtracting the equation $(\mathcal{A}u + \mathcal{N}u - \vartheta)(x) = 0$ from (9.2). Thus, because of (9.4), $u \in R$ satisfies also the differential equation of problem (2.1).

That each solution $u \in \mathcal{R}$ of problem (8.1) belongs also to R can be seen by formal calculations.

□

For obtaining existence statements and bounds, one may now apply the methods of sections 2,3 and 4, to problem (8.1), as described in section 8.

We will explain this application in some more detail for

Case 1: $n = 1$, $\psi \in C_0[0,1]$. In this case, where $L = L^*$, the EB-algorithm may be carried out as briefly described in the following. For simplicity, we assume here that all bounds obtained with procedure **E** are constant on $[0,1]$. (In a similar way, for example, piecewise constant bounds may be used.)

Applying procedure **A** one calculates α, β such that

$$\alpha \in R \quad , \quad L[\alpha] \approx 0 \quad , \quad \alpha(0) = 0 \quad , \quad \alpha(1) = 1 \quad ,$$
$$\beta \in R \quad , \quad L[\beta] \approx 0 \quad , \quad \beta(0) = 1 \quad , \quad \beta(1) = 0 \quad .$$

(The boundary conditions $\alpha(1) = 1$, $\beta(0) = 1$ may here be replaced by other requirements.) Then one checks whether

$$W(x) \neq 0 \quad \text{for} \quad 0 \leq x \leq 1 \quad , \quad \text{with} \quad W = \alpha\beta' - \alpha'\beta \quad . \tag{9.5}$$

For this purpose, one observes that $W' = L[\alpha]\beta - L[\beta]\alpha$ usually is "small", so that W is "nearly constant", and hence $W(x) \neq 0$ need, in general, only be checked for a few points x. For example, if

$$|\alpha| \leq \rho_0 \quad , \quad |\beta| \leq \rho_0 \quad , \quad |L[\alpha]| \leq \delta_L \quad , \quad |L[\beta]| \leq \delta_L \quad \text{on} \quad [0,1] \tag{9.6}$$

with constants ρ_0, δ_L, then $|W(x)| \geq |W(\frac{1}{2})| - \rho_0\delta_L$ for $0 \leq x \leq 1$. Such constants may be *calculated with procedure* **E** .

If (9.5) holds, one may, for example, choose $\Phi = -W_0^{-1}\alpha$, $\Psi = -W_0^{-1}\beta$ with a constant W_0 which is a good approximation of the "nearly constant" function $W(x)$ (say, $W_0 = W(\frac{1}{2})$). Then $\sigma_{00} = 0$, and $\sigma_{01} = W_0^{-1}W$ on $[0,1]$.

For veryfying the equivalence condition (9.4) one may differentiate the integral equation in the premise of (9.4) twice. Then one obtains for $0 \leq x \leq 1$

$$L[\beta](x) \int_0^x \alpha(x)r(s)\, ds + L[\alpha] \int_x^1 \beta(s)r(s)\, ds = W(x)r(x) \quad . \tag{9.7}$$

From this relation one can, in general, easily derive that $r \equiv 0$, since $L[\alpha]$ and $L[\beta]$ are "small" and W is bounded away from 0. For example, $r \equiv 0$ follows, if $|W(\frac{1}{2})| > 2\rho_0\delta_L$.

Condition (L) holds here with $\delta_0 = \Delta\delta$ if $|W_0^{-1}W| \geq 1 - \Delta$, and $\delta \leq \psi_0$, where the bound δ of $|\,\vartheta|$ can be obtained by applying procedure **E**. For example, $|\,\vartheta| \leq \rho_1\rho_0\bar{\delta}$ with $|\Phi| \leq \rho_1, |\Psi| \leq \rho_1, |d| \leq \bar{\delta}$, and ρ_0 as in (9.6). In a similar way, the bound k can be calculated.

Now let us briefly return to the general case $n \geq 1$. Here, it is not as easy to calculate Φ and Ψ such that $\sigma_{00} = \Psi\alpha^T - \Phi\beta^T \equiv 0$ (or even $\sigma_{00} \equiv 0$, $\sigma_{01} \equiv I$). Of course, one need not know Φ and Ψ explicitly, but only suitable bounds for these quantities. In general, this requires to know bounds of $\Phi(x)$ and $\Psi(x)$ at (finitely many, but) *a rather large set of point* x. Here, we will not discuss corresponding possibilities in detail, but only make two remarks.

Notice first that in applying the theory of the following section 10 only a few quantities Φ_j and Ψ_j have to be calculated, which have properties analogous to those of $\Phi(\xi_j)$ and $\Psi(\xi_j)$ for certain points ξ_j.

Another possibility is to modify the above approach, and, at the same time, extend its range of applicability.

Differentiating the equation in (9.2) yields

$$(\sigma_{11}u - \sigma_{10}u')(x) + \sigma_{00}(x)L[u](x) - \int_0^1 g_x(x,s)L[u](x)\,ds + \mathcal{L}_x u(x) = 0 \quad (9.8)$$

where

$$\sigma_{10} = \Psi'\alpha^T - \Phi'\beta^T \quad , \quad \sigma_{11} = \Psi'(\alpha')^T - \Phi'(\beta')^T \quad ,$$

and $\mathcal{L}_x u(x)$ is obtained, when in formula (9.3) (for $\mathcal{L}u(x)$) the terms $\Phi(x)$, $\Psi(x)$ are replaced by $\Phi'(x)$, $\Psi'(x)$.

By inserting $L[u] = \bar{d} - f[u]$ in both equations, (9.2) and (9.8), one obtains an integral equation for the vector $\begin{pmatrix} u \\ u' \end{pmatrix}$ with values in $\mathbb{R}^{2n}$. This system has a form analogous to (8.1) where, for example,

$$\mathcal{A}\begin{pmatrix} u \\ u' \end{pmatrix} = \begin{pmatrix} \sigma_{01} & -\sigma_{00} \\ \sigma_{11} & -\sigma_{10} \end{pmatrix}\begin{pmatrix} u \\ u' \end{pmatrix} \quad .$$

Now one may determine Φ, Ψ such that $\sigma_{01} \approx I$, $\sigma_{00} \approx 0$, $\sigma_{11} \approx 0$, $\sigma_{10} \approx -I$, by using collocation (interpolation). That means, one requires, for example, that $\sigma_{00}(x_k) = 0$ for suitable x_k. (Observe that $\sigma_{01} \equiv -\sigma_{10}$, and $\sigma_{11} \equiv 0$, if $\sigma_{00} \equiv 0$, $L^*[\alpha] \equiv 0$, and $L^*[\beta] \equiv 0$.)

We remark, without going into details, that *this approach can also be applied if the given differential equation has the form*

$$-U''(x) + F(x, U(x), U'(x)) = 0 \quad (0 < x < 1) \quad .$$

10. Transformation by using breakpoints

In each of the cases 1 through 4 discussed in section 7, we had to require that the operator A (in case 1) or some majorant A_0 (in cases 2, 3, 4) is inverse-positive. As the theory of inverse-positive differential operators shows, the inverse-positivity of such an operator depends in a certain sense on the length of the interval at whose ends the boundary conditions are given (see [21], p.70). This fact can be used to produce suitable transformed problems by introducing additional "boundary conditions" at certain points ξ_j $(j = 1, 2, \ldots, m)$ with $0 =: \xi_0 < \xi_1 < \ldots < \xi_m < \xi_{m+1} := 1$. The transformed operator $\mathcal{A}$ obtained in this way will be inverse-positive, in case 1, and it will have an inverse-positive majorant $\mathcal{A}_0$ in the cases 2, 3, and 4.

We will here only describe the construction of such a transformed problem without further discussing the various cases. The methods are essentially the same as described in section 7. The programmed algorithm in [6] is based on such a transformation.

To obtain the transformed problem, one replaces the conditions $(Au + Nu)(\xi_j) = d(\xi_j)$ $(j = 1, 2, \ldots, m)$ contained in (2.1) by integral equations at these points ξ_j, while for all other $x \in [0, 1]$ the equation in (2.1) remains unchanged. In particular, one obtains

$$\mathcal{A}u(\xi_j) = u(\xi_j) \quad , \quad \text{and} \quad \mathcal{A}u(x) = Au(x) \quad \text{for} \quad x \neq \xi_j \quad .$$

In some more detail, one proceeds as follows. Using procedure **A** one calculates matrix-valued functions $g_j : [0, 1] \to \mathbf{R}^{n,n}$ $(j = 1, 2, \ldots, m)$ such that

$$\begin{aligned} &L^*[g_j] \approx 0 \quad , \quad g_j(0) = g_j(1) = 0 \quad , \\ &g_j(\xi_j - 0) = g_j(\xi_j + 0) \quad , \quad g_j'(\xi_j - 0) - g_j'(\xi_j + 0) = I \quad . \end{aligned}$$

(Except this discontinuity of g_j' at ξ_j, the functions g_j shall be sufficiently smooth.) In the case $d(0) = d(1) = 0$, the transformed problem at $x = \xi_j$ is then given by

$$\begin{aligned} &u(\xi_j) + \int_0^1 g_j(s) f[u](s)\, ds + \mathcal{L}_j u(x) = \vartheta_j \quad , \\ &\mathcal{L}_j u(x) = \int_0^1 L^*[g_j](s) u(s)\, ds \quad , \; \vartheta_j = \int_0^1 g_j(s) \bar{d}(s) \quad . \end{aligned}$$

The quantities g_j may be calculated directly, or one may first calculate α, β as in section 9 and then determine $\Phi_j, \Psi_j \in \mathbf{R}^{n,n}$ such that

$$g_j(s) = \begin{cases} \Phi_j \beta^T(s) & \text{for} \quad \xi_j \leq s \\ \Psi_j \alpha^T(s) & \text{for} \quad \xi_j > s \end{cases}$$

has the desired properties, i.e.,

$$\Psi_j \alpha^T(\xi_j) - \Phi_j \beta^T(\xi_j) = 0 \quad , \quad \Psi_j(\alpha'(\xi_j))^T - \Phi_j(\beta'(\xi_j))^T = I \quad .$$

The transformed problem is posed on the space $\mathcal{R} = S^n_{0,2}[\xi_0, \xi_1, \xi_2, \ldots, \xi_m, \xi_{m+1}]$ which consists of all continuous $u : [0, 1] \to \mathbf{R}^n$ with the following property: The restriction of each component u_i to any of the subintervals $[\xi_j, \xi_{j+1}]$ has a continuous first derivative, and a piecewise continuous second derivative. This means, for example, that the bound functions ψ considered may have breaks (jumps in the first derivative) at the points $\xi_j \in (0, 1)$. Therefore, we speak of *breakpoints* ξ_j and *breakpoint functions* g_j. Such breakpoints have first been used by Küpper [11,12].

With the algorithm in [6] we calculated examples of type (1.1) with $F(x, U) = -\mu U + U^3 - \lambda x(1 - x)$ for $\mu = 25, 65, 125$: $F(x, U) = -65U + U^3 - \lambda x$, and also a problem with a nonlinearity $F(x, U, U') = -\lambda U[1 - \frac{1}{2}(U')^2 - \frac{1}{8}(U')^4]$. There, we had no essential difficulties in properly choosing the breakpoints.

In general, however, one will want to incorporate an algorithm which *produces* suitable breakpoints (or indicates that no breakpoint is needed). For constructing such an algorithm, one may exploit one of the various properties of inverse-positive differential operators (see [21]). Such methods are discussed in [6]. A (formally) very simple method consists in starting with no breakpoint, and adding more and more breakpoints, checking in each step whether the corresponding function ψ_0 satisfies $\psi_0 \geq 0$.

References

1. Ascher, U., Christiansen, J., Russell, R.D.: A Collocation Solver for Mixed Order Systems of Boundary Value Problems. Math. Comp. 33, 659-679 (1979).
2. Collatz, L.: Aufgaben monotoner Art. Arch. Math. 3, 366-376 (1952).
3. Collatz, L.: The Numerical Treatment of Differential Equations. Springer-Verlag 1960.
4. Ehlich, H., Zeller, K.: Schwankung von Polynomen zwischen Gitterpunkten. Math. Z. 86, 41-44 (1964).
5. Gärtel, U.: Fehlerabschätzungen für vektorwertige Randwertaufgaben zweiter Ordnung, insbesondere für Probleme aus der chemischen Reaktions-Diffusions-Theorie. Dissertation Köln 1987.
6. Göhlen, M., Plum, M., Schröder, J.: A programmed algorithm for existence proofs for two-point boundary value problems. Computing 44, 16-57 (1990).
7. Großmann, Ch., Krätzschmar, M., Roos, H.-G.: Gleichmäßig einschließende Diskretisierungsverfahren für schwach nichtlineare Randwertaufgaben. Numer. Math. 49, 95-110 (1986).
8. Großmann, Ch., Al-Zanaidi, M.: Monotone Iteration Discretization Algorithm for BVP's. Computing 41, 59-74 (1989).
9. Kaucher, E.W., Miranker, W.L.: Self-Validating Numerics for Function Space Problems. Academic Press, New York 1984.
10. Kedem, G.: A Posteriori Error Bounds for Two-Point Boundary Value Problems. SIAM J. Numer. Anal. 18, 431-448 (1981).
11. Küpper, T.: Einschließungsaussagen für gewöhnliche Differentialoperatoren. Numer. Math. 25, 201-214 (1976).
12. Küpper, T.: Einschließungsaussagen bei Differentialoperatoren zweiter Ordnung durch punktweise Ungleichungen. Numer. Math. 30, 93-101 (1978).

13. Lohner, R., Adams, E.: Einschließung der Lösung gewöhnlicher Anfangs- und Randwertaufgaben. ZAMM 64, T295-T297 (1984).

14. Lohner, R.: Einschließung der Lösung gewöhnlicher Anfangs- und Randwertaufgaben und Anwendungen. Dissertation Universität Karlsruhe 1988.

15. McCarthy, M.A., Tapia, R.A.: Computable a posteriori L_∞-Error Bounds for the Approximate Solution of Two-Point Boundary Value Problems. SIAM J. Numer. Anal. 12, 919-937 (1975).

16. Nakao, M.T.: A numerical verification method for the existence of solutions for nonlinear boundary value problems. This volume.

17. Nickel, K.: The Construction of a priori Bounds for the Solution of a Two-Point Boundary Value Problem with Finite Elements I. Computing 23, 247-265 (1979).

18. Plum, M.: Eigenvalue inclusions for second-order ordinary differential operators by a numerical homotopy method. ZAMP 41 (1990).

19. Plum, M.: Verified Existence and Inclusion Results for two-point Boundary Value Problems. This volume.

20. Ruttmann, B.: Untersuchungen zur Fehlerabschätzung von polynomialen Näherungslösungen bei Randwertaufgaben gewöhnlicher Differentialgleichungen. Dissertation Universität Köln 1982.

21. Schröder, J.: Operator Inequalities. Academic Press 1980.

22. Schröder, J.: Existence proofs for boundary value problems by numerical algorithms. Report Univ. Cologne 1986.

23. Schröder, J.: A method for producing verified results for two-point boundary value problems. Computing, Suppl. 6, 9-22 (1988).

24. Schröder, J.: Operator Inequalities and Applications. To appear in "Inequalities: Fifty years on from Hardy, Littlewood and Polya", Conference Birmingham 1987.

Aspects of Self-Validating Numerics in Banach Spaces

E. Kaucher
Institut für Angewandte Mathematik
Universität Karlsruhe
Federal Republic of Germany

C. Schulz-Rinne
Department of Mathematics
University of Maryland
USA

Abstract: Self-validating methods for two typical classes of partial differential equations are discussed.
In the first part implicit formulas for solutions to systems of first order quasilinear PDEs with initial and boundary conditions are developed. The goal is to find expressions suitable for highly parallel computation, in connection with E-verification. The achievable degree of parallelism depends essentially on the type of the PDEs. The systems of PDEs are transformed into a system of algebraic and/or ordinary differential equations which are expanded in the space coordinates. Hence, the original problem may be solved numerically or functionally in parallel for different space coordinates. Furthermore, there exist powerful E-Methods for algebraic systems as well as for functoid systems and systems of ODEs. Therefore, the E-verified numerical solution of the PDEs treated in this paper is possible.
In the second part, we discuss some iterative functoid processes for the validated computation of parabolic PDEs with initial and boundary conditions. Theoretical background are the W-principles referred to in Walter [25] which together with interval functoids deliver so-called exclusion sets for solutions.

Computer Arithmetic and Self-Validating Numerical Methods

ISBN 0-12-708245-X

1. Introduction

The authors decided to restrict the representation of this paper on a few but essential aspects of self-validating numerics in Banach spaces such as Fredholm Integral Equations, hyperbolic and parabolic PDE. Since the first author has published a paper "Self-validating computations of linear and nonlinear Fredholm integral equations of the second kind" together with coauthor Dobner in a second proceedings volume, we will refer to that volume and restrict ourselves here to partial differential equations.

We suppose the reader is familiar with the ideas of Function Space Arithmetics ([11] to [16]). The main idea is to compute functions on a computer similarly to numbers. The data are function elements represented generally by polynomial series, Fourier series, spline functions and the like. The aim is to formulate function space problems in terms of fixed point equations

$$x = f(x) \tag{1.1}$$

and to <u>validate</u> numerically the existence and inclusion by use of Banach or Schauder fixed point theorems. Therefore, a set-valued arithmetical environment has to be defined and implemented on a computer which then leads to so-called Interval Functoids.
Interval Functoids are represented e.g. by expansions as noted above but with coefficients in an interval space, for example $I\mathbb{R}^n$, $I\mathbb{C}^n$ or by a representation of interval functions F(x) by means of lower and upper functions on a domain D

$$F(x) = [u(x), v(x)] \tag{1.2}$$

where $f \in F$ then means $f(x) \in F(x)$, i.e. $u(x) \leq f(x) \leq v(x)$ for all $x \in D$, [15]. Such numerical validating methods are called E-Methods or validation methods. E-Methods compute an approximate solution, provide tight error bounds and simultaneously validate the inclusion or exclusion of the solution of the problem as defined by the input data. I n c l u s i o n means the computation of a very small set X and an automatic verification of the existence of a solution $x \in X$. E x c l u s i o n means the computation of a set X for which it is automatically verified that

"if a solution x of the problem exists, then $x \in X$"
or in other words
"in the complement of X a solution never exists".

If possible, they also confirm the local uniqueness of the solution. For partial differential equations properties proving the existence of a solution are often missing.

This paper consists of two parts. The first part deals with the solution to systems of first order nonlinear partial differential equations.
We consider the following class of systems of PDEs:

$$u_t + \sum_{i=1}^{n} \varphi_i(t,x,u)u_{x_i} = \alpha(t,x,u) \qquad (1.3)$$

for a vector x and vector-valued functions u and α.
In the case where u is a real-valued function, we also give some hints how to treat the general equation

$$h(t,x,u,u_t,u_x) = 0, \qquad (1.4)$$

where x may be a vector. Besides some general implicit formulas for solutions we also discuss the problems (1.3) and (1.4) in connection with the initial condition

$$u(0,x) = f(x) \quad , \quad x \in \mathbb{R}^n, \quad t \geq 0 \qquad (1.5)$$

for a given function f. In a final section we suggest how to treat boundary conditions.

Most numerical solution methods for these classes lead to finite element methods, boundary element methods, discretization methods or methods of characteristics. In general, however, the first three methods have the disadvantage that they are not easily parallelizable. In addition, in most cases it is impossible to provide error bounds.

In this paper we list several transformations of the classes of problems mentioned above into a system of algebraic equations and/or a system of ordinary differential equations or integral equations. These transformations are closely related to the method of characteristics, but we propose a new and more direct approach which seems to be more appropriate for these classes of problems.

Since there exist powerful E-Methods for algebraic systems as well as for systems of ordinary differential equations ([1], [2], [3], [6], [18] to [22], [24]), it is possible to validate the results numerically.

A further main goal is to find expressions and E-Methods suitable for highly parallel computation. In advance, the achievable degree of parallelism depends essentially on the type of the PDEs and on the type of the processor of the supercomputer. In chapter 5, we give a brief schedule for parallelly validated computations. A more detailed discussion will be published in a separate paper. In fact, there are types of problems where it is possible to achieve perfect parallelism. Thus, for example 3-D transport equations of Euler-type

$$\begin{aligned} u_t + \varphi(u)u_x &= A(u) \\ \rho_t + (\varphi(u)\rho)_x &= 0 \end{aligned} \tag{1.6}$$

and with some boundary conditions can be solved by validation.

The second part gives a rather general approach for a validated computation of a general class of parabolic PDE systems with initial and boundary conditions. On the background of the book by Walter [25] which developed some monotonic principles (here denoted as W-principles) based on the Nagumo Lemma we formulate some theorems and algorithms which give us a tool for validated computations of exclusion sets in functoids. We thereby extend ideas in [7] in so far, as we open the applicability of general functoids and roundings to those methods.

A. Aspects of Parallel and Validated Computations of a Class of Hyperbolic Systems

Throughout this part of the paper we define $I := \{1,...,n\}$, $K := \{1,...,N\}$ where $n, N \in \mathbb{N}$. $\varphi, \psi, \alpha, f, F$ represent continuously differentiable functions. Let $T \in \mathbb{R}$, $T > 0$. Then we define the set $D \in [0,T] \times \mathbb{R}^n$, where D is connected and $D \cap \{0\} \times \mathbb{R}^n$ is nonempty and compact. $C^1(D)$ denotes the set of all contiuously differentiable functions on D. Defining $x := (x_1,...,x_n)^T$, (t,x) denotes an element of $\mathbb{R} \times \mathbb{R}^n$.

Generally, let us assume that differentiation implies the existence of all the derivatives, and their continuity, if necessary. Furthermore, we only consider a solution $u(t,x)$ in a domain D, where some implicit natural conditions are satisfied. This will be explained in detail in Remark 2.1.

2. One Autonomous Nonlinear PDE

In this chapter first order quasilinear partial differential equations are considered whose coefficient functions $\varphi_i(u)$, $i \in I$ and the right-hand side $\alpha(u)$ are given. A real-valued function $u(t,x)$ solving the partial differential equation

$$u_t + \sum_{i=1}^{n} \varphi_i(u) \cdot u_{x_i} = \alpha(u) \quad , \quad (t,x) \in D \tag{2.1}$$

is desired, where t represents the time coordinate and $x = (x_1, \ldots, x_n)$ the space coordinates. Furthermore, let

$$u_t := \frac{\delta u}{\delta t} \text{ and } u_{x_i} := \frac{\delta u}{\delta x_i}, \quad i \in I \tag{2.2}$$

denote the partial derivatives of the function u.
Besides the solution u(t,x)of the equation (2.1), the corresponding initial value problem (IVP) is considered , where at t = 0

$$u(0,x) = f(x) \qquad \text{for all } x \in \mathbb{R}^n \tag{2.3}$$

is given.

The cases $\alpha(u) = 0$ and $\alpha(u) \neq 0$ are best treated separately. They are considered in the following two sections.

In this chapter we use the following notations:

$$\varphi(u) := (\varphi_1(u), \ldots, \varphi_n(u))^T$$

$$L(v) := v_t + \sum_{i=1}^{n} \varphi_i(u) \cdot v_{x_i} \quad \text{for } v(t,x) : \mathbb{R}^{n+1} \to \mathbb{R} \tag{2.4a}$$

$$F'(v,w) := F_v + F_w \cdot w_v \quad \text{for } F(v,w) : \mathbb{R} \times \mathbb{R}^n \to \mathbb{R} \tag{2.4b}$$

L is a linear differential operator of the form

$$L = \frac{\delta}{\delta t} + \sum_{i=1}^{n} \varphi_i(u) \cdot \frac{\delta}{\delta x_i}, \text{ i.e. L depends on the solution u.}$$

Now the equation (2.1) reads $L(u) = \alpha(u)$.

Analoguous to the notation introduced in (2.2), the partial

derivatives of a function are denoted by subscripts indicating the variable of differentiation.

Note that for the implicit solution u(x) of the equation g(u,x) = 0, u is an element of $C^1(X)$, $X \subseteq \mathbb{R}^n$ if $g \in C^1(\mathbb{R}^n \times X)$ and $g_u \neq 0$ in $\mathbb{R}^n \times X$ for $g: \mathbb{R}^n \times X \to \mathbb{R}^n$.

2.1 Homogeneous Differential Equation

In this section we consider the case $\alpha(u) = 0$ in the equation (2.1); then the differential equation has the form

$$u_t + \sum_{i=1}^{n} \varphi_i(u) \cdot u_{x_i} = 0. \tag{2.5}$$

Let us define the functions

$$\begin{aligned} w_i &:= x_i - t \cdot \varphi_i(u) \quad \text{for } i \in I, \\ w &:= (w_1, \ldots, w_n)^T \end{aligned} \tag{2.6}$$

and for short $w = x - t \cdot \varphi(u)$, $w: D \times \mathbb{R} \to \mathbb{R}^n$.

We now provide a transformation of (2.5) into a nonlinear algebraic system of equations defining solutions of (2.5).

Theorem 2.1:

Let the homogeneous differential equation (2.5) be given with a continuously differentiable function $\varphi(u)$. Let $F(u,w): \mathbb{R} \times \mathbb{R}^n \longrightarrow \mathbb{R}$ be a continuously differentiable function satisfying $F'(u,w) \neq 0$.
If the function $u(t,x) \in C^1(D)$ is a solution of

$$F(u,w) = F(u,x_1 - t \cdot \varphi_1(u),\ldots, x_n - t \cdot \varphi_n(u)) = 0 \qquad (2.7)$$

then u is also a solution of (2.5).

<u>Proof:</u>

Let u(t,x) be a solution of (2.7). Then we partially differentiate (2.7) with respect to t and $x_1,\ldots,x_n$:

$$\frac{\delta F}{\delta t} := F_u \cdot u_t - \sum_{i=1}^{n} F_{w_i} \cdot (\varphi_i(u) + t \cdot \varphi_i'(u) \cdot u_t) = 0 \qquad (2.8)$$

and

$$\frac{\delta F}{\delta x_j} := F_u \cdot u_{x_j} - \sum_{i=1}^{n} F_{w_i} \cdot t \cdot \varphi_i'(u) \cdot u_{x_j} + F_{w_j} = 0 \qquad (2.9)$$

for $j \in I$

The sum $\frac{\delta F}{\delta t} + \sum_{j=1}^{n} \varphi_j(u) \cdot \frac{\delta F}{\delta x_j}$ gives us

$$\begin{aligned} &F_u \cdot (u_t + \sum_{j=1}^{n} \varphi_j(u) \cdot u_{x_j}) - \sum_{i=1}^{n} F_{w_i} \cdot \varphi_i(u) \\ &- \sum_{i=1}^{n} F_{w_i} \cdot t \cdot \varphi_i'(u) \cdot (u_t + \sum_{j=1}^{n} \varphi_j(u) \cdot u_{x_j}) \\ &+ \sum_{j=1}^{n} \varphi_j(u) \cdot F_{w_j} = 0. \end{aligned}$$

Now with the notations (2.4) we finally have

$$F'(u,w) \cdot L(u) = 0. \qquad (2.10)$$

Since $F' \neq 0$, the equation (2.10) is equivalent to $L(u) = 0$. Hence u(t,x) is a solution of (2.5). □

Now we consider the initial value problem (IVP)

$$u_t + \sum_{i=1}^{n} \varphi_i(u) \cdot u_{x_i} = 0, \quad u(0,x) = f(x) \text{ for } (t,x) \in D \tag{2.11}$$

for an arbitrary once continuously differentiable function $f: \mathbb{R}^n \to \mathbb{R}$.

Theorem 2.2:
Let the IVP (2.11) with $f(w): \mathbb{R}^n \to \mathbb{R}$ be given.
If the function $u(t,x): D \to \mathbb{R}$ is a solution of

$$u - f(w) = 0, \text{ i.e. } u - f(x_1 - t \cdot \varphi_1(u), \ldots, x_n - t \cdot \varphi_n(u)) = 0 \tag{2.12}$$

then u is also a solution of (2.11).

Proof:
We define $F(u,w) := u - f(x - t \cdot \varphi(u)) = u - f(w)$.
Then by Theorem 2.1 a solution of (2.12) is also a solution of the differential equation in (2.11). For $t = 0$ the equation (2.12) simplifies to $u(0,x) - f(x) = 0$. Hence $u(t,x)$ solves the problem (2.11). □

Remark 2.1:
In the proof we have to assume that

$$F'(u,w) = 1 + \sum_{i=1}^{n} f_{w_i} \cdot t \cdot \varphi_i'(u) \neq 0, \text{ for } u \in U \text{ and } w \in W.$$

We take this as a natural definition of the domain D, where all assumptions are satisfied.

Let $X \subseteq \mathbb{R}^n$ and $[0,T] \subseteq \mathbb{R}$ such that

$$\{x - t\,\varphi(u) \mid x \in X,\ t \in [0,T],\ u \in U\} \subseteq W \tag{2.13}$$

then it is possible to choose $D = [0,T] \times X$.
D may be a bounded or even an unbounded set. Parts of the boundary

of D coincide with shock fronts of the solutions u. A more detailed discussion of this aspect and a computational method for shock fronts will be published in a separate paper. Throughout this paper, we imply the restrictions (2.13).

2.2 Inhomogeneous Differential Equation

In this section we consider the equation (2.1) with $\alpha(u) \neq 0$ and $\lambda(u) := 1/\alpha(u)$. Then the differential equation has the following form:

$$u_t + \sum_{i=1}^{n} \varphi_i(u) \cdot u_{x_i} = \frac{1}{\lambda(u)} \tag{2.14}$$

We define

$$\psi_i := \lambda(u) \cdot \varphi_i(u) \quad \text{for } i \in I,$$

$$\psi := (\psi_1, \dots, \psi_n)^T \text{ and for short } \psi = \lambda(u) \cdot \varphi(u)$$

and $w = w(x,u,R)$ with $w := x - \int_R^u \lambda(\tau) \cdot \varphi(\tau) d\tau = x - \int_R^u \psi(\tau) d\tau$ (2.15).

Theorem 2.3:

Let the inhomogeneous differential equation (2.14) be given with $\lambda(u) \neq 0$. Let $F(R,w)$: $\mathbb{R} \times \mathbb{R}^n \to \mathbb{R}$ be an arbitrary continuously differentiable function with $F(R,w) = 0$ and $F'(R,w) \neq 0$.

If functions $u(t,x)$, $R(t,x)$: $\mathbb{R}^{n+1} \to \mathbb{R}$ solve simultaneously the following nonlinear algebraic system of equations

$$t = \int_R^u \lambda(\tau)\, d\tau, \tag{2.16}$$

$$F(R,w) = F(R, x - \int_R^u \psi(\tau) d\tau) = 0, \tag{2.17}$$

then $u(t,x)$ is a solution of (2.14) and R is a solution of the adjoint homogeneous equation $L(R) = 0$ (see 2.4a).

Proof:

Let $u(t,x)$ and $R(t,x)$ be solutions of (2.16) and (2.17). Partially differentiation of (2.16) with respect to R yields

$$0 = \lambda(u) \cdot \frac{du}{dR} - \lambda(R). \tag{2.18}$$

Doing the same with (2.15) gives us

$$w_R = \varphi(R) \cdot L(R) - \varphi(u) \cdot \lambda(u) \cdot \frac{du}{dR}$$

and eliminating $\frac{du}{dR}$ from (2.18)

$$w_R = \lambda(R) \cdot (\varphi(R) - \varphi(u)). \tag{2.19}$$

Inserting the result (2.19) into (2.4b) we find

$$F'(R,w) = F_R + \lambda(R) \cdot \sum_{i=1}^{n} F_{w_i} \cdot (\varphi_i(R) - \varphi_i(u)). \tag{2.20}$$

We tabulate the first partial derivatives of (2.16) and (2.17) with respect to t and $x_1, \ldots, x_n$:

$$1 = \lambda(u) \cdot u_t - \lambda(R) \cdot R_t, \tag{2.21a}$$

$$0 = \lambda(u) \cdot u_{x_j} - \lambda(R) \cdot R_{x_j} \quad \text{for } j \in I, \tag{2.21b}$$

$$\frac{\delta F}{\delta t} := F_R \cdot R_t + \sum_{i=1}^{n} F_{w_i} \cdot (\psi_i(R) \cdot R_t - \psi_i(u) \cdot u_t) = 0 \tag{2.22a}$$

$$\frac{\delta F}{\delta x_j} := F_R \cdot R_{x_j} + \sum_{i=1}^{n} F_{w_i} \cdot (\psi_i(R) \cdot R_{x_j} - \psi_i(u) \cdot u_{x_j}) + F_{w_j} = 0$$

for $j \in I$. (2.22b)

Now we multiply (2.21b) by $\varphi_j(u)$ and add (2.21a):

$$\lambda(u) \cdot (L(u) - 1/\lambda(u)) - \lambda(R) \cdot L(R) = 0 \qquad (2.23).$$

Analogously, we multiply (2.22b) by $\varphi_j(u)$ and add (2.22a):

$$(-\sum_{i=1}^{n} F_{w_i} \cdot \psi_i(u)) \cdot L(u) + \sum_{j=1}^{n} F_{w_j} \cdot \varphi_j(u)$$
$$+ (F_R + \sum_{i=1}^{n} F_{w_i} \cdot \psi_i(R)) \cdot L(R) = 0$$

Now using $\psi = \lambda\varphi$ and reordering gives us

$$[-\lambda(u) \cdot \sum_{i=1}^{n} F_{w_i} \cdot \varphi_i(u)] \cdot (L(u) - 1/\lambda(u))$$
$$+ [F_R + \lambda(R) \cdot \sum_{i=1}^{n} F_{w_i} \cdot \varphi_i(R)] \cdot L(R) = 0. \qquad (2.24)$$

The equations (2.23) and (2.24) may be regarded as a two dimensional homogeneous linear system of equations with the unknowns L(R) and $L(u) - 1/\lambda(u)$. Now let A denote the matrix of that system and compute the determinant of A using (2.20):

$$\begin{aligned} \det(A) = \ & \lambda(u) \cdot (F_R + \lambda(R) \cdot \sum_{i=1}^{n} F_{w_i} \cdot \varphi_i(R)) \\ & - \lambda(R) \cdot (\lambda(u) \cdot \sum_{i=1}^{n} F_{w_i} \cdot \varphi_i(u)) \\ = \ & \lambda(u) \cdot F'(R,w). \end{aligned}$$

But $\lambda(u) \neq 0$ and $F'(R,w) \neq 0$. Hence A is regular and there is only the trivial solution $L(R) = 0$ and $L(u) - 1/\lambda(u) = 0$ and thus u(t,x) solves the inhomogeneous equation (2.14) and R solves the homogeneous equation (2.5).

□

Now we consider again the IVP for $(t,x) \in D$

$$u_t + \sum_{i=1}^{n} \varphi_i(u) \cdot u_{x_i} = \frac{1}{\lambda(u)}, \quad u(0,x) = f(x). \tag{2.25}$$

We define $F(R,w) := R - f(x - \int_R^u \psi(\tau)d\tau)$. By Theorem 2.3 we then know that (2.16) and (2.17) provide a solution $u(t,x)$ of (2.14). Furthermore, at $t = 0$ (2.16) becomes $0 = \int_R^u \lambda(\tau)d\tau$ and $\lambda \neq 0$ implies $u(0,x) = R(0,x)$. On the other hand, equation (2.17) simplifies at $t = 0$ to $R(0,x) - f(x) = 0$. Summarizing both we have $u(0,x) = f(x)$. Thus, $u(t,x)$ solves the IVP (2.25).

3. Systems of First Order Autonomous Nonlinear PDEs

Based on the results of chapter 2 for the transformation of first order quasilinear partial differential equations, in this chapter systems of such equations are considered. Supposed we are given an N-dimensional system of equations in the (n+1)-dimensional (t,x)-space with the coefficient functions $\varphi_i(u_1, \dots, u_N)$, $i \in I$ and the functions of the righthand side of the equation $\alpha_k(u_1, \dots, u_N)$, $k \in K$. We are looking for N real-valued functions $u_k(t,x)$, $k \in K$, which solve the N partial differential equations

$$
\begin{aligned}
(u_1)_t + \sum_{i=1}^{n} \varphi_i(u_1,\ldots,u_N) \cdot (u_1)_{x_i} &= \alpha_1(u_1,\ldots,u_N) \\
&\;\;\vdots \\
(u_N)_t + \sum_{i=1}^{n} \varphi_i(u_1,\ldots,u_N) \cdot (u_N)_{x_i} &= \alpha_N(u_1,\ldots,u_N).
\end{aligned}
\tag{3.1}
$$

Apart from the solution of the system (3.1), the corresponding IVP is considered at $t = 0$:

$$
\begin{aligned}
u_1(0,x) &= f_1(x) \\
&\;\;\vdots \\
u_N(0,x) &= f_N(x)
\end{aligned}
\tag{3.2}
$$

In this chapter we use the following notations:

$$
\begin{aligned}
u &:= (u_1,\ldots,u_N)^T \quad : D \to \mathbb{R}^N \\
v &:= (v_1,\ldots v_N)^T \quad : D \to \mathbb{R}^N \\
\varphi(u) &:= (\varphi_1(u),\ldots,\varphi_n(u))^T \\
\alpha(u) &:= (\alpha_1(u),\ldots,\alpha_N(u))^T \\
f(x) &:= (f_1(x),\ldots,f_N(x))^T \\
L(z) &:= z_t + \sum_{i=1}^{n} \varphi_i(u) \cdot z_{x_i} \quad \text{for } z : D \to \mathbb{R}
\end{aligned}
\tag{3.3}
$$

and the corresponding vector version:

$$
L(v) := v_t + \sum_{i=1}^{n} \varphi_i(u) \cdot v_{x_i} \quad \text{for } v : D \to \mathbb{R}^N. \tag{3.4}
$$

$$
F(v,w) := (F_1(v,w),\ldots,F_N(v,w))^T : \mathbb{R}^N \times \mathbb{R}^n \to \mathbb{R}^N
$$

$$F'(v,w) := \begin{bmatrix} \frac{\delta F_1}{\delta v_1} & \frac{\delta F_1}{\delta v_N} \\ \cdot & \cdot \\ \cdot & \cdot \\ \cdot & \cdot \\ \frac{\delta F_N}{\delta v_1} & \frac{\delta F_N}{\delta v_N} \end{bmatrix} + \begin{bmatrix} \frac{\delta F_1}{\delta w_1} & \frac{\delta F_1}{\delta w_N} \\ \cdot & \cdot \\ \cdot & \cdot \\ \cdot & \cdot \\ \frac{\delta F_N}{\delta w_1} & \frac{\delta F_N}{\delta w_N} \end{bmatrix} \cdot \begin{bmatrix} \frac{\delta w_1}{\delta v_1} & \frac{\delta w_1}{\delta v_N} \\ \cdot & \cdot \\ \cdot & \cdot \\ \cdot & \cdot \\ \frac{\delta w_n}{\delta v_1} & \frac{\delta w_n}{\delta v_N} \end{bmatrix}$$

$$= F_v + F_w \cdot w_v \qquad (3.5)$$

Now the system (3.1) reads simply $L(u_k) = \alpha_k(u)$, $k \in K$ or

$$L(u) = \alpha(u). \qquad (3.6)$$

Furthermore, the remarks at the end of section 2.1 apply.

First, we consider the homogeneous system (3.1) with $\alpha(u) = 0$ so that the system for $u(t,x)$ has the form

$$u_t + \sum_{i=1}^{n} \varphi_i(u) \cdot u_{x_i} = 0. \qquad (3.7)$$

Theorem 3.1:

Let the homogeneous system of differential equations (3.7) be given. Let $F(u,w)$: $\mathbb{R}^{n+1} \to \mathbb{R}^N$ be an arbitrary continuously differentiable vector function with $F'(u,w)^{-1}$ existing in a neighbourhood and with $w = w(u)$ as given in (2.6).

If the function $u(t,x)$: $\mathbb{R}^{n+1} \to \mathbb{R}^N$ is a solution of

$$F(u,w) = (u,\ x - t \cdot \varphi(u)) = 0 \qquad (3.8)$$

then u is also a solution of (3.7).

Proof:

Let $u(t,x)$ be a solution of (3.8). Using the notation (3.5) we find

$$F'(u,w) = F_u - t \cdot F_w \cdot \varphi'(u). \quad (3.9)$$

Partial differentiation of (3.8) with respect to t and $x_1, \ldots x_n$ yields

$$\frac{\delta(F)}{\delta t} := F_u \cdot u_t - F_w \cdot (\varphi(u) + t \cdot \varphi'(u) \cdot u_t) = 0 \qquad (3.10)$$

and

$$\frac{\delta(F)}{\delta x_j} := F_u \cdot u_{x_j} + F_w \cdot (e_j - t \cdot \varphi'(u) \cdot u_{x_j}) = 0 \qquad (3.11)$$

for $j \in I$.

Here e_j denotes the j-th identity vector in $\mathbb{R}^n$.

Now we compute the sum $\frac{\delta(F)}{\delta t} + \sum_{j=1}^{n} \varphi_j(u) \cdot \frac{\delta(F)}{\delta x_j}$:

$$F_u \cdot (u_t + \sum_{j=1}^{n} \varphi_j(u) \cdot u_{x_j}) - F_w \cdot \varphi(u)$$
$$-t \cdot F_w \cdot \varphi'(u) \cdot (u_t + \sum_{j=1}^{n} \varphi_j(u) \cdot u_{x_j}) + F_w \cdot \underbrace{\sum_{j=1}^{n} \varphi_j(u) \cdot e_j}_{= \varphi(u)} = 0$$

Referring to (3.9) and the notation (3.4) this expression can be shortened to

$$F'(u,w) \cdot L(u) = 0. \qquad (3.12)$$

Due to the assumption on F, the equation (3.12) is equivalent to $L(u) = 0$. Thus u(t,x) satisfies the homogeneous system (3.7). □

We now consider the IVP for $(t,x) \in D$

$$u_t + \sum_{j=1}^{n} \varphi_j(u) \cdot u_{x_j} = 0, \qquad u(0,x) = f(x). \qquad (3.13)$$

We define

$$F(u,w) := u - f(x - t \cdot \varphi(u)). \tag{3.14}$$

By Theorem 3.1 we then know that (3.8) provides a solution $u(t,x)$ of (3.7). Furthermore, for $t = 0$ the system of equations (3.8) simplifies to $u(0,x) - f(x) = 0$. Thus $u(x,t)$ solves the IVP problem (3.13). Here, $F'(u,w) = I + t \cdot f_w \cdot \varphi'(u) \neq 0$ is assumed where I denotes the N-dimensional identity matrix.

In the case where the system (3.1) is considered to be inhomogeneous with $\alpha_1(u) \neq 0$ and $\alpha_k(u) = 0$, $k = 2,\dots,N$ we come to similar results as in 2.2. (This is no restriction, since by linear combination of the equation a general inhomogenity $(\alpha_1(u),\dots,\alpha_N(u))$ can be transformed in this natural form. For the general case see also chapter 4). With $\lambda(u) := 1/\alpha_1(u)$ the system for $u(t,x)$ has the following form:

$$u_t + \sum_{i=1}^{n} \varphi_i(u) \cdot u_{x_i} = \begin{bmatrix} 1/\lambda(u) \\ 0 \\ \cdot \\ \cdot \\ \cdot \\ 0 \end{bmatrix} \tag{3.15}$$

In the following Theorem we formulate without proof an algebraic system of equations which is related to the system (3.15).

We use the following notations:

$$\begin{aligned}
\psi &:= (\psi_1,\dots,\psi_n)^T \quad \text{with } \psi = \lambda(u) \cdot \varphi(u) \\
R &:= (R_1, u_2,\dots,u_N)^T : D \to \mathbb{R}^N \\
w_i &:= x_i - \int_{R_1}^{u_1} \psi_i(\tau, u_2,\dots,u_N)d\tau, \quad i \in I \qquad (3.16) \\
w &:= (w_1,\dots,w_n)^T \quad \text{with } w = x - \int_{R_1}^{u_1} \psi(\tau, u_2,\dots,u_N)d\tau
\end{aligned}$$

Theorem 3.2:

Let the inhomogeneous system (3.15) be given with $\lambda(u) \neq 0$. Let $F(R,w): \mathbb{R}^{n+N} \to \mathbb{R}^N$ be a continuously differentiable function. Furthermore, assume the matrix A (see (3.23)) to be invertible.
If the functions $u(t,x) : D \to \mathbb{R}^N$ and $R_1(t,x): D \to \mathbb{R}$ are solutions of the algebraic system

$$t = \int_{R_1}^{u_1} \lambda(\tau, u_2, \ldots, u_N)\, d\tau \quad \text{and} \tag{3.17}$$

$$F(R,w) = F(R_1, u_2, \ldots, u_N, x - \int_{R_1}^{u_1} \psi(\tau, u_2, \ldots, u_N)\, d\tau) = 0, \tag{3.18}$$

then $u(t,x)$ also solves (3.15).

Considering the IVP for $(t,x) \in D$

$$u_t + \sum_{i=1}^{n} \varphi_i(u) \cdot u_{x_i} = \begin{bmatrix} 1/\lambda(u) \\ 0 \\ \cdot \\ \cdot \\ \cdot \\ 0 \end{bmatrix}, \quad u(0,x) = f(x). \tag{3.19}$$

We define

$$F(R,w) := R - f(x - \int_{R_1}^{u_1} \psi(\tau, u_2, \ldots u_N)\, d\tau). \tag{3.20}$$

By Theorem 3.2 we then know that (3.17) and (3.18) provide a solution $u(t,x)$ of (3.15). At $t = 0$ (3.17) becomes
$0 = \int_{R_1}^{u_1} \lambda(\tau, u_2, \ldots u_N) d\tau$ for all $x \in \mathbb{R}^n$. Due to $\lambda \neq 0$ this implies $u_1(0,x) = R_1(0,x)$. Now the system of equations (3.18) reads $u(0,x) - f(x) = 0$. Thus, $u(t,x)$ solves the IVP (3.19).

4. Generalized Hyperbolic Nonlinear Systems

First, we consider the general IVP according to the notations (3.1) to (3.3):

$$u_t + \sum_{i=1}^{n} \varphi_i(t,x,u) \cdot u_{x_i} = \alpha(t,x,u), \qquad u(0,x) = f(x) \tag{4.1}$$

Using the notations

$$\begin{aligned} R(t,x) &:= (R_1(t,x),\dots, R_n(t,x))^T : D \to \mathbb{R}^n \\ T(t,x) &:= (T_1(t,x),\dots, T_n(t,x))^T : D \to \mathbb{R}^n \end{aligned} \tag{4.2}$$

we state a Theorem (without proof) that defines n+1 ordinary differential equations with initial conditions and n equations whose solution gives one of (4.1).

Theorem 4.3:
Let the inhomogeneous IVP (4.1) with $f(x)\colon \mathbb{R}^n \to \mathbb{R}$ be given.
If the functions $T(t,\cdot)\colon \mathbb{R}^{n+1} \to \mathbb{R}^n$, $S(t,\cdot)\colon \mathbb{R}^{n+1} \to \mathbb{R}$ and $R(t,x)\colon \mathbb{R}^{n+1} \to \mathbb{R}^n$ are solutions of

$$T_i(t,\cdot)_t = \varphi_i(t,T,S), \qquad T_i(0,R) = R_i, \quad i \in I \tag{4.3}$$

$$S(t,\cdot)_t = \alpha(t,T,S), \qquad S(0,R) = f(R), \tag{4.4}$$

$$x = T(t,R(t,x)) \tag{4.5}$$

and if

$$u(t,x) := S(t,R(t,x)) \tag{4.6}$$

then $u(t,x)$ solves the IVP (4.1) if the Jacobian $T_R(t,R)$ is regular.

A similar result holds in the same symbolic notation for systems of quasilinear inhomogeneous equations. Replace the scalar function S by a vector function S.

Even in the most general case where we have the totally nonlinear IVP

$$h(t,x,u,u_t,u_{x_1}\dots,u_{x_n}) = 0 \tag{4.7}$$

and the nonlinear initial condition

$$f(x,u(0,x)) = 0 \tag{4.8}$$

we can transform this problem to a system of type (4.1) with nonlinear initial conditions and thus can formulate a corresponding ODE-system solving (4.7), (4.8). More details will be published in a separate paper.

Further generalization go into the direction of Euler-type PDE. We then add to a homogeneous (or inhomogeneous) autonomous systems (3.1), (3.2) a conservation law

$$\rho_t + \sum_{i=1}^{n} [\varphi_i(u_1,\dots,u_N) \cdot \rho]_{x_i} = 0 \tag{4.9}$$

and an initial condition

$$\rho(0,x_1,\dots,x_n) = g(x_1,\dots,x_n). \tag{4.10}$$

From Theorem 3.1 we may easily derive that the system (3.12)

$$L(v) = 0 \tag{4.11}$$

is solved by each v satisfying

$$F(v,w_1,\dots,w_n) = 0 \tag{4.12}$$

for some F. Hence, also the function $v := \rho \cdot d$ satisfies the equation $L(\rho \cdot d) = 0$. But one can prove (for more details we refer to a paper appearing separately) that for $d := \det\left(\frac{\delta F(u,w)}{\delta u}\right)$

$$L(\rho d) = d \cdot L(\rho) + \rho L(d)$$

and

$$L(d) = d \cdot \sum_{i=1}^{n} [\varphi_i(u)]_{x_i}.$$

Thus, we have

$$\begin{aligned} L(\rho d) = \; & d \cdot \{L(\rho) + \rho \sum_{i=1}^{n} [\varphi_i(u)]_{x_i}\} = \\ & d \cdot \{\rho_t + \sum_{i=1}^{n} [\varphi_i(u) \cdot \rho]_{x_i}\}. \end{aligned} \tag{4.13}$$

and $L(\rho d) = 0$ implies for $d \neq 0$ that ρ satisfies (4.9). Hence applying (4.12) to

$$F(\rho \cdot d, w) := \rho \cdot d - g(x - t\varphi(u)) = 0 \tag{4.14}$$

we have found, together with the results of chapter 3, an implicit algebraic system for the solutions $\rho(t,x)$ and $u(t,x)$.

5. Remarks on Boundary Conditions and Parallel Computations

We have seen in chapters 2 to 4 that a rather large class of hyperbolic PDE systems can be treated by adjoint algebraic systems. In particular, further results like those in (4.9) to (4.14) show the wide-spread applicability in gas dynamics and inviscous fluid dynamics. There, boundary conditions play a very important rule. Let us consider a simple case (N = 1, u = 1):

$$u_t + \varphi(u)u_x = \alpha(u) \tag{5.1}$$

with the boundary condition

boundary	condition
$\Gamma_1 := \{(t,x) \mid t = 0,\ 0 \leq x \leq 1\}$	$u(0,x) = f(x)$
$\Gamma_2 := \{(t,x) \mid x = 0,\ 0 \leq t\}$	$u(t,0) = g(t)$
$\Gamma_3 := \{(t,x) \mid x = 1,\ 0 \leq t\}$	$u(t,1) = h(t)$

(5.2)

Then the following algorithm delivers an Ω^* and an exclusion set $Y(t,x) \subseteq M$ for $(t,x) \in \Omega^*$:

<u>Algorithm 5.1:</u>

(1) Define

for $0 < x - \int_{R_1}^{u} \frac{\varphi(\tau)}{\alpha(\tau)}\, d\tau < 1$

$$\phi_1(t,x,u,R_1) := \begin{bmatrix} t - \int_{R_1}^{u} \frac{\varphi(\tau)}{\alpha(\tau)}\, d\tau \\ R_1 - f(x - \int_{R_1}^{u} \frac{\varphi(\tau)}{\alpha(\tau)}\, d\tau) \end{bmatrix} \tag{5.3a}$$

for $t \geq \int_{R_2}^{u} \frac{\varphi(\tau)}{\alpha(\tau)} > 0$

$$\phi_2(t,x,u,R_2) := \begin{bmatrix} x - 1 - \int_{R_2}^{u} \frac{\varphi(\tau)}{\alpha(\tau)}\, d\tau \\ R_2 - g(t - \int_{R_2}^{u} \frac{\varphi(\tau)}{\alpha(\tau)}) \end{bmatrix} \tag{5.3b}$$

for $t \geq \int_{R_3}^{u} \frac{\varphi(\tau)}{\alpha(\tau)}$

$$\phi_3(t,x,u,R_3) := \begin{bmatrix} x - 1 - \int_{R_3}^{u} \frac{\varphi(\tau)}{\alpha(\tau)}\, d\tau \\ \\ R_3 - h(t - \int_{R_3}^{u} \frac{\varphi(\tau)}{\alpha(\tau)}) \end{bmatrix} . \tag{5.3c}$$

(2) Compute for i = 1,2,3 a connected region Ω_i and function sets Y_i, Q_i such that the algebraic system

$$\phi_i(t,x,u,R_i) = 0 \tag{5.4}$$

(i) has a E-verified unique solution pair $(u,R_i) \in (Y_i, Q_i)$ for all $(t,x) \in \Omega_i$,

(ii) such that due to (5.2) $\Gamma_i \subseteq \Omega_i$ holds

and $f(x) \in Y_1(0,x)$ on Γ_1

$g(t) \in Y_2(t,0)$ on Γ_2

$h(x) \in Y_3(t,1)$ on Γ_3

Then we have the following E-verification:

(1) If for an $i,j \in \{1,2,3\}$ there holds $\Omega_{ij} := \Omega_i \cap \Omega_i \neq \phi$ and $Y_i(t,x) \cap Y_j(t,x) = \phi$ for a $(t,x) \in \Omega_{ij}$ then on Ω_{ij} does not exist a solution of the given problem.

(2) Let $\hat{\Omega}$ be the union of all sets where (1) hold. Then the union

$$Y := \bigcup_{i=1}^{3} Y_i \text{ on } \bigcup_{i=1}^{3} \Omega_i \setminus \hat{\Omega} \tag{5.5}$$

is a verified exclusion set.

A natural Boundary for the Ω_i are discontinuities of the functions f, g, h or singularity points of the corresponding Jacobian $\frac{\delta F(u,w)}{\delta u}$ which define shock fronts of the problem. Thus, it is possible to E-validate computationally those shock fronts.

For computations note that the algorithm 5.1 is a system of nonlinear algebraic equations (or generally ODEs). Thus, any appropriate finite element decomposition of Ω locally defines equation systems of the form (5.3) which are independent of one another. Hence, the solutions of these local solutions can be computed with validation almost separately and independently.
Therefore, a parallel processor system with ρ nodes can compute such problems with a parallel effectivity of almost 100%, that is a speed up of performance of almost a factor ρ.

B. Validation Aspects for Parabolic Systems

In this part we consider parabolic systems. Validation means the computation of so-called exclusion sets X due to a given problem such that one of the following properties holds for X:

(i) If there exists a solution x of the given problem then x is contained in X.

(ii) In the complement of X a solution of the given problem never exists.

To control rounding errors and truncation errors we are looking for a set Y which contains X, but Y can be computed in finite steps, say for example in an Interval Functoid. Sets X or Y with property (i) or (ii) are called shortly exclusion sets.

6. <u>A First Approach to the Problem</u>

We consider the boundary value problem (BVP)

$$u_t = f(t,x,u,u_x,u_{xx}) \text{ in } G \tag{6.1}$$

$$u|_{\delta G} = \eta \text{ on the boundary,}$$

where $u = u(t,x) \in \mathbb{R}^m$, $\overline{G} = G + \delta G$ (δG the boundary of G), e.g. G may be cylindrical, then $G = (0,T) \times D$ with bounded $D \subset \mathbb{R}^n$.

Let f be defined on $(0,T) \times D \times \mathbb{R}^{n \cdot m} \times \mathbb{R}^{n \cdot m \cdot m}$.

We say $u \in \mathbf{M}$ if u, u_z, u_x and u_{xx} are continuous on G.

$$Pu := u_t - f(t,x,u,u_x,u_{xx}) \text{ is the defect of u.} \tag{6.2}$$

Furthermore, we refer here to Theorems in [25] which are shortly denoted as W-principle.

We now consider the iteration process

$$u_{i+1} := u_i - QPu_i \tag{6.3a}$$

$$v_{i+1} := v_i - QPv_i$$

with $Q > 0$ on G and $Q = 0$ on δG (6.3b)

<u>Theorem 6.1</u>

Let for an $i \in N$, $u_i, v_i \in Z$,

Pv_i and Pu_i be bounded on G, $u_{i+1}|_{\delta G} = \eta$, $v_{i+1}|_{\delta G} = \eta$,

$$u_{i+1} > u_i \text{ and } v_{i+1} < v_i \tag{6.4a}$$

then $\mathbf{W} := [u_i, v_i]$ is an exclusion set with

$$u_i|_{\delta G} = \eta \ , \ v_i|_{\delta G} = \eta \tag{6.4b}$$

<u>Proof:</u>

From $u_{i+1} > u_i$ we have $u_i < u_{i+1} = u_i - QPu_i$ and from (6.3b) we follow $Pu_i < 0$ on G. This and the boundness of Pu_i on G and $Q|_{\delta G} = 0$ implies $u_i|_{\delta G} = \eta$ and due to W-principle in [25] there exist no solutions u of (6.1) with $u < u_i$.

A similar result holds for v_i and hence (6.4) is valid. □

Remark 6.1:

Primarily, the same Theorem holds if the number Q is replaced by inverse monotonic matrices for systems or by general inverse monotonic operators. In order to construct explicitely a Q which satisfies (6.3b) and which allows local convergence of the sequences in (6.3) we might define

$$Q := \frac{v - u}{Pv - Pu} \tag{6.5}$$

(for systems the matrix components of Q may be the divided differences), where u and v are arbitrary functions with

$$u|_{\delta G} = v|_{\delta G} = \eta \quad \text{and} \quad u < v \text{ on } G \tag{6.6}$$

such that $Pu \neq Pv$.

7. Some Algorithmic Aspects

Now let (M,Ω) be an ordered Bananch space with operations $\Omega = \{x,+,-,\cdot,/,\frac{\delta}{\delta t}, \frac{\delta}{\delta x}, \int,\ldots\}$, $(SM, S\Omega\}$ be a functoid of (M,Ω), in particular $(S^{\wedge}M, S^{\wedge}\Omega)$ the upper and $(S_{\vee}M, S_{\vee}\Omega)$ the lower functoid such that for all $a,b \in SM$ and all $\circ \in \Omega$

$$a \underset{\vee}{\circ} b \leq a \circ b \leq a \hat{\circ} b \tag{7.1a}$$

holds. In particular, this means for example for the differential operator $\delta_t = \frac{\delta}{\delta t}$ the property $\delta_{\vee t}\, a \leq a_t$.

The directed functional roundings $S^{\wedge}|_{\eta}$ and $S_{\vee}|_{\eta}$ are roundings dependent on boundary conditions, i.e.

$S^{\wedge}|_{\eta}u \geq u$ and $S^{\wedge}|_{\eta}u\ |_{\delta G} \geq \eta$ (7.1b).

$S_{\vee}|_{\eta}u \geq u$ and $S_{\vee}|_{\eta}u\ |_{\delta G} \leq \eta$ (7.1c).

These are so-called roundings with constraints (cf. [15]). It is clear that S has to be a good approximation with respect to some measures, for example, $\|Su - u\|$ has to be as small as possible.

With (ISM, $\Diamond\Omega$) we denote the interval functoid with the property that for all A,B $\in$ ISM and for all $\circ \in \Omega$

$$A \circ B \subseteq A \Diamond B \qquad (7.2a)$$

and for all $a \in M$ $a \in ISa$ (7.2b)

In particular, if we write IS(P)u for a function $u \in M$, then due to (6.2) we mean for an $u \in SM$

$$IS(Pu) := IS(u_t) \ominus f(t,x,IS(u),IS(u_x),IS(u_{xx}),IS\Omega) \qquad (7.3)$$

We now try to avoid the very restrictive conditions (6.6) on u and v for Q in (6.5) by constructing an iteration process, such that Q can be rather arbitrary in principal.

Theorem and Algorithm 7:

Under the conditions of the problem (6.1) let two arbitrary starting values u_0, $v_0 \in SM$ be given and define

$$A_0 := S^{\wedge}P(u_0)$$
$$B_0 := S_{\vee}P(v_0) \qquad (7.4a)$$

<u>repeat</u> $Q_i := S\,\dfrac{v_i - u_i}{B_i - A_i + \epsilon_i}$, ($\epsilon_i$ small number, see following Remark)

$$u_{i+1} := S|_{\eta} [(i_i - Q_i \cdot A_i) \sqcap (v_i - Q_i \cdot B_i)]$$

$$v_{i+1} := S|_{\eta} [(i_i - Q_i \cdot A_i) \sqcup (v_i - Q_i \cdot B_i)] \qquad (7.4b)$$

($\sqcup$, $\sqcap$ denote inf, sup-operation, respectively)

$$A_{i+1} := S\hat{}P(u_{i+1})$$

$$B_{i+1} := S\ P(v_{i+1})$$

<u>until</u> $A_{i+1} < 0 < B_{i+1}$ (7.4c)

If this process terminates successfully then we have the validated result:

$X := [u_i, v_i]$ is an exclusion set (7.4d)

<u>Proof:</u>
(7.4b) implies $u_{i+1}|_{\delta G} \leq \eta$ and $v_{i+1}|_{\delta G} \geq \eta$ and from (7.4c) we know $P(u_{i+1}) < A_{i+1} < 0 < B_{i+1} < P(v_{i+1})$. Hence, all conditions of the W-principle in [25] are satisfied and thus (7.4d) holds. □

<u>Remark:</u>

1. ϵ_i is a small arbitrary number which serves to avoid zeros of the denominator (7.4b). The problem to avoid zeros in the denominator is also a question for an appropriate decomposition of G such that $B_i - A_i$ has no zeros. This may be a natural strategy for a finite-element approach.
2. A functoid implementation of the operators $\sqcup$ or $\sqcap$ can be done along the following lines: Let be u.v ∈ SM with

$$u = \sum_{k=0}^{N} a_k \varphi_k = \phi * a$$

$$v = \sum_{k=0}^{M} b_k \varphi_k = \phi * b \qquad (7.5)$$

and basis $\phi = (\varphi_1 \ldots, \varphi_N)$ with bounded φ_i and $\varphi_1 > 0$ on $\overline{G}$ then there exist regular transformation matrices T such that

$$\psi := \phi T^{-1} \geq 0 \text{ componentwise on } G \tag{7.6}$$

Now we define
$\phi * a = \phi T^{-1} * Ta = \psi * Ta$ and $\phi * b = \phi T^{-1} * Tb = \psi * Tb$.
Then $W := u \sqcap v$ can be approximated by

$$w := \sum_{k=0}^{N} C_k \varphi_k = \phi * C \tag{7.7}$$

with $C := T^{-1}(Ta \sqcap Tb)$ componentwise, and analogously $u \sqcup v$.

8. Conclusions

The last algorithm can be used in a rather general class of problem whenever the properties of the W-principle in [25] are given. Note that the iteration can be done in directed functoids which can be very generally accomodated by the choice of the bases, roundings and arithmetics to the given problems. Any finite element or discretization principle can be applied and it is thus possible to satisfy locally all conditions of the W-principle. It is obvious that instead of a one-parametric relaxation step (6.3) any further generalized acceleration principle can be applied such as an n step method of the form

$$u_{i+1} = [u_i - \sum_{j=0}^{n} Q_j P(u_{i-j})] \sqcap [v_i - \sum_{j=0}^{n} Q_j P(v_{i-j})]$$
$$v_{i+1} = [u_i - \sum_{j=0}^{n} Q_j P(u_{i-j})] \sqcup [v_i - \sum_{j=0}^{n} Q_j P(v_{i-j})]$$

e.g. instead of (7.4b).

References

[1] Alefeld, G., Herzberger, J.: Einführung in die Intervallrechnung. Bibliographisches Institut, Mannheim, (1974).

[2] Böhm, H., Rump, S.M., Schumacher, G.: E-Methods for Nonlinear Problems, in: Kaucher, E., Kulisch, U., Ullrich, Ch. (eds.): Computerarithmetic: Scientific Computation and Programming Languages, Proceedings, B.G. Teubner, Stuttgart (1987).

[3] Corliss, G.F.: Computing Narrow Inclusions for Definit Integrals. In [9], pp. 150-169.

[4] Dobner, H.-J.: Einschließungsalgorithmen für hyperbolische Differentialgleichungen, Dissertation, Universität Karlsruhe (1986)

[5] Dobner, H.-J., Kaucher, E.: Solving Characteristic Initial Value Problems with Guaranteed Errorbounds. In [9]. pp. 170-185.

[6] FORTRAN-SC, A Study of a FORTRAN Extension for Engineering/Scientific Computation with Access to ACRITH, Language Reference and User's Guide. IBM Development Laboratory Böblingen, Germany (1988).

[7] Faaß, E.: Beliebig genaue numerische Schranken für die Lösung parabolischer Randwertaufgaben, Doctor Thesis, University of Karlsruhe (1975).

[8] IBM High-Accuracy Arithmetic Subroutine Library (ACRITH). Program Description and User's Guide, SC 33-6164-02, 3rd Edition (1986).

[9] Kaucher, E., Kulisch, U., Ullrich, Ch. (Hrg): Computer Arithmetic - Scientific Computation and Programming Languages. Teubner, Stuttgart (1987).

[10] Kaucher, E., Rump, S. M.: E-Methods for Fixed Point Equation $f(x) = x$. Computing 28, pp. 31-42 (1982).

[11] Kaucher, E.: Methoden zur Lösung von Integral- und Differentialgleichungen. In: Kulisch, U. (ed.): Wissenschaftliches Rechnen mit Ergebnisverifikation. Vieweg, Braunschweig, pp. 225-238 (1989).

[12] Kaucher, E., Miranker, W. L.: Residual Correction and

Validation in Functoids. Computing Suppl. 5, pp.169-192 (1984).

[13] Kaucher, E.: Self-Validating Computation of Ordinary and Partial Differential Equations. In: [9], pp. 221-254.

[14] Kaucher, E.: Self-Validating Numerical Methods. Proceedings of the AMSE Conference on Modelling & Simulation, Vol. 3.1, Athen (1984).

[15] Kaucher, E., Miranker, W. L.: Self-Validating Numerics for Function Space Problems. Academic Press, New York (1984).

[16] Kaucher, E.: Solving Function Space Problems with Guaranteed, Close Bounds. In: Kulisch, U., Miranker, W. L.: A New Approach to Scientific Computation, pp. 139-164, Academic Press, New York (1983).

[17] Kulisch, U., Miranker, W. L.: Computer Arithmetic in Theory and Practice. Academic Press, New York (1981).

[18] Kulisch, U., Stetter, H. J. (eds.): Scientific Computation with Automatic Result Verification. Computing Suppl. 6, Springer, Wien (1988).

[19] Kulisch, U. (ed.): PASCAL-SC: A Pascal Extension for Scientific Computation, Information Manual and Floppy Disks, Version IBM PC. B. G. Teubner, Stuttgart, John Wiley & Sons, Chichester (1987).

[20] Kulisch, U. (ed.): PASCAL-SC: A Pascal Extension for Sientific Computation, Information Manual and Floppy Disks, Version ATARI ST, B. G. Teubner, Stuttgart (1987).

[21] Lohner, R.: Einschließung der Lösung bei gewöhnlichen Anfangs- und Randwertaufgaben und Anwendungen. Doctor Thesis at the University of Karlsruhe (1988).

[22] Moore, R. E. (ed.): Reliability in Computing - The Role of Interval Methods in Scientific Computing. Academic Press, San Diego (1988).

[23] Rall. L. B.: Optimal Implementation of Differentiation Arithmetic, In [7], pp. 287-295.

[24] Schumacher, G.. Genauigkeitsfragen bei algebraisch-numerischen Algorithmen auf Skalar- und Vektorrechnern. Doctor Thesis at the University of Karlsruhe (1989).

[25] Walter, W.: Differential and Integral Inequalities, Springer, (1964).

INTERNATIONAL ASSOCIATION FOR MATHEMATICS AND COMPUTERS IN SIMULATION (IMACS)

GESELLSCHAFT FÜR ANGEWANDTE MATHEMATIK UND MECHANIK (GAMM)

Foreword to a resolution concerning computer arithmetic

The following IMACS-GAMM Resolution on Computer Arithmetic is based on unpleasant experiences made with existing vector processors. Most vector processors provide what is called elementary compound operations. They are heavily pipelined and make the computation really fast. Usually they are automatically inserted into a user's algorithm by the vectorizing compiler. If these operations are not carefully implemented the user looses complete control of his computation. The IMACS-GAMM Resolution is supposed to influence and put pressure on manufacturers to implement these operations with extreme care.

The "Resolution" was initiated by GAMM, and was unanimously approved by the GAMM Vorstandsrat (Steering Committee) at the GAMM meeting in 1987.

Its proponents felt that more weight would be given to it if it were also supported and approved by IMACS: It was thus submitted in 1988 to the IMACS Board of Directors who also approved it unanimously.

The text of this resolution which is thus an official IMACS-GAMM document follows:

IMACS-GAMM+) Resolution on Computer Arithmetic

The elementary floating-point operations +, -, *, / in electronic computers are currently required to be of highest machine accuracy: For any choice of operands, the computed result must coincide with the rounded exact result of the operation, rounded according to the rounding mode in use (if no overflow occurs). For reference, see the IEEE Arithmetic Standards 754 (binary floating-point arithmetic) and 854 (general floating-point arithmetic).

+) IMACS = International Association for Mathematics and Computers in Simulation
GAMM = Gesellschaft für Angewandte Mathematik und Mechanik

In recent years there has been a significant shift of numerical computation from general purpose computers towards vector and parallel computers - so-called supercomputers. Along with the 4 elementary operations +, -, *, /, these computers usually offer compound operations as additional elementary operations. This leads to an increase of several orders of magnitude in computing power. Some of these elementary compound operations are:

- multiply and add: a * b + c
- multiply and subtract: a * b - c
- accumulate: computes the sum of the components of a vector
- multiply and accumulate: computes the inner (or scalar) product of two vectors

and others.

IMACS and GAMM require that all elementary compound operations be implemented by the manufacturer in such a way that guaranteed bounds are delivered for the deviation of the floating-point result from the exact result. It is desirable and usually achievable that for all possible data the computed result of such a compound floating-point operation agrees with the result that would be obtained if the exact result were computed and then rounded by the rounding in use (if no overflow occurs). In this case no explicit error bounds need be delivered. The user should not be obliged to perform an error analysis every time an elementary compound operation, predefined by the manufacturer, is employed.

All elementary compound operations should also be provided with directed roundings, a feature needed both for fast computation of reliable and narrow bounds in numerical algorithms and for verification of the correctness of computed results. It must be ensured that the final floating-point result can differ from the exact result only in the direction defined by the rounding in use. This is already required of the elementary floating-point operations by the arithmetic standards mentioned above.

NOTES AND REPORTS IN MATHEMATICS IN SCIENCE AND ENGINEERING

Edited by William F. Ames, *Georgia Institute of Technology*

Vol. 1 John S. Gero, editor, *Design Optimization*

Vol. 2 Michael F. Barnsley and Stephen G. Demko, *Chaotic Dynamics and Fractals*

Vol. 3 Sigeru Mizohata, *On the Cauchy Problem*

Vol. 4 Heinz W. Engl and C. W. Groetsch, editors, *Inverse and Ill-Posed Problems*

Vol. 5. Dajun Guo and V. Lakshmikantham, *Nonlinear Problems in Abstract Cones*

Vol. 6 G. Fusco, M. Iannelli, and L. Savadori, editors, *Advanced Topics in the Theory of Dynamical Systems*

Vol. 7 Christian Ullrich, editor, *Computer Arithmetic and Self-Validating Numerical Methods*